Applications of Emerging Nanomaterials and Nanotechnology

Edited by

N.B. Singh[1], Md. Abu Bin Hasan Susan[2] and Ratiram Gomaji Chaudhary[3]

[1]Department of Chemistry and Biochemistry, SBSR, Sharda University, Greater Noida, India

[2]Department of Chemistry, University of Dhaka
Dhaka 1000, Bangladesh

[3]Department of Chemistry, Seth Kesarimal Porwal College of Arts, Science and Commerce, Kamptee, India

Published by **Materials Research Forum LLC**
Millersville, PA 17551, USA

Published as part of the book series
Materials Research Foundations
Volume 148 (2023)
ISSN 2471-8890 (Print)
ISSN 2471-8904 (Online)

Print ISBN 978-1-64490-254-7
eBook ISBN 978-1-64490-255-4

Distributed worldwide by

Materials Research Forum LLC
105 Springdale Lane
Millersville, PA 17551
USA
https://www.mrforum.com

Manufactured in the United States of America
10 9 8 7 6 5 4 3 2 1

Table of Contents

Preface

Nanotechnology is an interdisciplinary science that integrates several disciplines including biology, chemical, mechanical and electronics engineering to comprehend, manipulate and build devices/systems with remarkable functionalities and qualities at the atomic/molecular/supramolecular levels. It has revolutionized science and consumer products for several decades.

In recent years nanotechnology has been used as a significant tool for growth in different fields. Nanotechnology has become a principal tool in developing a variety of industries. It has the potential to produce a wide range of novel products. Keeping the importance of nanomaterials and nanotechnology in mind, this book briefly discusses applications of nanomaterials and nanotechnology. In this book, there are 11 chapters written by experts in the area. The book will be useful for students and researchers alike.

Prof. N.B. Singh
Prof. Md. Abu Bin Hasan Susan
Dr. Ratiram Gomaji Chaudhary

Applications of Emerging Nanomaterials and Nanotechnology
Materials Research Foundations 148 (2023) 1-26

Materials Research Forum LLC
https://doi.org/10.21741/9781644902554-1

Chapter 1

Overview of Applications of Nanotechnology

Richa Tomar[1*] and N.B. Singh[1,2]

[1] Department of Chemistry and Biochemistry, Sharda University, Greater Noida, India

[2] Research Development Cell, Sharda University, Greater Noida, India

*richa.tomar@sharda.ac.in

Abstract

Nanotechnology has changed science and technology. Every discipline is now making use of nanomaterials (NMs) and nanotechnology. NMs have size below 100 nm. Variety of nanoparticles (NPs) with different size and functions are available in the nature. Nowadays, NPs are being synthesized under controlled conditions in laboratories and are called engineered nanoparticles. Nanotechnology is an interdisciplinary discipline and used in different industries such as agriculture, food, biotechnology, medicine, textile, electronics, drugs, etc. In Layman's language, nanotechnology is the science of organizing, manipulating, manufacturing, and engineering products and materials at the nanoscale. This article gives an overview of nanotechnology applications in different areas.

Keywords

Nanomaterials, Nanotechnology, Agriculture, Food, Textile, Construction

Contents

1.   Introductions

The term ''nano'' has been taken from a Greek word ''nanos'' meaning ''dwarf''. Particles with size less than 100 nm are known as nanomaterials. Properties of materials at this scale are entirely different as compared to that of bulk materials. The study of science of such materials is termed as nanoscience. Nanotechnology is a branch of science and engineering, which concerns with designing and producing devices using nanomaterials. Nanotechnology is an "atomically precise technology" or "engineering with atomic precision" [1]. It relates to systems and materials, in which the components and structures possess novel physical, chemical, and biological properties, because of their nanoscale size. Nanotechnology is also defined as "the design, characterization, manufacture and shape and size-controlled application of matters in the nanoscale" [1]. Nanotechnology also plays an important role in producing innovative methods and products, which perform better in different sectors [2]. Nanotechnology also gives solution to many existing problems in different sectors [3]. Classification of nanotechnology can be made as given in Fig.1 [3].

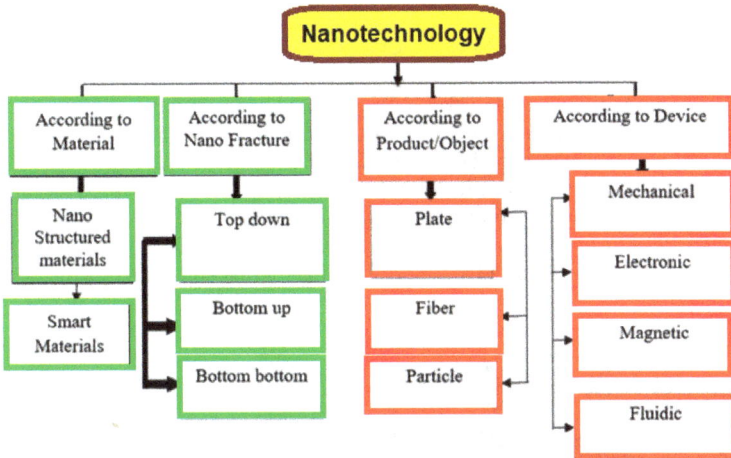

Figure 1 General classification of nanotechnology

Seeing the importance of nanomaterials and nanotechnology, this chapter gives an overview of applications of nanotechnology in different sectors.

2.   Applications of nanotechnology

Nanotechnology is used for various applications in different sectors (Fig.2). Some of the applications of nanotechnology are discussed.

Figure 2 Applications of Nanotechnology

2.1   Agriculture

Many agrochemicals are used for protecting the plants from diseases and increasing production of crops, and improving the quality and yield. Agrochemicals also protect plants from damage and infections. However, only a limited portion of agrochemicals are taken up by plants and other portion becomes just a waste. Many a time, agrochemicals are dangerous. If nano-enabled agrochemicals are used, they offer controlled effectiveness and reduce the toxicity of agrochemicals. Nano-enabled agrochemicals increase uptake of agro- nutrients, improve stability and solubility and control disease of crops. Nanomaterials improve the growth of plants, quality of seeds and flowers, quality of soil, etc. The use of nanomaterials and their role for different functions in agriculture are shown in Fig.3[4,5].

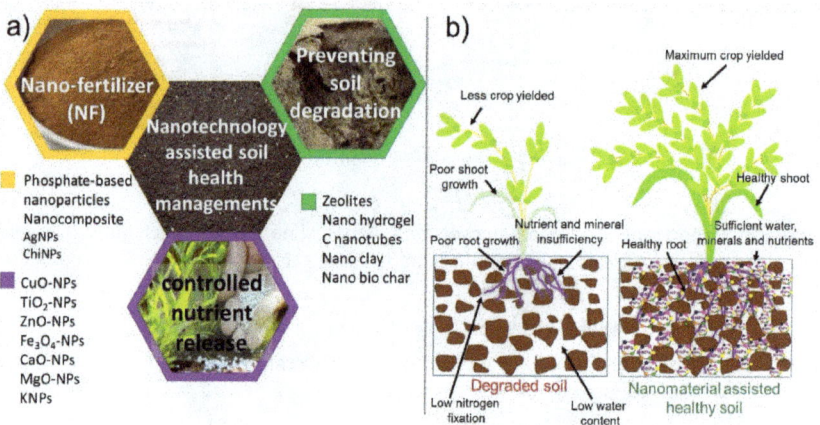

Fig. 3. (a) Soil health and nanotechnology (b) Plant growth in two types of soil.

Nanotechnology has increased agricultural productivity and proficiency with less waste and lower cost. Nanotechnology protects plant from different diseases as shown in Fig.4 [4].

Materials Research Forum LLC
https://doi.org/10.21741/9781644902554-1

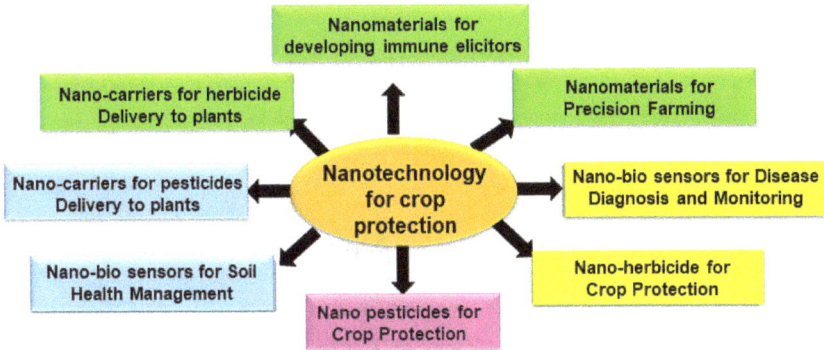

Fig. 4. Nanotechnology in plant disease management [4]

2.2   Food sector

It is believed that nanotechnology may have greater advantage in the food sector in the 21st century [6,7]. Nanotechnology is being used in different industries including agriculture to food processing, packaging, safety, and nutrient delivery (Fig.5) [8,9]. Use of nanotechnology in food systems has generated novel products with better food quality such as taste, texture, sensory properties, stability, etc.

Food processing and packaging are the most important aspects of nanotechnology. Food processing combines different steps such as procuring suitable raw materials, palatability enhancement, removal of toxin, enzymes deactivation, spoilage organisms, minimization of pathogens, fortification and enrichment with micronutrients packaging, storage, and transportation. Since number of steps are involved, there is huge possibility of intervention of nanotechnology-based applications [9]. Nanomaterials are generally used as fillers in food and additives in improving durability and mechanical strength of packaging materials. Use of different type of nanomaterials in food packaging serves the purpose [9]. Applications of nanotechnology in food sector is still in early stages of development and depends on cost-effectiveness [7].

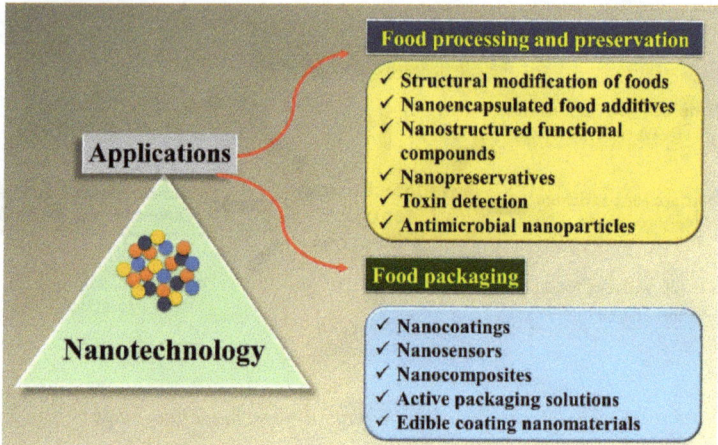

Fig.5 Applications of nanotechnology in Food sector

2.3   Cosmetics

Technically, the word 'cosmetics' refers to the products that amplify the "appearance of the skin, intensify the cleansing, and promote skin beauty". Although, the use of 'cosmetics' word backs to the time of Egyptians around 4000 BCE, but this term was first coined in 1961 by the "founding members of US Society of Cosmetic Chemist, Raymond Reed". The technology of using nanoparticles started early 1960s when the use of liposome technology was applied to moisturizers and skin creams. On using liposomes, the optical properties of the products, solubility and absorption increased. Number of cosmetics such as lotion, sunscreens, haircare products, anti-aging creams, etc. have been using NPs since long and even today nanotechnology has a brighter future in cosmetic industry [10, 11]. Nanocosmetics are categorized depending upon the characteristics of the system, method of preparation, and involved components [10]. Figure 7 gives general classification about nanomaterial's applications in cosmetics [12].

Nanotechnology based cosmetics have been developed but some regulations should also be developed for toxicity.

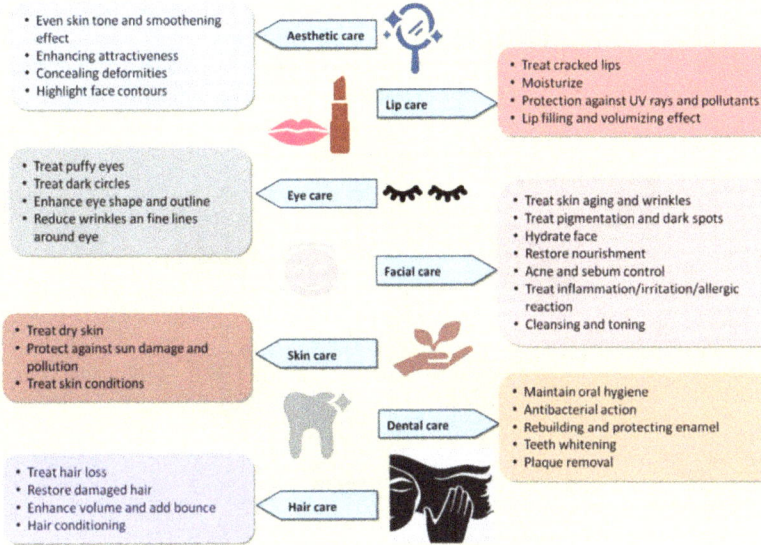

Fig.6 Applications of Nanocosmetics

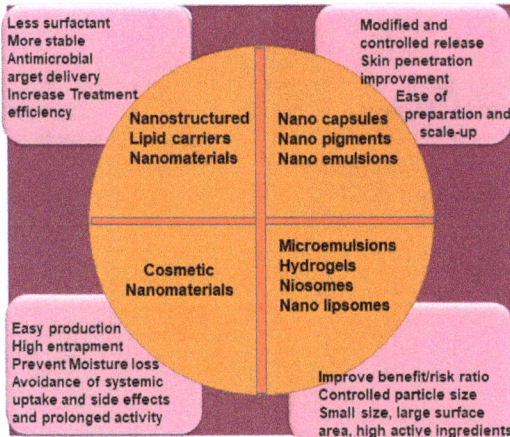

Fig.7 Types of nanocosmetic formulations based on type of nanocarrier system or nanomaterial employed [10]

2.4   Construction

In construction industry ordinary Portland cement (OPC) concrete and geopolymer cement and concrete are the most important binding materials. NMs, when added in cement and concrete, modify different properties as shown in Fig.8 [13].

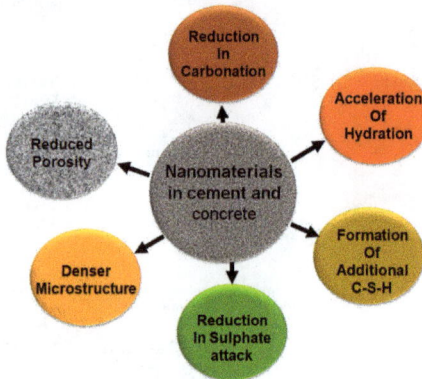

Fig.8 Effect of NMs in cement and concrete [13]

Effect of nano silica (NS), nano alumina, nano $CaCO_3$, nano TiO_2, nano Fe_2O_3, nano ZnO, nano clay, carbon nanotube, graphene, etc. on the properties of cement and concrete have been studied extensively and all of them improve the properties. For example NS accelerates the heat evolution and improves the mechanical properties (Fig.9,10) [14,15]. Improvements of properties are due to additional nucleation sites provided by NMs. For examples the presence of NS in cement, due to pozzolanic activity, consumes $Ca(OH)_2$, converting it to calcium silicate hydrate (C-S-H), an additional binding material. As a result, the hydration is accelerated. The overall mechanism of hydration of cement in presence of NS is given in Fig.11 [16].

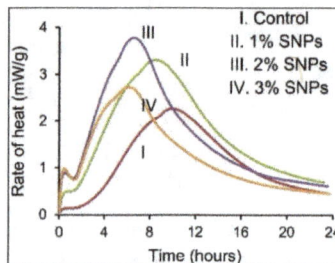

Fig.9 Effect of NS on heat evolution of cement concrete [13]

Materials Research Forum LLC
https://doi.org/10.21741/9781644902554-1

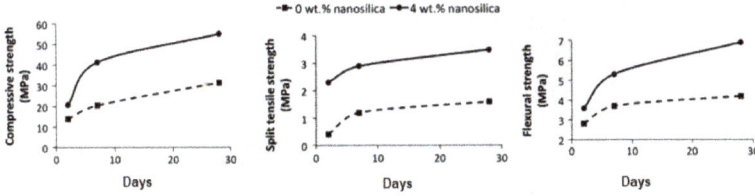

Fig.10 Effect of NS on the mechanical properties of cement

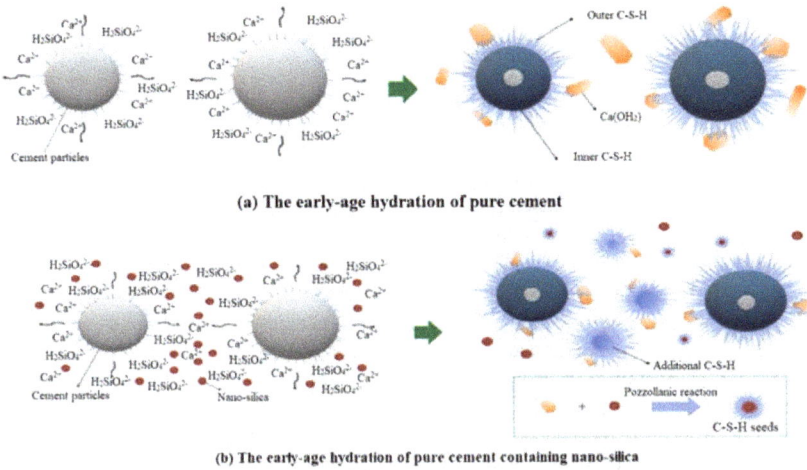

(a) The early-age hydration of pure cement

(b) The early-age hydration of pure cement containing nano-silica

Fig.11 Mechanism of action of NS on early-age hydration [16]

When CNTs are mixed in concrete, it acts as a bridge during crack formation and protects the structure (Fig.12) [15].

Applications of Emerging Nanomaterials and Nanotechnology   Materials Research Forum LLC
Materials Research Foundations 148 (2023) 1-26   https://doi.org/10.21741/9781644902554-1

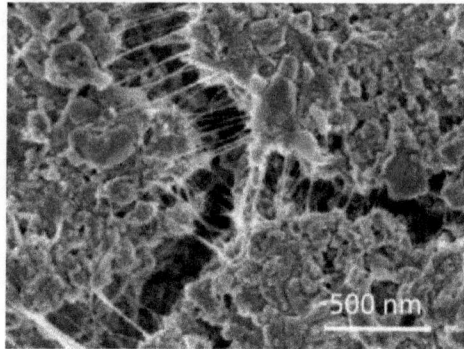

Fig.12 Crack bridging in OPC-CNT composite [15]

2.5 Geopolymer cement and concrete

Geopolymers are new type of binders and can replace OPC composites. The production of geopolymer composites has lower carbon footprint and utilises less energy as compared to OPC manufacture. In recent years, attempts have been made to add various kinds of nanoparticles (NPs) to geopolymer composites in order to improve the properties and performance [17]. Switching off one material from the macro scale to the nanoscale, resulted in significant changes in their chemical reactivity, mechanical properties, surface energy, shape, conductivity, and optical properties. Numerous studies have been done using geopolymer concrete and NMs to examine the new composite's performance in terms of chemical durability, mechanical and structural performance, fresh and microstructural properties, physico-mechanical properties, fire resistance and permeability. The filler effect, which occurs when nanoparticles are added to cement composites, is the result of free water becoming immobile because the NMs fill the pores and spaces among cement particles. Additionally, the NMs contribute to the pozzolanic reaction's formation of fresh calcium silicate hydrate (C–S–H) gel, which improves the interfacial transition zone between the aggregates and binder pastes and, as a result, improves the bond strength properties of the mixture. Geopolymer composites have low compressive strength when cured in an ambient condition. In order to overcome this problem, NMs are mixed within geopolymer concrete composites. This is achieved by atomically modifying the geopolymer concrete's microstructure, which significantly improves the material's fresh and hardened state features as well as its structural properties. Numerous NMs such as nano-silica slurry (NS), nano-zinc oxide (NZ), waste glass nano powder (WGNP), colloidal nano-silica (CNS), nano-calcium carbonate (NC), nano-clay platelets (NCP), multi-wall carbon nanotubes (MWCNTs), nano-titanium oxide (NT), carbon nanotubes (CNTs), etc. have been used to enhance different properties of geopolymer composites. However, NS is most used NM in geopolymer composites (Fig. 13) [17]. In the presence of different NMs, polymerization, mechanical properties and resistance against aggressive atmosphere are increased and microstructures become dens.

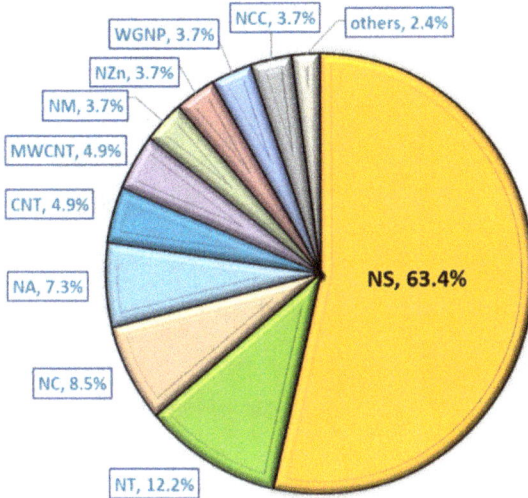

Figure 13. Uses of NPs in geopolymer composites [17]

2.6   Textiles

In recent years, smart fabrics are being designed using nanotechnology. NMs when used in textile industry, generate fabrics with ultraviolet resistant, antimicrobial, optical, electrically conductive, hydrophobic, and flame-retardant properties. Various functions such as sensing, drug release, energy harvesting and storage, and optics are exhibited when NMs based smart devices are integrated with the textiles. This type of technology has now applications in the fashion industry and are being used in healthcare and on-body energy harnessing applications and defence. The incorporation of nanotechnology in textiles show innovative applications in different areas (Fig. 14)[18].

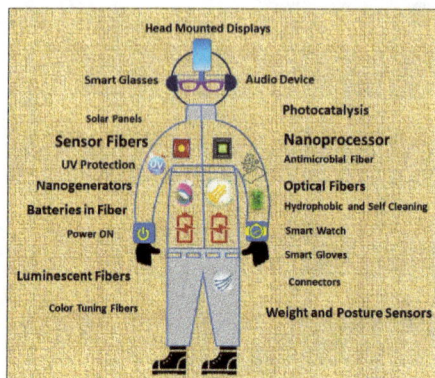

Fig.14 Smart clothing made from nanomaterial [18]

2.7 Sports

Different NMs with different benefits are being used in sports industry (Table 1)[19]

Table 1 NMs in different sports with their benefits [19]

Sports	Nanomaterials	Benefits
Tennis/Badminton	Carbon nanotubes	Increase consistency, stiffness, impact, durability, resiliency, repulsion power and vibration control of rackets
Golf		Lower torque/spin of clubs and reduce weight
Kayaking		Easy padding in kayaks and enhance abrasion/crack resistance
Archery		Better vibration control in arrows
Tennis/Badminton	Silica nanoparticle	Increase power, durability and stability of rackets
Skiing		Facilitate transition in skis and decrease torsion index
Fly fishing		Enhance hoop and flex strength of rods
Tennis/Badminton	Fullerenes	Reduce weight and twisting of racket frames
Golf		Facilitate flexible club whipping
Bowling		Reduce chipping and cracking of balls
Carbon nanofibers	Cycling	Increase stiffness of bicycles and reduce weight
Watercraft	Nano clay	Enhance speed of water-boats and reduce weight
Tennis/Golf		Increase bounce of balls and residency
Tennis/Badminton	Nano-titanium	Transmit more power to ball /shuttlecock, increase strength and durability of rackets, resist deformation of rackets
Golf	Nano nickel	Increase stability of clubs and moment of inertia

2.8   Energy storage

Metal organic framework based templates are receiving more attention for the porous metal oxides and nanocomposites preparation because metal oxides have some limitations, such as decreased capacitive performance because of continuous faradic reactions from electrolyte to electrode. They can increase their capacity for electrochemical storage by preparing Metal organic framework based oxides in conjunction with carbon-based materials including graphene, carbon nanotubes and reduced graphene oxide. Due to their high specific surface area, high surface to volume ratio, and high ion mobility during intercalation, nanomaterial-based devices are advantageous in the conversion and storage of energy.

As nitride-based nanomaterials bear good specific capacitance value, they are gaining greater attention in energy devices together with other graphene based nanomaterials, metals or metal oxides, and graphene quantum dots-based nanomaterials. NiO electrodes with particle sizes of 70 nm can show enhanced capacitance value to 132 F/g at 10mV/s of scan rate. Cobalt (Co), with its high theoretical capacitance value of 3560 F/g, has received a lot of attention for its usage as an electrode in super capacitors and for its strong electrochemical performance. Polymers like polyaniline vary their behaviour during the charging and discharging process as a result of ion doping and dedoping. To solve this problem, a coating of carbon or metal oxide can be deposited on a polyaniline film, improving its stability and capacitance behaviour.

Nanomaterials based on carbon are more appropriate as electrode materials as they offer high electrical conductivity, excellent chemical stability, high specific surface area and bear strong electrochemical property. While having less specific surface area than activated carbon, multiwall carbon nanotubes' mesopores nevertheless allow ions of electrolyte to easily pass across the electrode electrolyte interface. It has been shown that multiwall carbon nanotubes can be carbonised to increase their specific surface area, which in turn improves capacitance with a decent rate of charge and discharge as well as good retention after several cycles. An increase in capacitance value was seen when polymer, metal oxide, or metal nanoparticles were combined with graphene or graphene oxide.

In comparison to individual carbon nanotubes, hybrid structures made of carbon nanotubes and gold nanowire has been shown to have a greater power density. Hybrid nanomaterial-based electrodes were used in lithium ion batteries in addition to supercapacitors [20]. The potential hydrogen storage materials include a variety of nanomaterials like carbon-based materials, metal organic framework-based nanomaterials, boron nitride nanotubes, nano-Mg-based hydrides, TiS_2/MoS_2 nanocomposite, and polymer-based monohybrid.

Hydrogen is an energy carrier which can store a huge amount of energy. Fuel cells using hydrogen may produce power, energy, and heat. The nanomaterial modifies the diffusion length and rate, which has an impact on the thermodynamics and kinetics of hydrogen sorption and absorption. It has been found that alloys based on magnesium make excellent hydrogen storage precursors because of their superior mechanical properties. Mg2NiH4 has been used to encapsulate MgO nanoparticles, giving them exceptional hydrogen sorption capabilities. They are also resistant to further oxidation by lingering oxygen.

2.9 Defense

Modern armed forces are utilising new technologies in an effort to outperform rivals on a qualitative level. By incorporating many cutting-edge technologies, it is possible to shift from "mass and mobility" to unconventional methods of boosting comparative combat capabilities. We are forced to adopt revolutionary combat strategies by technologies like stealth technology, remote sensing, night vision, sensors, precision-guided missiles, image processing,   and, most essential, computer networks and digital communications. Aeronautics, nautical, shielding, electronics, automotive, electromagnetic interference (EMI), energy storage and constructions are among the industries involved in the defence industry. Nanotechnology has the potential to have a significant impact on the defence industry in a number of areas, specifically increasing weapon viability, surveillance, securing communication and protecting targets like strategic assets, soldiers and equipment, and creating lightweight aerial naval, and ground platforms.

The development of defensive technologies from the Stone Age to the present can be used to illustrate the significance of material science and technology. In addition to functionality, the choice of material for military applications must be durable in adverse conditions, such as temperature changes, sandstorms, rain, corrosion due to humid and salty environments, and structural stability. The material should also be light in weight and affordable. Submarines, aircraft, naval ships, combat vehicles, military personnel, sailors, mariners, and pilots also need modern material components that enable significant changes in manoeuvrability, safety from chemical, nuclear, and biological weapons, protective cover from explosives through signature step-down and heavy military force, and involvement with highly focused firepower and robust logistics. The military and defence industries have witnessed the spectacular expansion of nanotechnology over the years as it greatly raises the calibre of military equipment and improves worker comfort and safety.

The most practical way to develop advanced defence technologies is to use carbon-based composites that incorporate different carbonaceous materials like black carbon, coke, activated carbon, char, carbon fibre along with other carbon based nanomaterials like graphite, carbon nanofibres, carbon nanotubes and graphene. PNCs are used to create military infrastructure and equipment that is lightweight, relatively tiny, less expensive, more accurate, creative, and robust.

Fig: 15 Applications of nanotechnology in the defense sector.

2.9.1 Smart military uniforms

For the protection of soldiers, PNCs are used to create footwear, helmets, gloves, intelligent materials, and bulletproof vests. When supplemented with nanomaterials like carbon nanotube (CNT), Kevlar, graphene etc., polymer matrix permits the fabrication of exceptionally robust, smart, and lightest high-tech fighting suits. When silica nanopowder-containing shear thickening fluids are added, body armour becomes more compact, durable, and extremely flexible. The wearer of these body armour vests can move more freely, protect themselves from poisonous substances, endure the impact of high-speed gunshots, and avoid getting hurt by blunt objects like bars, stones, and sticks.

Lightweight military platforms and armors

Steel-made heavyweight military platforms in concern to their speed, high fuel consumption, thick armours to defend against explosive and ballistic attacks and mobility in the fighting effectiveness represent a big issue. Composites made of hybrid nanoparticles and fibres exhibit good mechanical and fatigue properties as well as superior scratch and impact energy attenuation. Composites can transmit electricity when conductive nanoparticles like graphite, carbon black (CB), carbon nanotubes, graphene, or metals are added.

2.9.2 Acoustics absorption

Designing military and aerospace vehicles takes into account acoustics damping in order to achieve comfort and structural stability. Nowadays, nanofillers like silica are added to PU foams, which are already widely utilised as sound-absorbing materials, to create polymer matrix composites for ballistic protection. The tensile, flexural, and shear strengths of glass fibre with 6, 7, and 8 wt% SiO_2 were improved, along with the component adhesion and surface damage area. The tensile impact and flexural strength of aramid fibre are improved by adding 0.15, 0.3, and 0.5 weight percent of multiwall carbon nanotube. Barium sulfate, calcium carbonate, or fibreglass are added to advanced polymer nanocomposites to increase the level of acoustic absorbency. Sung et al. looked into the PU/nanosilica foams' ability to absorb sound and discovered a considerable reduction in acoustic wave. Baferani et al. created PU/CNT foam nanocomposites that can effectively absorb sound waves with a wide frequency range (400–3600 Hz). [21].

2.9.3 Camouflage printed fabric

With advancements in long-range munitions and picture identification technology and equipment, soldiers and military infrastructure can now be completely invisible, giving commanders incredible control over the combat. The electromagnetic spectrum's many wavelength ranges must be covered by the current military materials, particularly the thermal or far-infrared region (35 and 814 m), the near-infrared region (NIR) (750-1200 nm), and the visible region (400-800 nm) [23]. Karpagam et al. used thermochromic colorants, which showed reversible colour changes in response to electrical power and temperature change, to create a camouflage printed fabric with a jungle theme. According to Alehi et al., adding activated carbon nanoparticles and carbon black to viscose/polyester fabrics significantly reduced the textiles' IR reflectivity for both the green and black textures.

In order to tailor the fabrics' NIR concealment and visibility for a desert environment, Mehrizi et al. created nylon/cotton textiles with varying concentrations of CB nanoparticles and colors. Additionally, Mehrizi et al. looked at how adding multiwalled CNT particles (MWCNTs) to a fabric made of 50% nylon and 50% cotton might lower the NIR reflectance criteria for printed textiles in light brown, olive green, and dark brown [22]. Similar to this, Siadat and Mokhtari demonstrated the impact of coated cotton/nylon textiles with cerium dioxide (CeO_2) and zirconium dioxide (ZrO_2) nanoparticles and cerium dioxide (CeO_2) nanoparticles in desert and forest patterns, as well as coated nylon/cotton textiles with ZrO_2 and magnesium oxide (MgO) in desert and jungle patterns [23].

2.9.4 Corrosion protection

Metal corrosion has now become a significant problem that has cost the military sector a lot of money and poses a big risk to the public. Since impairment or decline in quality cannot be entirely stopped, it is restricted and hampered in order to minimise financial losses. Because of the huge surface area to volume ratio of the nanoparticle, water molecules cannot diffuse to the twisted channel, providing better surface protection than bulk counterparts. It seems like an intriguing idea to create a PNC coating to protect metal surfaces against corrosion. There have been many nanoparticle-based coatings found, including TiO_2, Al_2O_3, SiO_2, ZnO, and Fe_2O_3.

2.9.5 Wound care for soldiers

The contamination of war wounds is frequently severe, and in the event of bomb assaults, contamination with foreign human material can make things even more difficult. Combat wounds are challenging to dress because to their size and exudate. There are antimicrobial components in many dressings. While some dressings collect wound exudate and employ the antibiotic there to limit microbial development, other dressings discharge the antimicrobial straight into the wound. Dressings with nanomaterial coatings provide regulated protein and medication release over a predetermined period of time. Additionally, the distinct feature of nanoparticle aggregate wound dressing makes it feasible for the exciting prospects of controlling the quantities of growth factors and other active components to hasten wound healing [24].

2.9.6 Sensory applications

Innovative nanomaterial-based sensor technologies are in demand to intensify military intelligence aggregation by military personnel on the battleground. Nanosensors, for instance, can recognise hazardous substances. Chemical sensors built on graphene (G) and CNT-infused polymer composites are being researched for a number of applications, such as gas sensing, biosensing, and chemical sensing based on optical and electrochemical recognition methods. Chemical sensors based on individual single-walled nanotubes were developed by Kong et al. They discovered that the electrical resistance of semiconducting single walled nanotubes significantly changed when it was exposed to gas molecules like NH_3 and NO_2 [25].

Electronic pollution in the environment

Electromagnetic (EM) radiations or waves, electronic noise, radiofrequency interference (RFI) and electromagnetic interference (EMI) are a few examples of particular types of electronic pollution in the environment that have arisen due to the rapid development of technology and the substantial rise in the use of electronic and electrical devices. Polyacrylonitrile (PAN) solution was able to permeate CNT fibres made via floating catalytic chemical vapour deposition, according to Li et al. Such composite fibres demonstrated an almost 300% increment in breaking load, a 350% improvement in strength, and a roughly 700% rise in modulus when compared to pure CNT fibres. The materials' electrical characteristics have been improved by the addition of CNT as carbon nanoparticles.

2.10   Biotechnology

The expensive old manufacturing process will be replaced by less expensive, environmentally friendly goods that have flexible, precise, and long-lasting architecture. Using this technology, it is possible to create strong yet lightweight materials for things like sensors, microrobots, surgical instruments, electronic devices, and circuits. Another example in nanobiotechnological research is nanospheres coated with luminous polymers. Researchers are trying to develop polymer patterns that can dim fluorescent light when they interact with a certain class of molecules. DNA nanotechnology is a significant instance of bionanotechnology. Utilizing the intrinsic properties of nucleic acids (such as DNA) to create synthetic membranes and self-assembling proteins is the promising field of this modern technology.

The uses of the molecules made from DNA are being investigated due of their complexity. This is achievable if these structures are changed into colloidal superstructures when DNA is utilised to govern delivery, removing inorganic nanoparticles. A full superstructure-designed molecule's building blocks are nanoparticles that have the surface chemistry, size, and assembly architecture of the superstructure. The entire structure can interact with cell organelles in accordance with their inherent design and is later catabolized into components for exocytosis. This shortens the time that nanoparticles spend inside cells, which improves the accumulation of a tumour in vivo as well as the removal of the entire body.

This shortens the time that nanoparticles spend inside cells, which improves the accumulation of a tumour in vivo as well as the removal of the entire body. The superstructures could be used for imaging or as a carrier for therapeutic agents to prevent the deterioration of the enzymes. There should be some methods for creating nanostructures that could result in both multifunctional nanomedicines and compounds that are capable of biodegradation.

2.11   Biomedical

Innovative nanoparticles for medical applications were created using a variety of metal salts, including titanium, zinc, silver and other inorganic salts. Infection in the wound region is a common issue, regardless of the type of wound. The primary cause of wound complications, microorganisms, are progressively developing resistance to the widely used antibacterial medications. To keep the pathogenic bacteria under control, it is necessary to produce nanoparticles with effective antimicrobial potential. In general, there are four steps in which a wound heals: haemostasis, inflammation, proliferation, and maturation. In addition to being utilised topically as an ointment, these nanoparticles can also be incorporated into electrospun nanofibers, hydrogels or sponges.

According to research, inorganic nanoparticles like (CoO, CuO, AgO, TiO_2, ZnO, Ag) that are made using environmentally friendly procedures can treat skin sores. Due to their adaptable activity against a variety of pathogenic bacteria and their low or nonexistent toxicity towards mammalian cells, silver nanoparticles (AgNPs) have become one of them and have significantly increased in popularity in the biomedical field. Additionally, materials based on nanosilver are useful therapeutic and preventive agents for preventing microbial colonisation of wounds. Consequently, nanosilver has been utilised in a variety of applications, including contact lens coatings, antimicrobial gel formulations, orthopaedic applications, instruments, medical catheters, dressings for wound healing, implants, and 3D and 4D printing [26].

Another relevant inorganic substance is zinc oxide (ZnO), which has potential biocidal activity as a result of its photocatalytic properties. Nano ZnO and biomaterials derived from it have surfaces that produce free radicals when they come into contact with light. It has been shown that the active radicals produced in this way can suppress microorganisms. Additionally, they are biocompatible, non-toxic, affordable, transparent, and eco-friendly materials, making them perfect for cutting-edge medical applications. They are efficient adsorbents due to their large surface area to volume ratio and strong adsorption characteristics. Nanomaterials are used to regulate bone homeostasis and modulate cellular activity as therapeutic carriers or multifunctional platforms. In order to detect bone microfractures, malignancies, and metastases at early stage, biological imaging is made possible by rare earth, gold, and metal oxide nanomaterials and quantum dots.

Additionally, nanomaterials have amazing promise as biosensors for indicators of arthritis in peripheral blood and synovial fluids. Bone tissue regeneration is significantly impacted by carbon-based nanomaterials with effective cell proliferation and osteogenic differentiation. New therapeutic options for spinal cord injuries have been made possible by nanomaterials. The microenvironment of traumatic injury can be improved with the use of performance-based nanoparticles made from a variety of materials, and in some situations, this can encourage neuron regeneration. Functional nanoparticles' nanoscale size enables medications to pass across the blood-spinal cord barrier (BSCB) and collect in the location of the injury. Within 96 hours of an injury, 200 nm nanoparticles may extravasate into the cord parenchyma at the lesion site. Linking targeted groups can also be used to achieve active directed delivery.

2.12 Optical devices and electronics

Carbon atom sheets that form hexagonal patterns like benzene rings in each layer of the layered substance known as graphite. Graphene is the name for a monolayer of graphite [27]. It is a remarkable drug with a long list of accolades. It is both the thinnest substance in the universe and the toughest material ever researched. Future models could replace silicon with long, thin graphene strips called graphene nanoribbons (GNRs), which can widen the bandgap in the semimetal. In the realm of electronics, particularly in the creation of graphene and related nanomaterials and batteries are widely used. Reduced graphene and nickel sulphide were combined to create a nanocomposite that was employed as anode components in the manufacture of sodium-ion and potassium ion batteries. Additionally, solar cells employ graphene and related nanoparticles as a transparent conductive cathode electrode. The reduced graphene oxide can act as both a working electrode and a cathode electrode for the electrochemical deposition of material. Graphene nanoparticles are employed to construct solar cells because the cathode electrodes they use are so important for solar cell applications.

2.13 Water remediation

Nanomaterials (NMs)-based technologies have produced promising results for accessing cleansed water from various resources, and nanotechnology has emerged as a unique possibility for improving current breakthroughs in wastewater treatment. The majority of organic pollutants are chemically stable, making it difficult to clean water using traditional physiochemical approaches without suffering from one or more drawbacks. According to recent reports, nanotechnology may be able to effectively breakdown a variety of organic contaminants, including colours, into harmless byproducts. This process is known as photocatalysis. AgNPs were immobilised onto graphene oxide using green tea extract during an in-situ fast reduction process. The created composite showed 633 mg g^{-1} of methylene blue degradation.

The photocatalytic activity of green AgNPs against methylene blue, orange red, and 4-nitro phenol was reported. Ruellia tuberosa leaf extract was used as a green reducing and stabilising agent to create iron NPs, which were successful in the photocatalytic degradation of reactive black 5. The removal of dyes and heavy metal ions from water has also been reported to be accomplished by a range of nanomaterials, including Au, SiO_2, FeS, Fe_3O_4 nanoparticles, ZnO, CuO, FeO, TiO_2, SnO_2, Pd nanoparticles, Ag/Pt NPs, Ag-TiO_2 NCs, and ZnO/SnO_2 nanocomposite [28-29].

Bacterial nanotechnology (Bac-Nano), which is helping to advance wastewater treatment technologies, is focused on the intriguing interplay between bacterial cells and nanomaterials. Analyses of the economic viability, environmental risk, performance and possibility for commercial-scale evolution of Bac-Nano for wastewater treatment are conducted. Reactive oxygen species (ROS) production and the impact of metal cations were indicated as the main chemical and physical interactions in the majority of the documented processes. The cell membrane, membrane-linked enzymes, proteins, and nucleic acids are all targets of these mechanisms, making it difficult for bacterial cells to develop defences against all of these biological components. Blocking nutrition intake is another way to induce inhibition. For instance, it has been demonstrated that gallium ions (Ga^{3+}) can inhibit bacterial growth via a "trojan horse" mechanism in which the bacterial cells take up the Ga^{3+} along with the Fe^{3+} ions due to their similar chemical properties, and metabolic inhibition results as a result of the inability of the bacterial cells to reduce the Ga^{3+}.

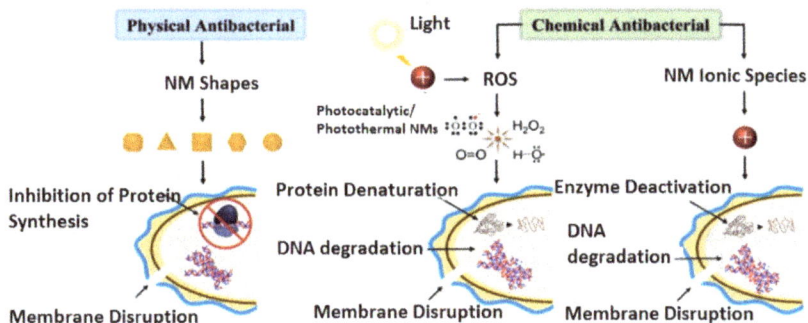

Fig. 16 Physical and chemical mode of actions of nanomaterials for different antibacterial mechanisms [30].

Some antibacterial NMs are primarily activated by light and magnetism, which results in a variety of modes of action. Specific light wavelengths activate photocatalytic and photothermal NMs, which restrict bacterial growth by producing ROS and concentrated temperature spikes, respectively. the use of an external magnetic field to increase the antibacterial activity of Fe_3O_4-ZnO nanocomposites against E. coli and S. aureus. After 15-20 minutes of modification with C60, MWCNT, and rGO, TiO_2 completely destroyed the bacterium.

2.14 Miscellaneous

2.14.1 Biosensors

With a linear range of 28 mM x mm(-2), Biocompatible-Graphene Oxide demonstrated good sensitivity (8.045 mA x cm(-2) x M(-1)) as a reliable glucose biosensor. The biocompatibility of as-synthesised GO-nanosheets with human cells was effectively proven in this work for the very first time. The electrode was altered with graphene oxide, and glucose oxidase was immobilised there. This increases the reproducibility and storage capacity of the biosensor. By creating a nanocomposite of Au nanocluster and graphene oxide through electrostatic interaction and sonication, Ge et al. showed an electrochemical biosensor that has the tendency to detect L-Cysteine. An electrochemical biosensor with 0.02 mol/L of LOD can detect L-Cysteine using graphene oxide and an Au nanocluster. Urea with a LOD of 5 g/mL may be detected using a nanocomposite of graphitized nanodiamond and graphene nanoplatelets. Colorimetric biosensors found effective in heavy metal detection due to its high sensitivity, simplicity, specialised detection capabilities without the need for costly equipment and low cost. Looking into the high extinction coefficients and size-dependent optical characteristics of AuNPs, they are frequently used as a colorimetric reporter. Recently developed artificial biomimetic enzymes known as nanozymes are a prospective candidate for use in the production of colorimetric sensors for heavy metal ions detection [31].

2.14.2 Lubricants

Everyday, lubricants are one of the most practical materials since they reduce equipment wear and friction. As lubricant additives, nanocomposites (Mn_3O_4/G)e were utilised. Utilizing the solvothermal method, Sun et al. also created the nanocomposites (graphene oxide/Fe_3O_4). As lubricant additives, this nanocomposite is employed. Due to their high dispersion stability in water, the nanocomposites in this study had excellent tribological properties. They also have enhanced friction coefficient and reduced wear scar diameter by 33.6% and 32.3% in comparison to base oil alone. Recently, employing electrophoretic deposition on a silicon substrate, Mia et al. created a nanocomposites film (Graphene oxide/Ti_3C_2) for the first time. This study's tribological test showed that the synthesised nanocomposites film was effectively deposited under 35 V and that wear resistance and the effect of graphene oxide and Ti_3C_2 synergy on friction rate and wear resistance were also demonstrated.

2.14.3 Humidity-sensor

Due to its transparency, flexibility, and extreme water permeability, GO was used in humidity sensors. Its attributes combined with water enable efficient humidity sensing with a response time of up to 30 ms. Numerous applications, such as touchless user interfaces that can be exhibited by whistling-based recognition detection, are made possible by such humidity-based sensors.

2.14.4 Gas transport

Metal organic frameworks (MOFs), covalent organic frameworks (COFs), and other 2D nanomaterials are emerging nanomaterials for gas separation and pervaporation that have recently gained a lot of attention. Metal organic frameworks (MOFs), a new class of inorganic materials, are crystalline porous substances produced by the self-assembly of metals and organic ligands. Due to its thinness of just one carbon atom, graphene is a potential material for membranes. Since it creates separation membranes by lowering transparent resistance and raising flow, it is ideal for membrane applications. A study claims that GO sheets are employed and created to achieve the requisite gas separation. The gas transportation is heavily dependent on the degree of interlocking in the stacking structure of graphene oxide and selective gas diffusion is achieved in this by manipulating the channels and pores of gas flow with the aid of various stacking methods. The desired gas was successfully separated since it has unique qualities that make it appropriate for membrane use. According to the proposed study, a membrane made of few-layered graphene oxide and graphene successfully transported a certain gas.

Conclusions

In this chapter, it is highlighted that nanotechnology plays an important role in producing innovative methods and products, which perform better in different sectors. Nanotechnology also gives solution to many existing problems. Classification of nanotechnology has been made. Applications of nanotechnology in different sectors such as Agriculture and food, Energy storage, Defence, Biotechnology, Biomedical, Optical devices and Electronics, Construction, Sports, Textiles, Cosmetics, Paints, Water remediation, etc. have been discussed. In the coming days, nanotechnology may be used in almost all disciplines.

References

[1] Mahmoud Nasrollahzadeh, S. Mohammad Sajadi, Mohaddeseh Sajjadi and Zahra Issaabadi Chapter 1 An Introduction to Nanotechnology (Chapter 1), Interface Science and Technology, Vol. 28. 2019 Elsevier Ltd. https://doi.org/10.1016/B978-0-12-813586-0.00001-8

[2] J. Lee, S. Mahendra , P.J.J. Alvarez. Nanomaterials in the construction industry: a review of their applications and environmental health and safety considerations. ACS Nano 4(7)(2010):3580-90 https://doi.org/10.1021/nn100866w

[3] Rakesh Kumar, Mohit Kumar, Gaurav Luthra, Fundamental approaches and applications of nanotechnology: A mini review, Materials Today: Proceedings, https://doi.org/10.1016/j.matpr.2022.12.172

[4] Mukta Rani Sarkar , Md. Harun-or Rashid , Aminur Rahman , Md. Abdul Kafi , Md. Ismail Hosen , Md. Shahidur Rahman , M. Nuruzzaman Khan, Recent advances in nanomaterials based sustainable agriculture: An overview, Environmental Nanotechnology, Monitoring & Management 18(2022)100687 https://doi.org/10.1016/j.enmm.2022.100687

[5] J. Sangeetha, D. Thangadurai, R. Hospet, E.R. Harish, P. Purushotham, M.A. Mujeeb, et al., Nanoagrotechnology for soil quality, crop performance and environmental management In: Nanotechnology: An Agricultural Paradigm. (2017), pp. 73-97 https://doi.org/10.1007/978-981-10-4573-8_5

[6] R.Ravichandran, Nanotechnology applications in food and food processing: Innovative green approaches, opportunities and uncertainties for global market, International Journal of Green Nanotechnology: Physics and Chemistry, 1 (2) (2010), pp. P72-P96, https://doi.org/10.1080/19430871003684440 https://doi.org/10.1080/19430871003684440

[7] Shibasini Murugana , Vijay Karuppiaha , Kavitha Thangavela, Sivasakthivelan Panneerselvam Applications of nanotechnology in food sector: Boons and banes in Nanotechnology Applications for Food Safety and Quality Monitoring, 2023, Pages 473-492. https://doi.org/10.1016/B978-0-323-85791-8.00009-4

[8] Rahul Biswas, Mahabub Alam , Animesh Sarkar, Md Ismail Haque, Md. Moinul Hasan, Mominul Hoque, Application of nanotechnology in food: processing, preservation, packaging and safety assessment, Heliyon 8(2022)e11795 https://doi.org/10.1016/j.heliyon.2022.e11795

[9] Monalisa Sahoo, Siddharth Vishwakarma, Chirasmita Panigrahi, Jayant Kumar, Nanotechnology: Current applications and future scope in food, Food Frontiers.2(2021)3-22 https://doi.org/10.1002/fft2.58

[10] Suman Singh, Satish Kumar Pandey and Neelam Vishwakarma, Functional nanomaterials for the cosmetics industry (Chapter-22) Handbook of Functionalized Nanomaterials for Industrial Applications, Micro and Nano Technologies, 2020, 717-730 https://doi.org/10.1016/B978-0-12-816787-8.00022-3

[11] Sunil Kumar Dubey, Anuradha Dey, Gautam Singhvi, Murali Manohar Pandey, Vanshikha Singh, Prashant Kesharwani, Emerging trends of nanotechnology in advanced cosmetics, Colloids and Surfaces B: Biointerfaces 214 (2022) 112440 https://doi.org/10.1016/j.colsurfb.2022.112440

[12] Khadijeh Khezri, Majid Saeedi, Solmaz Maleki Dizaj, Application of nanoparticles in percutaneous delivery of active ingredients in cosmetic preparations, Biomedicine & Pharmacotherapy, 106(2018)1499-1505 https://doi.org/10.1016/j.biopha.2018.07.084

[13] N.B.Singh, Properties of cement and concrete in presence of nanomaterials (Chapter-2) in Smart Nanoconcretes and Cement-Based Materials, 2020,9-39 https://doi.org/10.1016/B978-0-12-817854-6.00002-7

[14] Ali M. Onaizi , Ghasan Fahim Huseien , Nor Hasanah Abdul Shukor Lim , Mugahed Amran , Mostafa Samadi, Effect of nanomaterials inclusion on sustainability of cement-based concretes: A comprehensive review, Construction and Building Materials, 306(2021)124850 https://doi.org/10.1016/j.conbuildmat.2021.124850

[15] Monica J. Hanus , Andrew T. Harris, Nanotechnology innovations for the construction industry, Progress in Materials Science. 58 (2013) 1056-1102. https://doi.org/10.1016/j.pmatsci.2013.04.001

[16] Mohammad Tabish, Mohd Moonis Zaheer, Abdul Baqi, Effect of nano-silica on mechanical, microstructural and durability properties of cement-based materials: A review, Journal of Building Engineering 65 (2023) 105676 https://doi.org/10.1016/j.jobe.2022.105676

[17] Hemn Unis Ahmed , Azad A. Mohammed, Ahmed S. Mohammed, The role of nanomaterials in geopolymer concrete composites: A state-of-the-art review, Journal of Building Engineering 49 (2022) 104062. https://doi.org/10.1016/j.jobe.2022.104062

[18] Mudasir Akbar Shah, Bilal Masood Pirzada , Gareth Price , Abel L. Shibiru, Ahsanulhaq Qurashi, Applications of nanotechnology in smart textile industry: A

critical review, Journal of Advanced Research, 38 (2022) 55-75
https://doi.org/10.1016/j.jare.2022.01.008

[19] Luiz Pereirada Costa, Engineered nanomaterials in the sports industry (chapter-14) Handbook of Nanomaterials for Manufacturing Applications, Micro and Nano Technologies, 2020, 309-320 https://doi.org/10.1016/B978-0-12-821381-0.00014-4

[20] Vandenberg A, Hintennach A, A comparative microwave-assisted synthesis of carboncoated LiCoO2 and LiNiO2 for lithium-ion batteries. Russ J Electrochem 51(4) ((2015) 310-317. https://doi.org/10.1134/S102319351504014X https://doi.org/10.1134/S102319351504014X

[21] A.H. Baferani, A.A. Katbab, A.R. Ohadi, The role of sonication time upon acoustic wave absorption efficiency, microstructure, and viscoelastic behavior of flexible polyurethane/CNT nanocomposite foam, Eur. Polym. J. 90 (2017) 383391 https://doi.org/10.1016/j.eurpolymj.2017.03.042

[22] M. Khajeh Mehrizi, F. Bokaei, N. Jamshidi, Visible-near infrared concealment of cotton/nylon fabrics using colored pigments and multiwalled carbon nanotube particles (MWCNTs), Color. Res. Appl. 40 (2015) 9398. https://doi.org/10.1002/col.21852

[23] S.A. Siadat, J. Mokhtari, Influence of ceramic nano-powders and cross-linker on diffuse reflectance behavior of printed cotton/nylon blend fabrics in near infrared and shorts in near infrared and shortwave infrared spectral ranges, J. Text. Inst. (2020) https://doi.org/10.1080/00405000.2020.1802893

[24] M.M. Mihai, M.B. Dima, B. Dima, A.M. Holban, Nanomaterials for wound healing and infection control,Materials 12 (2019) 2176. https://doi.org/10.3390/ma12132176

[25] J. Kong, N.R. Franklin, C. Zhou, M.G. Chapline, S. Peng, K. Cho, et al., Nanotube molecular wires as chemical sensors, science 287 (2000) 622625. https://doi.org/10.1126/science.287.5453.622

[26] K. Varaprasad, C. Karthikeyan, M.M. Yallapu, R. Sadiku, The significance of biomacromolecule alginate for the 3D printing of hydrogels for biomedical applications, Int. J. Biol. Macromol. (2022 May 25). https://doi.org/10.1016/j.ijbiomac.2022.05.157

[27] M. Bandeira, B.S. Chee, R. Frassini, M. Nugent, M. Giovanela, M. Roesch-Ely, J.D. Crespo, D.M. Devine, Antimicrobial PAA/PAH electrospun fiber containing green synthesized zinc oxide nanoparticles for wound healing, Materials 14 (11) (2021) 2889. https://doi.org/10.3390/ma14112889

[28] A. John, A. Shaji, K. Vealyudhannair, M. Nidhin, G. Krishnamoorthy, Anti-bacterial and biocompatibility properties of green synthesized silver nanoparticles using Parkiabiglandulosa (Fabales: fabaceae) leaf extract, Curr. Res. Green and Sustain. Chem. (2021), 100112. https://doi.org/10.1016/j.crgsc.2021.100112

[29] S. Li, J. Zeng, D. Yin, P. Liao, S. Ding, P. Mao, Y. Liu, Synergic fabrication of titanium dioxide incorporation into heparin-polyvinyl alcohol nanocomposite: enhanced in vitro antibacterial activity and care of in vivo burn injury, Mater. Res. Express 8 (8) (2021), 085012. https://doi.org/10.1088/2053-1591/abe1fb

[30] Ahmed ElMekawy, Hanaa M. Hega, Habiba Alsafar, Ahmed F. Yousef, Fawzi Banat, Shadi W. Hasan. Bacterial nanotechnology: The intersection impact of bacteriology and nanotechnology on the wastewater treatment sector, Journal of Environmental Chemical Engineering 11 (2023) 109212 https://doi.org/10.1016/j.jece.2022.109212

[31] Wu, J., Wang, X., Wang, Q., Lou, Z., Li, S., Zhu, Y., Qin, L., & Wei, H. Nanomaterials with enzyme-like characteristics (nanozymes): Next-generation artificial enzymes (II). Chem. Soc. Rev, 48 (2019) 1004-1076. https://doi.org/10.1039/C8CS00457A

Applications of Emerging Nanomaterials and Nanotechnology   Materials Research Forum LLC
Materials Research Foundations 148 (2023) 27-62   https://doi.org/10.21741/9781644902554-2

Chapter 2

Fabrication of Nanomaterials via Ionic Liquids for Optoelectronics

Md. Arif Faisal and Md. Abu Bin Hasan Susan[*]

Department of Chemistry, University of Dhaka, Dhaka 1000, Bangladesh

*susan@du.ac.bd

Abstract

Nanomaterials can be fabricated with fascinating morphologies to develop smart optoelectronic devices based on nanotechnology. One of the best ways to fabricate materials of this kind is to follow the template-based strategy where the ionic liquids (ILs) offer great potential as the shape-controlling agent. Due to their unique physicochemical characteristics, ILs can act as a better template overpowering the conventional templates to synthesize nanomaterials with desirable morphology and characteristics. The exquisite morphologies of metals or metal oxides can be achieved using ILs, which may further advance the performance of optoelectronics. In this chapter, we have discussed optoelectronics, ILs and their classifications, controlled morphology of smart materials for optoelectronics, and fabrication of nanomaterials via ILs.

Keywords

Ionic Liquid, Morphology, Optoelectronics, Synthesis, Nanofabrication, Metal Oxide, Template

Contents

1.   Introduction

Nanomaterials such as metal and metal oxide nanoparticles (NPs) have been very effective in the field of electronics. Researchers all over the world are trying to enhance the efficacy of electronics by utilizing nanotechnology. The fabrication of the nanomaterials is crucial to attaining the best efficiency through nanotechnology [1-3]. Especially, in the case of optoelectronics, the structure of the NPs is important to advance the optoelectronic properties of the nanomaterials and hence the efficiency of the devices [4]. The structure-property relationship deeply depends on the architecture of the nanostructure of the material. Controlling the size and shape of NPs is crucial for optimizing the performance of new smart devices made possible by nanotechnology.

The challenge is to achieve control over the morphology of the nanomaterials. The bottom-up strategy is the best one to obtain NPs with fewer defects [5] and according to this, template-based synthesis of NPs has been vital for acquiring control over morphology [6,7]. There have been several types of templates used to produce NPs such as polymeric templates, molecular solvents, etc. [8-10]. All these are mainly distinguished into two parts, e.g. hard and soft templates [11]. Often the hard templates are difficult to remove from the product nanomaterials and they cause rupture on the surface of the NPs [12]. These consequences can highly degrade the performance of the nanostructures since the morphology is hampered. On the contrary, soft templates are greatly desired for nanomaterial synthesis because they eradicate all the difficulties caused by hard templates and introduce better pathways to tune the morphology of nanomaterials.

Ionic liquids (ILs) have recently gained significant attraction from researchers to synthesize NPs with controlled size and shape. Due to their unique properties, ILs could be utilized as soft templates. ILs can be superior to other soft templates used to synthesize metal and/or metal oxide NPs [13]. They exhibit supramolecular interactions in the aqueous system so that they can form a highly organized state acting like a self-directing agent for the NPs to grow in a certain direction [12-14]. Thus, ILs have been effective as a templating agent for the controlled growth of NPs to achieve the distinctive morphology of nanomaterials.

Metal oxides such as ZnO, TiO_2, CuO, transition metal oxides, etc. have been successfully synthesized by ILs with various methods e.g. chemical precipitation method, hydrothermal method, sol-gel method, etc. [15]. Such metal oxide NPs are better for the enhancement of semiconducting properties and hence the optoelectronic devices might advance with their performance. Due to the combination of high carrier mobility, large-area electrical uniformity, good optical transparency, mechanical flexibility, and straightforward synthetic routes metal oxides clearly offer advantages for optoelectronic applications compared to conventional materials [4]. These characteristics of the metal oxide nanomaterials are greatly influenced by the morphology of the NPs. Therefore, controlling the size and shape of the nanomaterials would be the key to advancing the performance of optoelectronic devices.

In this chapter, we reviewed metal oxide nanomaterials that enhance optoelectronics. Understanding the impact of morphology on the functioning of these electronic devices has been the focus. In this regard, the role of ILs as the shape-controlling agent for the synthesis of metal oxide NPs has been critically investigated. Finally, morphology-dependent characteristics of nanomaterials have been analyzed for their application to the development of optoelectronics.

2. Optoelectronics

The study and use of electronic systems and devices that can serve the purpose of location detection, and regulation of light are known as optoelectronics (also termed optronics). It is typically regarded as a branch of photonics. Light frequently includes not only visible light but also radiations such as ultraviolet, infrared, x-rays, and gamma rays. Examples of such devices include light-emitting diodes (LEDs), laser diodes, and superluminescent diodes, which convert electrical energy into light; photodetectors, which convert optical signals into electrical currents; optoisolators, which transmit analog or digital signals while maintaining isolation; imaging detectors, which are based on electronic image sensors; electro-optic modulators, which are used to control the power, phase, or polarization of light; and optical fiber communication; solar cells which convert light into electrical energy.

Optoelectronic devices are instruments that use optical-to-electrical or electrical-to-optical transducers in their operation [16]. The quantum-mechanical interactions of radiation with electronic materials, notably semiconductors, sometimes under an electric field, are the basis of optoelectronics [17]. Semiconducting materials have appropriate bandgap energies for absorbing light, such as visible and near-infrared light. These applications also require their electric conductivity, even if it isn't ideal. Aside from some photodetectors that make use of the external photoelectric effect, it would be difficult to use dielectrics in either case, whereas metals are primarily used as conductors. Therefore, in order to improve the performance of the devices, it may be essential to manipulate the bandgap of the semiconductors. Moreover, metal oxide semiconductors exhibit fascinating properties due to their simple fabrication pathways and tunable morphology at the nanoscale.

3.   Controlled morphology for optoelectronics

Since semiconductor materials are the basis of optoelectronics, it is certain that the synthetic route of such semiconductors is very important to ensure the perfection of optoelectronics. The specific synthetic procedure causes the morphology of the material to vary significantly. Moreover, the size of the materials also depends on the method of synthesis. Both the shape and size of the nanomaterials are crucial variables to tune the bandgap. Thus, controlling the morphology is the tool to develop fascinating characteristics for the advancement of smart devices.

The electronic and electrical properties resulting from quantum confinement effects with the reduction in the size of the materials are responsible for the enhanced semiconducting properties. NPs show better semiconducting properties than their bulk phase. So, it is desired that the particle size remains in the nanoscale. The size quantization (Q-size) effect manifests when the confinement of charge carriers causes the size of the semiconductor particle to decrease from its bulk to that of the Bohr radius, for example, in the first excitation state. As a result, electrons and holes in the quantum-sized semiconductor are contained in a potential well and do not experience the delocalization that occurs in the bulk phase. In light of this, as a semiconductor particle becomes ultra-fine with particle size below the band gap minimum, the bandgap widens and the effect becomes more prominent with a decrease in the size of particles. Numerous uses are envisaged for the particular size- and shape-dependent optoelectronic characteristics of nanostructured semiconductors.

The direct bandgap of ZnO colloids can be estimated from a plot of hv vs $(\alpha hv)^2$ where α is the absorption coefficient and is related to the bandgap, E_g by $(\alpha hv)^2 = k(hv - E_g)$, where hv refers to the energy of the incident light and k is a constant [18]. The E_g is produced by extrapolating the linear portion until it intersects the hv axis. The E_g is size-dependent, and as shown in Fig. 1, the bandgap of the semiconductor increases as the particle size decreases [19].

Applications of Emerging Nanomaterials and Nanotechnology   Materials Research Forum LLC
Materials Research Foundations 148 (2023) 27-62   https://doi.org/10.21741/9781644902554-2

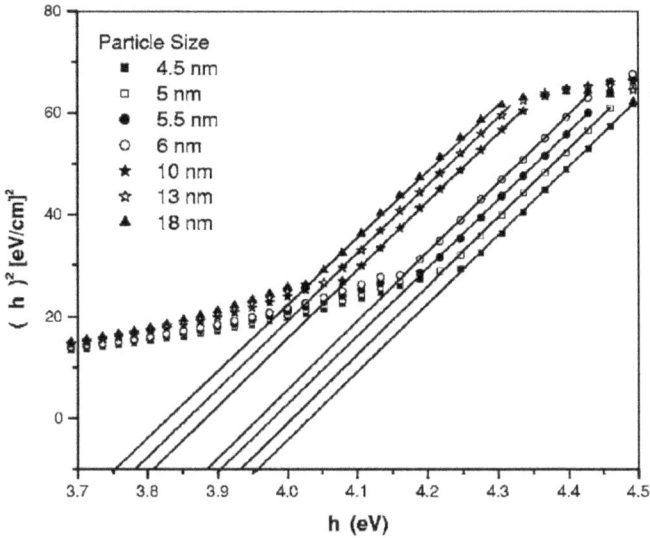

Figure 1. Size-dependent variation of the bandgap of ZnO NPs. Reproduced with permission [19].

Clearly, the E_g increased with a decrease in the size of the NPs to confirm that the size is indeed a key factor to exploit characteristics of semiconductor nanomaterials in optoelectronics. This phenomenon is due to quantum size effects [20, 21]. Every particle is only made up of a very small number of atoms or molecules as the particle size decreases, which also results in a decrease in the number of orbitals or energy levels that overlap. As a result, the energy gap between the conduction and valence bands will widen. This explains why the energy gap in NPs is greater than in the corresponding bulk material. The bandgap is the region where electrons are not allowed to exist. The larger this forbidden region, the greater the restriction on electron movement. So, compared to the bulk, NPs have lower electrical conductivity. As a result, the absorption spectrum is moving toward the blue or lower wavelength region. Therefore, ZnO exhibits low optical bandgap energy as particle size increases.

Similarly, the shape of the nanostructures is also significant in controlling the bandgap as well as other important features of the nanomaterials. For instance, in the case of CuO NPs with a tubular shape, they exhibit an increase in vibration absorption due to the decreased compactness of the tubular nanostructure [22]. Hence, the shape of the particles indirectly influences aggregation behavior and consequently affects the bandgap. The photocatalytic performance is strongly dependent on the size, morphology, and surface area of metal oxide

NPs. Although small particle size and large surface area are advantageous for increasing photocatalytic activity, they also make separation difficult and recycling less efficient, especially in large-scale industrial applications. Hollow structure materials, on the other hand, may be able to resolve the issue due to their unique characteristics, such as low density, good permeation, large surface area, and distinct optical properties. For this purpose, attempts have been made to produce hollow-shaped metal oxide NPs. Surfactant-assisted hydrothermal processes, etching ZnO rod arrays in aqueous solution, and thermal evaporation methods were used to generate three-dimensional center-hollow ZnO architectures with multigonal star-shapes [23], vertically aligned ZnO nanotubes [24], and hollow ZnO microspheres [25]. Furthermore, hollow TiO_2, ZrO_2 micro-shells, and hollow SnO_2 have been synthesized by polymeric templates [26, 27], carbon templates [28], and solid-state salt templates [29].

In fact, metal oxide semiconductors have exponentially gained popularity in optoelectronics due to their interesting morphological variation. Up to now, most of the suitable methods proposed for tunable morphology are focused on template-based synthesis [13,15]. The problem is that in most cases, the typical templates are toxic to the environment, hard to be removed from the product, and non-recyclable. Therefore, the necessity of ILs as green, recyclable, soft template come to the scenario for nanomaterial fabrication, which can provide the optimum conditions to control the morphology of metal oxide NPs.

4.   Template-based approach for fabrication of nanomaterials

Precursors in the gaseous, solid, and liquid phases are considered in NP and nanomaterial assembly and synthesis techniques. The building blocks of nanostructures are incorporated into the final material structure using either chemical reactivity or physical compaction [5]. Synthesis of nanomaterials relies mainly on two approaches:

- The "top-down" approach, in which larger particles are reduced to nanoscale size while retaining their original properties. It starts with a suitable starting material and "sculpts" the functionality out of it. Mechanical attrition, etching, lithography, etc. are the ways to sculpt nanostructures from bulk material.

- The "bottom-up" approach, in which materials are engineered by assembling or self-assembling atoms or molecular components. In this case, first, the formation of nanostructured building blocks occurs and then these blocks are assembled to produce the desired nanomaterial. Powder/aerosol compaction and chemical synthesis are the methods to carry out the bottom-up strategy for NP production.

Both of these strategies are useful, however, the surface of the NPs or the morphology gets hampered in the case of a top-down approach. The bottom-up strategy promises an improved possibility of creating nanostructures with fewer flaws, better compositional uniformity, and superior ordering both in the short- and long-range. This is so that the nanomaterials created using this method remain in a condition that is closer to the thermodynamic equilibrium state since the lowering of Gibbs free energy is the primary driving force for the bottom-up approach [5]. On the other hand, a top-down approach is likely to add contaminations, internal stress, and surface defects. Therefore, the bottom-up strategy is favored over the top-down strategy for the synthesis of metal oxide nanomaterials. Additional subcategories of the bottom-up strategy include solution-, vapor-, and template-based methods [30]. When compared to vapor deposition and solution-based methods, the bottom-up strategy for controlled synthesis of nanostructures is advancing with the template-assisted synthesis method. It is a practical method to control the structure and morphology of NPs for specific applications [31].

There are several methods for producing metal oxide NPs, but not all of them are suitable for controlling the size and shape of nanostructures. Two important techniques for the fabrication of various metal oxide NPs are spontaneous growth and template-based synthesis. The process of fabricating without a template might be self-catalyzed or employing metal catalysts. A benefit of using metal catalysts, such as Au, is that they can promote aligned and selective area growth [32]. Without using a metal catalyst, aligned nano-rods may also be produced hydrothermally [33]. However, the spontaneous growth technique accompanies some drawbacks such as the aggregation of particles cannot be prohibited; uncontrolled growth may occur; uniformity of size and shape of the NPs may not be achieved; and overall the morphology is difficult to control. In contrast, the template-based synthesis offers superior control over the size and shape of the NPs. The main benefit is the synthesis of highly ordered mesoporous nanostructures [34]. It employs a template that serves as a scaffold for directing the growth of nanostructures to achieve various structures and morphologies [35].

Two major templates, hard and soft have so far been reported [36]. Mostly polymeric templates or molecular solvents have been utilized so far for the production of nanomaterial. Often the polymeric materials are found to be difficult to be removed from the surface of the synthesized NPs. It also causes ruptures in the shape of the nanostructures which damages the target NPs. Thus, IL was introduced for the synthesis of ZnO NPs as soft templates. Recently, the potential and success of IL medium in the preparation of interesting morphologies of NPs have been demonstrated. Before dipping into ILs and their advantages of being a template, we should understand what types of templates are there and why should we choose IL templates for the synthesis of morphology-controlled NPs.

4.1 Types of template-based synthesis

The main component of template-assisted synthesis is the "template," which can be any entity with nanostructured properties. Size, morphology, and charge dissemination all have an impact on the properties that guide structure [37]. Physical techniques like surface coating and chemical techniques like elimination, addition, isomerization, or substitution reactions can be used to fabricate the desired material. Once the reaction is complete, the template can be removed using either a chemical method like calcination or a physical method like dissolution [36]. The main advantage of the method is that it allows for better control over the dimension, morphology, and structure of the final product. There are three types of template-assisted synthesis: (I) hard template methods; (II) soft template methods; and (III) colloidal template methods.

(I) Hard template

The method is often called "nano casting". A solid mold is adopted in this method which is well-organized with definitive pores and these pores are then filled with precursor molecules to produce the final product. After the completion of the reaction, the mold can be removed retaining the desired properties of the final product [38]. A variety of hard templates, including in situ formed templates and macroscopic structures, can be used during the synthesis. In situ, hard templates are those that are created when a compound is physically or chemically altered while still in its precursors. Salt, carbon, and ice crystals are a few typical examples [39, 40]. The components of macroscopic structures can be films, powders, or fibers. Anodized aluminum oxide [41], tissues and biomolecules of biological origins [42], silica [43], and polymeric microspheres [44] are common examples. A careful choice of template is essential for the preparation of desired products since the structural and morphological characteristics of the product heavily depend on the attributes and quality of the template and the means of interaction with the precursors [45]. Although synthesis assisted by a hard template is a popular method for producing crystalline oxides, it has some drawbacks that make it difficult for effective use. These include things like their relatively less controllable pore size, lower yield due to the uncontrollable nucleation taking place outside the pore, and their time-consuming, expensive processes [46].

(II) Soft template

The utilization of non-rigid flexible nanostructures as templates for the purpose of getting NPs, is known as the soft template method. These flexible soft materials exhibit intermolecular interactions to infer the templating property. Such soft templates could be flexible organic polymers, surfactants, and block copolymers [47]. These templates interact with the precursors via supramolecular interactions e.g. hydrogen bonding, van der Waals, or electrostatic interactions [48]. Due to these weak yet significant interactions with the precursors, soft templates are easily removable from the end product. Thus, compared to hard templates, soft templates are far better for use in the synthesis of metal and metal oxide NPs. One of the most important advantages of soft templates is the ease with which they can be removed from the product surface, resulting in the safety and enhanced stability of the morphology of the nanostructures. Also, these templates offer better control over the size, morphology, and overall structure of the NPs compared to hard templates.

The incorporation of the precursor molecule into the soft template can bring about significant changes in the self-assembly process, resulting in the formation of a micelle or liquid crystal for the sol-gel process [49]. The crucial step in soft template methods is considered to be the control of the sol-gel conversion of the precursors during self-assembly. The first step is the simultaneous assembly of precursors and templates. The process, occurring in many reactions, is known as "cooperative assembly" and involves the formation of small groups of precursors and soft matter in the solution that are held together by the supramolecular interactions and then combine to form larger structures [50]. In the end, this causes the precursors to form a liquid-crystalline phase that eventually precipitates out of the solution. The desired product can be obtained by carefully controlling this step of cooperative assembly. In another pathway, the soft template undergoes self-assembly in the reaction media and becomes crucial for the synthesis of nanomaterials. In this case, initially, the soft template creates a highly ordered structure in the solution, and then the incorporation of precursor along with reactant cause the nucleation and growth of the desired product within a certain direction. Therefore, a certain size or morphology could be achieved via the soft template method. There are several ways to remove the soft templates, including calcination, combustion, depolymerization through thermal treatment, or low-temperature chemical treatment [36].

The method based on soft templates is preferred over the hard template one for the synthesis of nanomaterials with fewer defects. Template removal is easier in the case of the soft template method and this is comparatively cost-effective. However, the recovery, reusability, and low thermal stability of the soft templates remain a challenge. Here comes the exquisite role of ILs as a template.

(III) Colloidal template

The colloidal template method was developed in response to the need to improve the thermal stability of polymeric soft templates while limiting crystal growth. In this method, an inorganic core material is incorporated into the polymer-based template to prepare the colloidal template [51]. Here, the core inorganic part acts as a hard template, while the polymer tails act as a soft template. So, basically, the colloidal template method is an attempt to achieve the synergistic features of the methods based on hard and soft templates. The thermal stability of the template is enhanced by the inorganic core while the polymeric tails serve to control the growth of the desired nanomaterials. Colloidal templates have been used to create ordered nanostructures with exceptional structural integrity and crystallinity and controlled pores that can withstand higher calcination temperatures [52]. Many transition metal oxide NPs and noble metal hybrids have been produced using this method for a variety of applications [53, 54]. However, the recyclability and reusability of the template still remain an issue. The synthesis of next-generation nanomaterials is in the need of a suitable method that would be inexpensive, simple, and effective. Thus, modification of these conventional methods is ongoing and ILs are a promising candidate in this regard, which possess the advantageous characteristics of these template-based approaches for the synthesis of nanomaterials.

Fig.2 schematically represents all of the above-mentioned template methods.

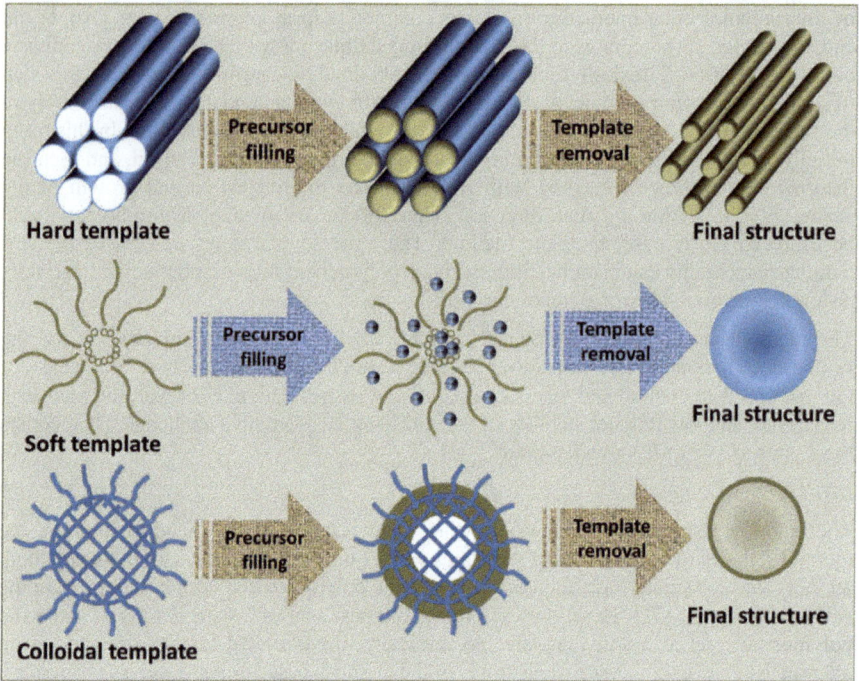

Figure 2. Different templates used for the synthesis of nanomaterials [13]. (Reprinted from [13]).

From the viewpoint of the limitations of these template-based methods, it should be realized that the main challenges are: enhancing the stability of the template; removing the template after the synthesis with easier processes; recyclability, and reusability of the template. Therefore, the standard template will be something that has blending properties of the three types of templates mentioned above. ILs could be a prominent solution to the difficulties observed till now. Hence, the exceptional properties of IL would be discussed critically since they make them very attractive for use as a template to synthesize nanostructured materials of metals and metal oxides.

5. Ionic liquids

ILs are salts that only contain ions and have weak inter-ionic interactions, allowing them to be liquid at ambient temperature. The melting temperatures of salts of this kind are below 100 °C, but the values can go down far below room temperature. These salts are usually composed of organic cations and inorganic anions and their properties are heavily influenced by the constituent cation and anion species. The structure of different ILs are shown in Scheme 1. Chemical and thermal stability, high ionic conductivity, negligible volatility, low melting temperature, inflammability, moderate viscosity, high polarity, and high solubilization capacity with many compounds make them appealing for a wide range of applications [55-61].

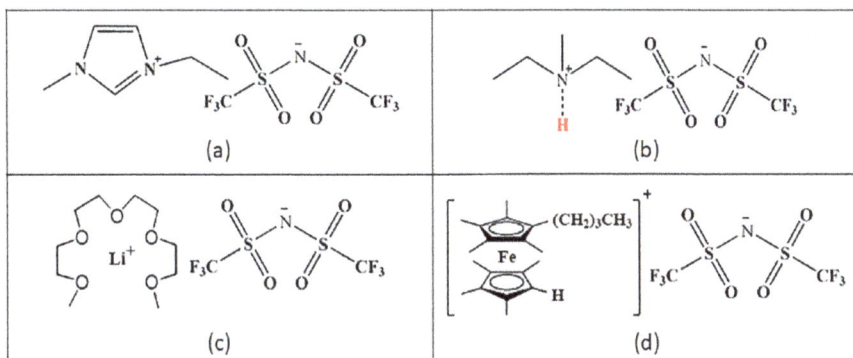

Scheme 1. Structure of different ILs: (a) [C₂mim](CF₃SO₂)₂N, (b) [dema](CF₃SO₂)₂N, (c) [Li(G4)](CF₃SO₂)₂N, and (d) [C₄Fc](CF₃SO₂)₂N. Reproduced with permission [61]. Copyright 2022, Elsevier.

The ILs typically comprise combinations of organic cations, such as ammonium, imidazolium, pyrrolidinium, pyridinium, phosphonium, and sulfonium derivatives, and bulky and soft anions, such as $CF_3SO_3^-$, PF_6^-, BF_4^-, and $(CF_3SO_2)_2N^-$ [62]. ILs are referred to as "designer solvents" because their physicochemical properties can be easily modified by changing the structure of the component anion [63] or cation [64]. Even the change in the alkyl chain length of the imidazolium cation can be used to tune the physicochemical properties of the ILs systematically [65]. Hence, the careful design of the IL structures for a specific task is very important to acquire the best possible performance. Consequently, a series of ILs with a wide variation in structures have so far been developed and they are now classified into different categories.

5.1 Classification of ILs

Based on proton activity, ILs may be both protic or aprotic. As a whole, ILs are of four categories: aprotic, protic, inorganic, and solvate [66]. Regarding the hydrophilic nature, ILs are of two kinds: hydrophilic and hydrophobic ILs.

5.1.1 Aprotic ILs

ILs that do not contain any dissociable hydrogen, are known as aprotic ILs. They are typical combinations of cations, such as ammonium, imidazolium, pyrrolidinium, pyridinium, piperidinium, and phosphonium derivatives, and bulky and soft anions, such as $CF_3SO_3^-$, PF_6^-, BF_4^-, and $(CF_3SO_2)_2N^-$. The synthesis of aprotic ILs is induced by the transfer of any group with a complexity higher than the proton. The structure of a typical aprotic IL is shown in Scheme 1(a). Watanabe et al. presented the fascinating tunable physicochemical characteristics of aprotic ILs via extensive research on the design and manipulation of the component ions of the ILs and laid the groundwork for their exquisite application in various fields [62-65].

5.1.2 Protic ILs

A protic IL is a salt formed by the proton transfer reaction between a Brønsted acid and a Brønsted base. Protic ILs are simpler to produce because the acid and base only need to be combined, in contrast to other types which require a series of steps for synthesis. The first protic IL, hydroxyethylammonium nitrate, with a melting temperature (T_m) of 52–55 °C, was reported by Gabriel in 1888 [67]. Since then, the development of research about protic ILs has increased exponentially. Till now reported protic ILs are mostly based on ammonium [68] or phosphonium [69] cations with anions similar to their aprotic analogs. Protic and aprotic ILs differ significantly from one another in that protic ILs show proton activities that result from the addition of H^+, whereas their R^+ counterparts are considered to be aprotic ILs. Since the proton transfer reaction is reversible, the equilibrium between the reactants and products can shift depending on the conditions. The presence of neutral acid and base species in the solution affects the properties of protic ILs significantly. The ΔpK_a value of the constituent acid and base has been a governing factor for protic ILs and their physicochemical properties [70]. Protic ILs exhibit superior performances in the fields of fuel cells [71], proton conductors [72], etc. due to their suitable proton transfer characteristics. The structure of a protic IL, [dema]$(CF_3SO_2)_2N$ i.e., diethylmethylammonium bis(trifluoromethylsulfonyl) imide, is shown in Scheme 1(b).

5.1.3 Inorganic ILs

By utilizing the same packing issues that result in low-melting ILs of the organic cation type, inorganic ILs can be obtained in both aprotic and protic forms. Aprotic examples include silver trifluoromethanesulfonate ($AgCF_3SO_3$) and protic examples include hydrazinium nitrate. Scheme 1(d) depicts the structure of a ferrocene-based inorganic IL, $[C_4Fc](CF_3SO_2)_2N$.

5.1.4 Solvate ILs

Solvate ILs are a relatively new class of ILs made up of a coordinating solvent and salt that results in a chelate complex with IL-like properties [73]. The dissolution of lithium bis(trifluoromethanesulfonyl) imide (Li$(CF_3SO_2)_2N$) in tetraethylene glycol dimethyl ether to yield [Li(G4)]$(CF_3SO_2)_2N$, as shown in Scheme 1(c), is the most well-known example. By sequestering a hard cation in an ethereal solvent, a cationic charge is diffused in most solvate ILs [74]. Molten salt hydrates, such as Ca$(NO_3)_2$.4H_2O, for instance, were found to be almost ideal mixtures but with a short lifetime. New cases have recently been reported in which the ligating groups all belong to the same molecule, ensuring a long lifetime. The $[Li(glyme)]^+$ complex cation has been similar to common anions of ILs such as PF_6^-, BF_4^-, and $AlCl_4^-$. Such ions are the adducts of a Lewis acid (Li^+ for the [Li(glyme)] cation and PF_5, BF_3, and $AlCl_3$ for the common anions) and Lewis base (glyme for the [Li(glyme)] cation and Cl^- and F^- for the common anions) [75].

5.1.5 Hydrophilic vs hydrophobic ILs

ILs are divided into two groups: hydrophilic and hydrophobic based on their affinity to water. The affinity of ILs with water depends on their structures. More specifically, the size and structure of the anion and/or cation of the IL determine its hydrophilicity. For instance, $[C_2mim](CF_3SO_2)_2N$ is hydrophobic but $[C_2mim]CH_3CO_2$ is hydrophilic; the greater the size of the anion makes $[C_2mim](CF_3SO_2)_2N$ hydrophobic. Chloride and iodide, two hydrophilic anions, produce ILs that are miscible with water in any proportion. The hydrophobicity increases with increasing alkyl chain length for a series of 1-alkyl-3-methylimidazolium cations, from butyl to hexyl to octyl [76].

6. Ionic liquid as a template for the synthesis of nanomaterials

ILs are undoubtedly a very promising material for the synthesis of metal oxide NPs. They exhibit some vital characteristics which make them very useful as a template. Among the three types of templates, IL possesses the best properties of all while having the solutions to the difficulties observed in these templates. ILs can serve as: (a) reaction media, (b) 'all in one' solvent, and (c) templates.

Firstly, ILs can be an excellent reaction medium. In order to produce nanostructures, a variety of molecular solvents are utilized, with organic solvents being the most popular. But organic solvents are more often associated with pollution issues due to high vapor pressure and there has been a surge of interest in environmentally benign solvents to minimize environmental pollution. ILs have the potential to be the green solvent to replace organic solvents [77]. ILs more often offer a great number of solvent properties even better than conventional organic solvents. ILs have been successfully employed as the reaction media to prepare inorganic nanomaterials via IL-assisted hydrothermal and microwave methods. Polymerized ILs have also been used as reaction media [78] by combining advantageous properties of polymers and ILs [79].

Secondly, ILs can be utilized as the 'all in one' solvent. IL can work as a multipurpose solvent in addition to being a solvent or template. It is possible to properly adjust the anions and cations of the ILs to enable dual functionality as a capping agent for the nanostructures as well as a reaction medium. Highly organized ILs function as a tailoring agent for nanocrystal formation while providing a stable medium for the effective growth of NPs. Long alkyl side chains in the cations of some ILs can separate into nonpolar domains while forming polar domains in other areas of the IL [80]. ILs are capable of facilitating the dissolution of a wide range of substances due to their highly hydrogen-bonded networks and the nanoscale domains for which the solutes have a greater affinity [81,82].

Thirdly and most importantly, as a whole, ILs can be suitable templates for nanomaterial synthesis. The unique properties of ILs make them exceptional for use as a template. Since ILs are entirely composed of ions and yet remain liquid, they are different from conventional liquids. Their ions typically have delocalized electrostatic charges and are asymmetric [83]. ILs have more complex molecular forces and interactions than traditional salts do. Strong Coulombic interactions combined with weak directional interactions facilitate the development of nano-scale structures in ILs and IL/molecular solvent or IL/solute mixtures [84, 85]. Due to these supramolecular interactions, the ILs can form a highly ordered structure in the reaction medium so that the growth of the nanostructures can be directed in a certain direction [12, 14]. It is well known that the rate of nucleation and growth of the product affect the particle size. IL can play two different roles in this process: (i) it can prevent particle agglomeration, which means the reaction rate should be a little slower but still fast enough for the molecules to self-assemble; and (ii) it can promote preferential growth of the metal oxide crystal in a particular direction, serving as a soft template [86]. Since the growth rate can be controlled by ILs, a certain shape is obtainable by using ILs as a template. There have been a number of methods reported for the synthesis of metal oxides such as ZnO, CuO, SnO_2, TiO_2, etc. using ILs as templates. For the successful synthesis of metal oxide nanostructures, the use of ILs as a self-directing agent and templating material has so far been demonstrated to be effective.

7. Fabrication of nanomaterials via ionic liquids

By using a one-step low-temperature method, Wang et al. produced a series of ZnO nanostructures with controllable shapes in ILs [87]. Systematic research was performed to determine the influence of varying cations in the ILs on the mechanism of formation of ZnO nanostructures and consequently their shapes. Three different ILs with varying cations were used: 1-ethyl-3-methylimidazolium tetrafluoroborate, $[C_2mim]BF_4$; 1-butyl-3-methylimidazolium tetrafluoroborate, $[C_4mim]BF_4$; 1-butyl-2,3-dimethylimidazolium tetrafluoroborate, $[C_4dimim]BF_4$. The formation of hydrogen bonds between the hydrogen atom of the imidazole ring at position 2 and the oxygen atoms of O-Zn crystal cores may serve as a useful link between the generated ZnO nuclei and cations of ILs, which is essential for the directional growth of 1D nanocrystals. Furthermore, due to the steric hindrance effect, a longer alkyl chain at position 1 of the imidazole ring of the IL will prevent the 1D ZnO nanostructures from growing longer. Scheme 2 describes the role of ILs as the shape-controlling agent.

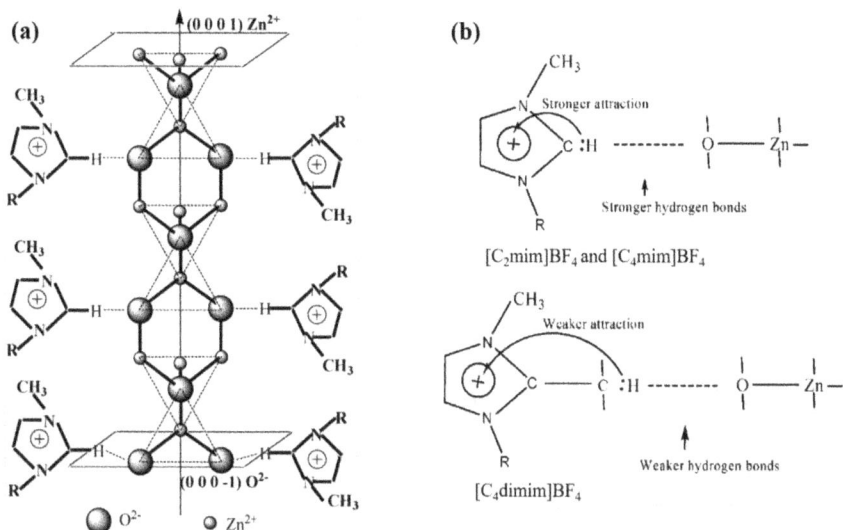

Scheme 2. (a) Directional growth of 1D ZnO nanostructures in IL; (b) Comparison of hydrogen bonds in ILs [C₂mim]BF₄, [C₄mim]BF₄, and [C₄dimim]BF₄. Reproduced with permission [87]. Copyright 2008, American Chemical Society.

Scheme 2 shows the hydrogen bonds between ZnO crystal and ILs. This is how the IL surrounds or directs the growth of the ZnO clusters in a particular direction and acts as a shape-controlling agent. When no IL was used in the synthesis, irregular shapes of ZnO particles were obtained. This shows that IL in the reaction system is crucial to the preferential growth of ZnO nanostructures. [C₂mim]BF₄ and [C₄mim]BF₄ provided nanorods and nanowires while [C₄dimim]BF₄ produced NPs with totally different morphology. The wurtzite-structured ZnO crystal is structurally defined as having a number of alternating planes made up of O^{2-} and Zn^{2+} ions that are alternately stacked along the c axis in 4-fold tetrahedrally coordinated arrangements. Zn^{2+}-terminated (0001) top and O^{2-}-terminated (0001) bottom surfaces come from the zinc and oxygen atoms being positioned alternately along the c-axis. ZnO also possesses additional typical side surfaces, such as O^{2-}-terminated (10Ī1) and (10Ī0) planes, in addition to these surfaces. According to Scheme 2(a), the cations of the ILs can be readily adsorbed on the surface of the O^{2-}-terminated facet by electrostatic force, and the hydrogen bond formed between the oxygen atoms of O-Zn and the hydrogen atom at position 2 of the imidazole ring may act as an effective bridge to connect the O^{2-}-terminated plane of the produced nuclei of metal oxide and the cations of the ILs. As a consequence, the Zn^{2+}-terminated (0001) facet, which has the highest initial growth rates, experiences preferential growth.

[C$_2$mim]BF$_4$, [C$_4$mim]BF$_4$, and ZnO crystal cores form strong hydrogen bonds because the imidazole ring is an electron-withdrawing group and may pull the electron pair shared by hydrogen and carbon in position 2 of the imidazole ring (Scheme 2(b)). However, the effects of hydrogen bonding are altered when [C$_4$mim]BF$_4$ is replaced by [C$_4$dimim]BF$_4$, which substitutes a methyl group for the hydrogen atoms at position 2 of the ring structure of imidazole. The hydrogen bonding between [C$_4$dimim] cations and ZnO crystalline cores become rather loose as the attraction of the imidazole ring to electron pairs weakens, allowing ZnO crystal cores to grow fairly freely. Hence, only ZnO NPs without any definitive morphology were produced.

Another interesting finding was established by Qi et al. where they produced hollow ZnO ring- and tube-like shapes by using 1-propyl-3-methylimidazolium bromide, [C$_3$mim]Br [12]. A simple hydrothermal route was followed to carry out the synthesis of a series of ZnO NPs prepared in varying concentrations of [C$_3$mim]Br. The hexagonal disk-like structure (obtained without IL) of ZnO transformed into hollow rings or tube-shaped NP due to the selective adsorption of [C$_3$mim]Br on certain facets of ZnO disk and promotion of [0001] facet. Scheme 3 represents the mechanism of the formation of different shapes and the effect of the concentration of the IL on the shapes.

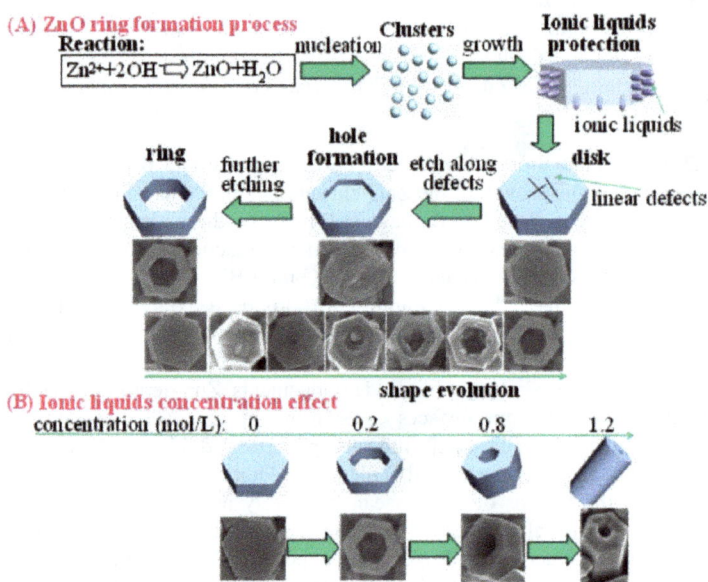

Scheme 3. (A) Process of formation of the ZnO rings and (B) effect of concentration of the [C$_3$mim]Br component and corresponding SEM images [12]. Reproduced with permission [12].

The typical reaction begins the nucleation process and eventually, cluster formation occurs. Then these clusters of ZnO start to grow. Initially, in the absence of IL in the reaction medium, the free growth of the clusters produced a disk shape. Introduction of [C$_3$mim]Br in the reaction medium caused this disk shape to transform as the IL protected the lateral (1010) facets via adsorption by hydrogen bonding, п-п stacking interaction, van der Waals interaction, etc. and promoted the growth along [0001] direction. First-principles calculations show how [C$_3$mim]Br affected the growth of ZnO crystals. The key points are: (i) [C$_3$mim]Br acted as a stabilizing agent and shows preferential adsorption on the ZnO (1010) facet, which significantly lowered the surface energy; (ii) the mutual interionic interactions (including [C$_3$mim]$^+$–[C$_3$mim]$^+$, [C$_3$mim]$^+$–Br$_2$ and Br$_2$–Br$_2$) made [C$_3$mim]Br array on the (1010) facet orderly to serve as a soft template. As a result, [C$_3$mim]Br protected the lateral (1010) facets of ZnO crystals and assisted in the formation of ring-like structures, and promoted growth along the [0001] direction to form tubes. On the other hand, the hydrothermal route used in this synthesis produced H$^+$ ion that initiated the etching along the surface defects of the disks. The [C$_3$mim]Br species, however, created a comparatively integrated layer for adsorption on the lateral (1010) facets, protecting them and reducing the rate of etching. In comparison to the (0001) facet with the higher surface energy, the (1010) facet is more stable. The etching rate of the metastable (0001) facets is quicker than that of the (1010) facets because the system aims to reduce total surface energy. As a result, (0001) facets of the as-grown disks were removed preferentially. The hollow ring-like structure in the final products was a result of selective etching of the ZnO crystals. [C$_3$mim]Br controlled the shape and size of the ZnO NPs and after synthesis, it could be easily removed from the product surface by washing with ethanol and distilled water. Overall, [C$_3$mim]Br worked perfectly like a soft template.

Akter et al. used a hydrophilic IL, 1-ethyl-3-methylimidazolium methylsulfate, [C$_2$mim]CH$_3$SO$_4$ to investigate the effect of IL concentration on the size and morphology of the ZnO NPs [14]. A simple chemical precipitation method was followed by which the size of the NPs acquired was in the range of 2-55 nm. The ZnO NPs varied in shape from spheres, capsules, flakes to rods depending on the concentrations of the IL. The IL may dissociate into constituent ions when [C$_2$mim]CH$_3$SO$_4$ is added to the reaction medium in low concentrations. Because of its ability to withdraw electrons by sharing an electron pair with hydrogen and carbon at position 2 of the imidazole ring, the [C$_2$mim]$^+$ may interact with the bulk (Scheme 4(I)). As the ZnO crystals form, [C$_2$mim]CH$_3$SO$_4$ is adsorbed on their surface, somewhat limiting the growth of the forming material. As a result, the crystal grows in an anisotropic manner. The IL encourages the development of various nanostructures via the hydrogen bonding-co-π-π* stacking mechanism. Ion clusters were formed in the reaction media when IL was added there in high concentrations. Instead of dissolving into single ions, positively or negatively charged ion clusters surrounded the NP surface and formed an electrical double-layer, exerting an electrostatic force to keep the NPs apart. In this instance, the nucleation rate can be greater than the growth rate. Consequently, the resultant morphology was reported to be spherical.

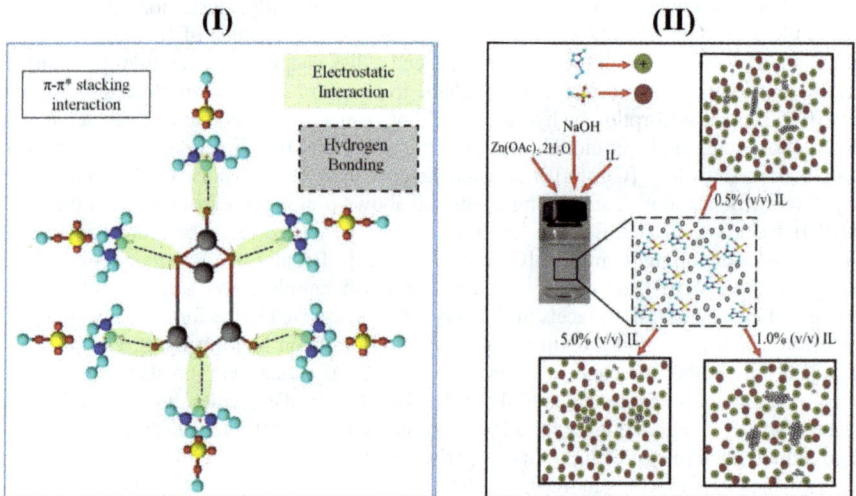

Scheme 4. (I) Interaction between [C₂mim]CH₃SO₄ and ZnO crystal and (II) difference in morphology due to the varying interaction [14]. Reproduced with permission [14].

The cationic species of the IL are electrostatically drawn to the surface of negatively charged NPs to form a layer of positive ion, followed by adherence of counter ions to the NPs to form a second layer. Therefore, the difference in how the IL interacts with the developing ZnO nuclei caused a variety of shapes, from capsules to spheres on the nano-scale. The mechanism is shown in Scheme 4(II).

Besides ZnO NPs, syntheses of many other metal oxides and metals involve the use of ILs as a template. Fig. 3 shows morphologies of different metal oxides and metals synthesized via different methods using ILs. Manjunath et al. synthesized spherical TiO_2 NPs (Fig. 3(a)) by a hydrothermal method where the IL, [C₄mim]Cl served as a template [88]. Sol-gel method was used for the production of sponge-like morphology of TiO_2 NPs (Fig. 3(b)) where [C₄mim]BF₄ acted like a solvent medium [89]. 1-hexadecane-3-methylimidazolium bromide, [C₁₆mim]Br-assisted hydrothermal synthesis of mesoporous ZrO_2–TiO_2 nanocomposite (Fig. 3(c)) was also reported [90]. The IL was used as a morphological template in this case. Therefore, the variation of the cation or anion can drastically change the morphology of the material. Also, not only a pure metal oxide but also a nanocomposite of two different metal oxides can be produced via the use of an IL system. Another interesting fact is that the mesoporous form of the nanocomposite was obtained due to the templating effect of the IL, which again proves the highly ordered structure possessed by the IL in the reaction medium. This organized structure acts like a hard template and so the pores of the final product are found to be precise. However, the soft-templating property of the ILs allows them to be easily removed after the synthesis to ensure the mesoporosity of the nanocomposite.

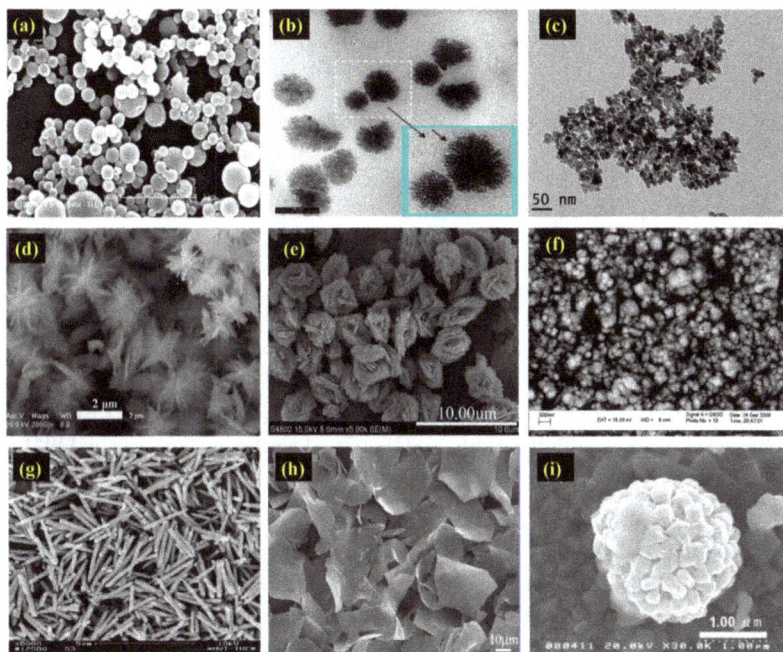

Figure 3. (a) SEM image of TiO$_2$ NPs (reprinted from [88]); (b) TEM image of sponge-like TiO$_2$, reproduced with permission from [89]. Copyright 2003, American Chemical Society; (c) TEM image of ZrO$_2$–TiO$_2$ nanocomposite, reproduced with permission [90]. Copyright 2013, Elsevier; (d) SEM image of flower-like CuO, reproduced with permission [91]. Copyright 2009, Elsevier; (e) SEM image of peach stone-like CuO architectures, reproduced with permission [92]. Copyright 2010, Elsevier; (f) SEM image of SnO$_2$ NPs, reproduced with permission [93]. Copyright 2009, Elsevier; (g) SEM image of Ag nanorods, reproduced with permission [94]. Copyright 2009, Elsevier; (h) SEM image of Au single-crystal NPs, reproduced with permission from [95]. Copyright 2005, American Chemical Society; (i) SEM image of Au NPs (reproduced with permission [96]).

Leaf- and flower-like CuO NPs (Fig. 3(d)) have been synthesized with the templating effect of [C$_8$mim]C$_2$F$_3$O$_2$ through the microwave heating method [91]. Highly crystalline products with thicknesses in the range of 50-60 nm were obtained. On the contrary, when a hydrothermal route was followed with $[C_8mim]C_2F_3O_2$, three-dimensional hierarchical peach stone-like (Fig. 3(e)) CuO architectures were obtained [92]. Therefore, the synthetic route along with the IL media are significant as shape-controlling agents. This is happening because the templating effect of the IL would work on the growth of the nanomaterials differently in different synthetic routes. The time or way of the addition of IL in the reaction medium could be different for various synthetic methods and hence variation is noted in controlling the nucleation and/or growth rate of the NPs. Sometimes this would be decisive for the role of the IL in the medium. That is whether it would act as a template or a solvent or a reaction medium. Although the templating effect plays a major role, other influences on the shape of the nanostructures cannot be overruled. So, it is ideal to fix the synthetic route along with a certain IL for the desired morphology of the nanomaterial for a specific application. Fig 3(f) represents SnO_2 NPs that have been produced by the sonochemical method and in the presence of $[C_2mim]C_2H_5SO_4$ [93]. A tetragonal rutile phase with a particle size of ca. 30 nm was synthesized where the IL acted as a solvent. Since the use of the sonochemical route would only allow the IL of the role as a solvent, it should be comprehended that the synthesis method and ILs work synergistically to produce distinctive morphologies of the nanomaterials.

Even metal NPs can also be synthesized via ILs used as a template. Fig. 3(g) depicts the Ag nanorods synthesized by $[C_2mim]CF_3SO_3$-assisted electrochemical method [94]. Here the IL played both roles of a stabilizer and a template for the metal nanorods. Moreover, By heating $HAuCl_4$ in $[C_4mim]BF_4$ using a microwave, large-size single-crystal gold nanosheets (Fig. 3(h)) have been successfully prepared without any additional templating agent [95]. In another report, $[C_4mim](CF_3SO_2)_2N$ has been used as a stabilizer and a template to synthesize Au NPs by electron beam irradiation technique [96]. These results show that the ILs could also take part to stabilize the final product as well as performing as the morphological template. Also, it shows the potential of ILs for the fabrication of both metal oxides and metals, there could not be any barriers that might hamper its advantageous effect on the development of optoelectronics.

No matter which synthetic route is followed, ILs can play their specific role to drive the morphology to a desired state. Highly crystalline nanostructures with precision and purity can be produced by IL templates. Often, surface functionalization is acquired by using ILs during synthesis. Thus, the soft templating property is in fact not the only characteristic of the IL. Moreover, the ILs can be recycled with certain precision and purity after the synthesis procedure [97, 98]. So, it can be used repeatedly for the synthesis of nanomaterials. Recent studies have provided dependable information that suggests that ILs could be a cost-effective template for the synthesis of nanomaterial. Hence, the fabrication of nanomaterials via ILs is a promising aspect to be exploited in the advancement of optoelectronics.

8.   Application in optoelectronics
ILs can fabricate prominent metal oxides such as ZnO, CuO, TiO_2, SnO_2, ZrO_2, etc. to use them as prospective components of optoelectronic devices. Among such devices, the most

common are p-n junction devices, random access memories (RAMs), solar cells, etc. The design of these devices is shown in Scheme 5 for a better understanding of the functions played by metal oxide NPs in them.

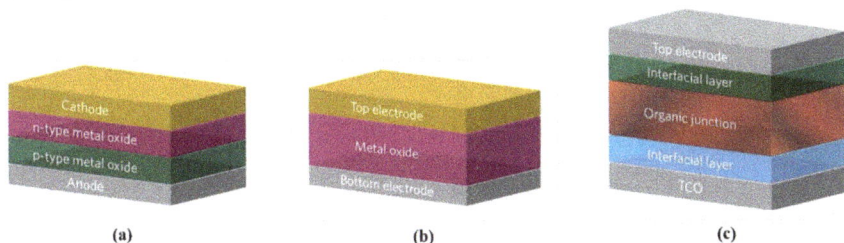

Scheme 5. Components of (a) p-n junction device, (b) RAM, and (c) solar cell.
Reproduced with permission from [4].

In the case of p-n junction diodes or transistors, both p- and n-type metal oxide NPs are necessary for the completion of successful production. Thin film transistor (TFT) channels have been employed using metal oxide nanoparticles. Oxide nanostructures exhibit high carrier mobilities, but it is difficult to manage their density and orientation on a wide scale. Fabrication of the morphology of the nanostructures via ILs is therefore a dominant step for advancing these devices. SnO_2, ZnO, and In_2O_3 nanowires and nanoribbons are used in the most effective TFTs [99-101]. These shapes have been frequently produced by different ILs as reported in the previous section.

Diodes can be developed through a combination of p- and n-type metal oxide layers to create p-n junctions (Scheme 5a). In order to achieve rectifying current-voltage characteristics, conducting anodes, semiconductor p-n junctions, and conducting cathodes are employed in devices like diode rectifiers, sensors, and solar cells. Diverse p-n junction architectures, such as semiconductor blends, core-shell, bilayer, and multilayer structures, have been used to study the usage of metal oxide diodes for gas sensors. Cu_2O-SnO_2 p-n multilayer heterostructures in resonant tunneling devices also display exceptional sensitivity to H_2S vapors at 25 °C [102]. All metal oxide diodes have also been suggested in some studies for photovoltaic cells. The power conversion efficiencies (PCEs) of the Al-doped thin-film heterojunction solar cells of ZnO(n-type)-Ga_2O_3(i)-Cu_2O(p-type) fabricated at 25 °C by pulsed-lased deposition were 5.38%, which was higher than the PCEs of the optimized Cu_2O solar cells [103]. It is now possible to fabricate practical nanostructured oxide heterojunction solar cells thanks to the fabrication of nanosized ZnO(n)-Cu_2O(p) and ZnO(n)-TiO_2(i)-Cu_2O(p) heterostructure diodes from vapor-phase-synthesized Cu_2O nanowires exhibiting a PCE of 0.39% [104].

RAMs are crucial elements of logic circuits for storing data, and metal oxides are excellent for floating-gate TFT flash memories and resistive RAMs (ReRAMs) (Scheme 5b).

ReRAMs have so far used a variety of metal oxides, sandwiching two electrodes with an oxide layer. Scaling down device sizes can improve performance because the size of the memory element has an impact on the ReRAM switching speeds as well. The most widely used flash memories contain charge carriers that have been trapped in a floating gate within a SiO_2 or SiN_x gate dielectric or in a non-conductive discrete trapping layer. Metal oxide nanoparticles are excellent for memory devices as non-conducting trapping layers in the TFT channel or dielectric [105].

Organic photovoltaic (OPV) cells are excitonic devices that use an organic semiconductor-based p-n junction photoactive layer to convert light into power. The components of the most studied OPV design are the bulk-heterojunction combination of the donor (p-type) and acceptor (n-type) organic semiconductors, charge-extracting layers, and a top metal electrode (Scheme 5c). To improve PCE, metal oxides are frequently used as the TCO (transparent conducting oxide) electrode or as interfacial layers (IFLs) for extracting charge. Electrical and optical characteristics of TCO are influenced by electron concentration (N) and carrier mobility, with N having an impact on both the short wavelength transmittance cut-off limit (bandgap) and the long wavelength transmittance cut-off (plasma wavelength) [106]. To increase the optical bandgap while maintaining $N <$ 10^{18} cm^{-3}, secondary dopant ions are added to wide bandgap metal oxides like In_2O_3, SnO_2, and ZnO (Burstein-Moss effect). For inverted organic solar cell (OSC) architectures, metal oxides can be used to change the function of ITO from hole- to electron-collecting, allowing for the extraction of electrons and the use of stable top-electrode metals with high work functions. On the ITO electrode, typical inverted OPVs use ZnO e-IFLs (electron interfacial layers) derived from sol-gels [107, 108]; however, Cs_2CO_3 [109] and TiO_2 [110] have also been reported. Once more, formulations based on NPs enable the simplification of processes, effectively eliminating high-temperature sol-gel annealing and extending the lifetime of the device [111]. Metal nanowire meshes, typically silver nanowire (AgNW) networks, have developed into efficient OSC electrodes.

Metal oxides are the key component or raw materials to perfectly develop optoelectronic devices. Hence, the fabrication of these raw materials to achieve interesting morphologies is crucial. ILs play a vital role by introducing their special templating effect for the production of unique morphologies. There are reported works where the ILs have been used as a template to obtain metal oxides with various fascinating morphologies that enhanced the photocatalytic activity, semiconducting property, and overall optoelectronic activity. The ILs are the main driver here to manipulate the photoluminescence activity due to their ability to variate the shape of the nanostructures. The photoluminescence spectra of ZnO NPs with various forms are depicted in Fig. 4. With shape, the photoluminescence response changes significantly either in intensity or in peak position. The disk-, tube- and ring-like morphologies are obtained from [C_3mim]Br-assisted hydrothermal synthesis and the nanorods are produced from [C_4mim](CF_3SO_2)$_2$N-assisted ultrasound synthesis.

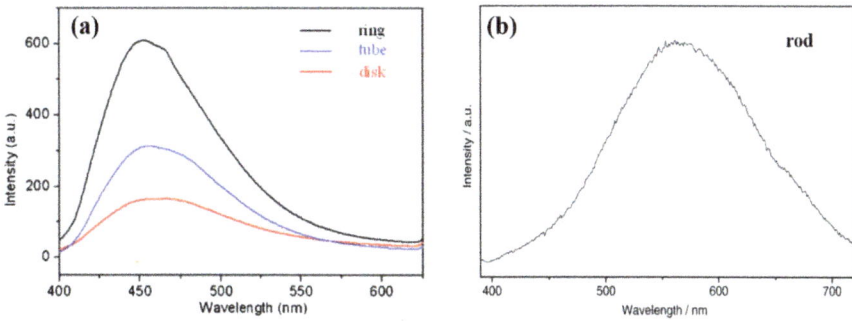

Figure 4. Photoluminescence spectra ZnO nanostructures of different shapes at room temperature. Reproduced with permission from [12] and [112].

A broad peak at 450 nm intensified as the shape changed to disk- to ring-like structure whereas rods-like architecture caused the strong broad peak at 563 nm. So, Fig. 4(a) and 4(b) correspond to blue emission and green-yellow emission, respectively. These different emissions could have occurred due to different defects present on the NP surface. However, the introduction of such defects onto the product surface can be controlled by the proper use of the ILs in the synthesis. Acquiring a certain shape or overall specific morphology can be possible only when templates like ILs are used. Either the use of different ILs or the manipulation of the concentration effect of the ILs can result in different morphologies. Thus, the defect concentration of the NPs can also be manipulated via the judicious use of the ILs. The next step is to apply such nanostructured materials to optoelectronic devices to achieve the enhancement of performance.

CuO NPs were fabricated via four different pyridinium-based ILs with a common anion (Br⁻) and varying cations e.g. N-pentylpyridinium, N-octylpyridinium, N-benzylpyridinium and pentylen dipyridinium [113]. Various morphologies such as nanosheet, leaf-, oval-, sword-like structures were produced and therefore they showed interesting shifts in the bandgap. The results showed that the ILs greatly affected the bandgap energy of the CuO samples. The bandgap of nano CuO with various morphology ranged from 1.67 - 1.83 eV, which are red shifts with regard to the bandgap for bulk CuO (3.25 eV). Therefore, these prepared nanomaterials exhibited fascinating potential in optoelectronic and photovoltaic applications. The advantages of ILs are thus exploited to control the shape as well as bandgap of the nanostructures to enhance the semiconducting and optoelectronic properties.

Min et al. reported an interesting finding where they established a synthetic route to produce ZnO quantum dots deposited onto graphene oxide (GO) using the IL-assisted solvothermal method [114]. The enhanced visible light response was achieved by the ZnO/graphene hybrid semiconductor with a p-n heterojunction. Fig. 5 depicts the photoelectronic properties of ZnO/graphene heterostructure photocatalysts by investigation of photocurrent density and incident PCE (IPCE).

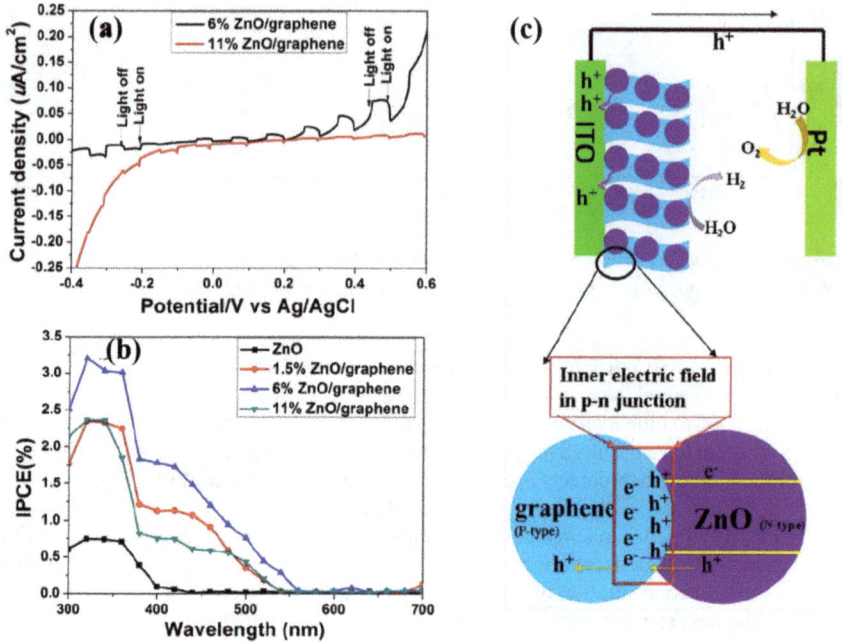

Figure 5. (a) Voltage-photocurrent functions of ZnO and ZnO/graphene under visible light, (b) IPCE spectra of ZnO and ZnO/graphene, (c) Electron transport in ZnO/graphene-based p-n hetero-junction for solar water splitting. Reproduced with permission [114]. Copyright 2012, Elsevier.

Fig. 5(a) confirmed the existence of a p-type semiconductor graphene and since the inner electric field was created in the region of p-n junction, it caused the region of p-type semiconductor graphene to have a negative charge while the ZnO region had a positive charge at equilibrium. The inner electric field caused the holes to fly into the negative field and the electrons to travel to the positive field. To analyze the optoelectronic properties of the nanocomposites of ZnO/graphene, IPCE spectra were investigated and 3-5 times enhancement of photoresponse occurred for the nanocomposite than their pure components. These improved physical and electrochemical characteristics were directly used in solar water splitting, where the positive and negative poles are produced by the inner electric field in the p-n heterojunction. The electrons may then travel to the positive field with the help of solar irradiation or an external potential basis, while the holes would go into the negative field. As a result, the electron-hole pairs will be successfully separated by the p-n junction created at the ZnO/graphene interface, increasing the optoelectronic characteristics. Now, these fascinating characteristics would not have been possible if quantum dots like ZnO was not deposited onto the graphene sheet. To achieve quantum dots, even the size of ZnO NPs has to be below 10 nm. Thus, the role of the IL is understood as it served as the controlling agent of nucleation and growth of the ZnO clusters. Owing to the electrostatic and other supramolecular interactions, ZnO NPs exhibited a controlled growth, and also the deposition of the produced dots was driven by the IL as the graphene was wrapped with it. Therefore, the placement of the ZnO dots was carried out by the supramolecular interactions of the IL and the enclosed system provided by the IL caused the proper formation of the ZnO/graphene nanocomposite system. Thereby, the IL served as both a template and an appropriate solvent for the fabrication of the nanocomposite.

Since, the fabrication of metal oxides and metals is very important for the manipulation of bandgap, the role of IL as a suitable templating agent for emerging applications, cannot be overlooked. So far, the key raw materials utilized in optoelectronics are mostly semiconductors and metals which provide enormous efficacy in performance according to their size and shape. ILs could be a significant tool in this regard, which can give the control of morphology in our hands. Size- and shape-dependent properties are the main roots for the successful production of a smart device and ILs could be the most powerful tools to develop the field of optoelectronics. In fact, ILs could give researchers the most demanding opportunity, which is the fabrication of smart materials at the nanoscale. All we need to do is tailoring the ILs by modification of the anion or cation structure to influence the shape and/or size of the desired materials. ILs can thus fabricate the materials at the nanoscale as they can control both the nucleation rate and the growth rate of the materials. ILs not only act as a soft template for the synthesis of nanomaterial but also provides all the advantages of the hard template method. Therefore, it may otherwise be called the "IL-template method" for the fabrication of nanomaterials. Furthermore, the already reported works provided the excellent and prominent vibe of a successful future of ILs for the fabricated smart materials used in the advancement of optoelectronics.

Conclusions

ILs have been proven useful and effective materials that can give us the change we need in nanostructures to be exploited in optoelectronics. Metal and metal oxide NPs can be fabricated by ILs for tunable morphology, irrespective of the synthetic route. Due to the supramolecular interactions, ILs act as a self-directing agent for controlled directional growth and provide a suitable templating characteristic that can be compared to a mixture of all the advantages of hard-, soft- and colloidal templates. They have unique physicochemical properties to overcome the limitations of conventional templates. They can offer a better environment for the production of materials that might need a high-temperature medium for synthesis. Even the ILs can be used for surface functionalization of the nanostructured materials to achieve task-specific applications. Thus, nanomaterial synthesis by the IL-template method would be more favored by future researchers since it has promising potential that could lead to innovative ways of nanofabrication. ILs are the rarest material that could simultaneously control both the shape and size of the materials. Moreover, they perform like a solvent, a reaction medium, and essentially as a template, for the nanofabrication of smart materials, which gives them the edge in the field of optoelectronics. The recyclability of ILs could pave the way for revolutionary change in the industrial synthesis of nanomaterials, making it cost-effective and consequently cheaper production of optoelectronic devices would be possible. It could be said that in the future, ILs would be the bridge between nanotechnology and optoelectronics, connecting the two emerging fields by developing the properties of advanced nanomaterials.

Acknowledgment

The authors acknowledge the Semiconductor Technology Research Center and the Centennial Research Grant from the University of Dhaka, Bangladesh for supporting the themed research.

Competing financial interests

There are no competing financial interests to declare.

References

[1] J. H. Myung, S. j. Park, A. Cha, S. Hong, Integration of biomimicry and nanotechnology for significantly improved detection of circulating tumor cells (CTCs), Adv. Drug Deliv. Rev., 125 (2018) 36–47. https://doi.org/10.1016/j.addr.2017.12.005

[2] Y. Liu, C. N. Ong, J. Xie, Emerging nanotechnology for environmental applications, Nanotechnol. Rev. 5 (2016) 1–2. https://doi.org/10.1515/ntrev-2015-0072

[3] Z. L. Wang, W. Wu, Nanotechnology-enabled energy harvesting for self-powered micro-/nanosystems, Angew. Chem., Int. Ed. 51 (2012) 11700–11721. https://doi.org/10.1002/anie.201201656

[4] X. Yu, T. J. Marks, A. Facchetti, Metal oxides for optoelectronic applications, Nature Mater 15 (2016) 383–396. https://doi.org/10.1038/nmat4599

[5] Cao, G. Nanostructures and Nanomaterials: Synthesis, Properties & Applications, 2nd ed.; Imperial College Press: UK, 2004

[6] A. K. Boal, F. Ilhan, J. E. DeRouchey, T. Thurn-Albrecht, T. P. Russell, V. M. Rotello, Self-assembly of nanoparticles into structured spherical and network aggregates, Nature, 404 (2000) 746–748. https://doi.org/10.1038/35008037

[7] T. Y. Zhai, L. Li, Y. Ma, M. Y. Liao, X. Wang, X. S. Fang, J. N. Yao, Y. Bando, D. Golberg, One-dimensional inorganic nanostructures: synthesis, field-emission and photodetection, Chem. Soc. Rev. 40 (2011) 2986–3004. https://doi.org/10.1039/C0CS00126K

[8] Z. L. Wang, Zinc oxide nanostructures: Growth, properties and applications, J. Phys. Condens. Matter. 16 (25) (2004) R829. https://doi.org/10.1088/0953-8984/16/25/R01

[9] X. Wang, Y. Ding, C. J. Summers, Z. L. Wang, Large-scale synthesis of six-nanometer-wide ZnO nanobelts. J. Phys. Chem. B 108 (26) (2004) 8773–8777. https://doi.org/10.1021/jp048482e

[10] A. H. Moharram, S. A. Mansour, M. A. Hussein, M. Rashad, Direct precipitation and characterization of ZnO nanoparticles. J. Nanomater. (2014) 2014. https://doi.org/10.1155/2014/716210

[11] Y. Peng, A. W. Xu, B. Deng, M. Antonietti and H. Co¨lfen, Polymer-controlled crystallization of zinc oxide hexagonal nanorings and disks, J. Phys. Chem. B, 110 (2006) 2988–2993. https://doi.org/10.1021/jp056246d

[12] K. Qi, J. Yang, J. Fu, G. Wang, L. Zhu, G. Liu, W. Zheng, Morphology-Controllable ZnO Rings: Ionic liquid-assisted hydrothermal synthesis, growth mechanism and photoluminescence properties. CrystEngComm, 15 (34) (2013) 6729–6735. https://doi.org/10.1039/C3CE27007F

[13] R. R. Poolakkandy, M. M. Menamparambath, Soft-template-assisted synthesis: a promising approach for the fabrication of transition metal oxides, Nanoscale Adv. 2 (2020) 5015-5045. https://doi.org/10.1039/D0NA00599A

[14] Akter, M.; Satter, S. S.; Singh, A. K.; Rahman, M. M.; Mollah, M. Y. A.; Susan, M. A. B. H. Hydrophilic Ionic Liquid-Assisted Control of the Size and Morphology of ZnO Nanoparticles Prepared by a Chemical Precipitation Method, RSC Adv. 6 (94) (2016) 92040–92047. https://doi.org/10.1039/C6RA14955C

[15] J. Łuczak, M. Paszkiewicz, A. Krukowska, A. Malankowska, A. Zaleska-Medynska, Ionic liquids for nano- and microstructures preparation. Part 2: Application in synthesis, Adv. Colloid Interface Sci., 227 (2016) 1-52. https://doi.org/10.1016/j.cis.2015.08.010

[16] N. Koch, Supramolecular Materials for Opto-Electronics, Royal Society of Chemistry 2014. https://doi.org/10.1039/9781782626947

[17] R. Paschotta, Optoelectronics in Encyclopedia of Laser Physics and Technology, Wiley, 2008.

[18] P. Kubelka, F. Munk, An Article on Optics of Paint Layers. Z. Tech. Phys, 12 (1931) 593–601.

[19] L. Irimpan, V. P. N. Nampoori, P. Radhakrishnan, A. Deepthy, B. Krishnan, Size dependent fluorescence spectroscopy of nanocolloids of ZnO. J. Appl. Phys. 6 (2007) 102. https://doi.org/10.1063/1.2778637

[20] R. Koole, E. Groeneveld, D. Vanmaekelbergh, A. Meijerink, C. de Mello Donegá, Size effects on semiconductor nanoparticles, Nanoparticles, (2014) 13-51. https://doi.org/10.1007/978-3-662-44823-6_2

[21] H. Lin, C. P. Huang, W. Li, C. Ni, S. I. Shah, Y. H. Tseng, Size dependency of nanocrystalline TiO$_2$ on its optical property and photocatalytic reactivity exemplified by 2-chlorophenol, Appl. Catal. B, 68 (2006) 1–11. https://doi.org/10.1016/j.apcatb.2006.07.018

[22] X. Wan, X. Liang, C. Zhang, X. Li, W. Liang, H. Xu, S. Lan, S. Tie, Morphology controlled syntheses of Cu-doped ZnO, tubular Zn (Cu) O and Ag decorated tubular Zn (Cu) O microcrystals for photocatalysis, Chem. Eng. J., 272 (2015) 58-68. https://doi.org/10.1016/j.cej.2015.02.089

[23] H.M. Hu, C.G. Deng, X.H. Huang, Hydrothermal growth of center-hollow multigonal star-shaped ZnO architectures assembled by hexagonal conic nanotubes, Mater. Chem. Phys. 121 (2010) 364–369. https://doi.org/10.1016/j.matchemphys.2010.01.044

[24] D.W. Chu, Y. Masuda, T. Ohji, K. Kato, Formation and photocatalytic application of ZnO nanotubes using aqueous solution, Langmuir 26 (2010) 2811–2815. https://doi.org/10.1021/la902866a

[25] Y. Tian, J.C. Li, H. Xiong, J.N. Dai, Controlled synthesis of ZnO hollow microspheres via precursor-template method and its gas sensing property, Appl. Surf. Sci. 258 (2012) 8431–8438. https://doi.org/10.1016/j.apsusc.2011.12.090

[26] Z. Jin, F. Wang, F. Wang, J. Wang, J. C. Yu, J. Wang, Metal nanocrystal-embedded hollow mesoporous TiO$_2$ and ZrO$_2$ microspheres prepared with polystyrene nanospheres as carriers and templates, Adv. Funct. Mater., 23(17) (2013) 2137-2144. https://doi.org/10.1002/adfm.201202600

[27] H. Lu, L. Zhang, W. Xing, H. Wang, N. Xu, Preparation of TiO$_2$ hollow fibers using poly (vinylidene fluoride) hollow fiber microfiltration membrane as a template, Mater. Chem. Phys.,94(2-3) (2005) 322-327. https://doi.org/10.1016/j.matchemphys.2005.05.008

[28] B. Réti, G. I. Kiss, T. Gyulavári, K. Baan, K. Magyari, K. Hernadi, Carbon sphere templates for TiO$_2$ hollow structures: Preparation, characterization and photocatalytic activity, Catal. Today, 284 (2017) 160-168. https://doi.org/10.1016/j.cattod.2016.11.038

[29] C. Zhang, Y. Shi, Z. Fu, A facile method for the fabrication of SiO$_2$ and SiO$_2$/TiO$_2$ hollow particles using Na$_2$SO$_4$ particles as templates, J. Sol-Gel Sci. Technol., 91(3) (2019) 431-440. https://doi.org/10.1007/s10971-019-05035-x

[30] G. Zhang, X. Xiao, B. Li, P. Gu, H. Xue, H. Pang, Transition metal oxides with one-dimensional/one-dimensional-analogue nanostructures for advanced supercapacitors, J. Mater. Chem. A, 5 (2017) 8155–8186. https://doi.org/10.1039/C7TA02454A

[31] C. M. Ghimbeu, J. M. Le Meins, C. Zlotea, L. Vidal, G. Schrodj, M. Latroche, C. Vix-Guterl, Controlled synthesis of NiCo nanoalloys embedded in ordered

porous carbon by a novel soft-template strategy, Carbon, 67 (2014) 260–272. https://doi.org/10.1016/j.carbon.2013.09.089

[32] P. Yang, H. Yan, S. Mao, R. Russo, J. Johnson, R. Saykally, N. Morris, J. Pham, R. He, H. J. Choi, Controlled Growth of ZnO Nanowires and Their Optical Properties, Adv. Func. Mater. (2002) 323–331. https://doi.org/10.1002/1616-3028(20020517)12:5%3C323::AID-ADFM323%3E3.0.CO;2-G

[33] L. E. Greene, M. Law, D. H. Tan, M. Montano, J. Goldberger, G. Somorjai, P. Yang, General Route to Vertical ZnO Nanowire Arrays Using Textured ZnO Seeds, Nano Lett. 5 (7) (2005) 1231–1236. https://doi.org/10.1021/nl050788p

[34] Y. Zhai, Y. Dou, X. Liu, S. S. Park, C. S. Ha, D. Zhao, Carbon, 49 (2011) 545–555. https://doi.org/10.1016/j.carbon.2010.09.055

[35] N. Pal, A. Bhaumik, Soft templating strategies for the synthesis of mesoporous materials: Inorganic, organic–inorganic hybrid and purely organic solids, Adv. Colloid Interface Sci. 189–190 (2013) 21–41. https://doi.org/10.1016/j.cis.2012.12.002

[36] N. D. Petkovich, A. Stein, Controlling macro-and mesostructures with hierarchical porosity through combined hard and soft templating, Chem. Soc. Rev. 42 (2013) 3721–3739. https://doi.org/10.1039/C2CS35308C

[37] V. M. Prida, V. Vega, J. Garc´ıa, L. Iglesias, B. Hernando, I. Minguez-Bacho, Electrochemical methods for template-assisted synthesis of nanostructured materials, (2015) Magnetic Nano- and Microwires: Design, Synthesis, Properties and Applications, pp. 3-39. https://doi.org/10.1016/B978-0-08-100164-6.00001-1

[38] X. Y. Liu, K. X. Wang, J. S. Chen, Template-directed metal oxides for electrochemical energy storage, Energy Storage Mater. 3 (2016) 1–17. https://doi.org/10.1016/j.ensm.2015.12.002

[39] A. Ahmed, R. Clowes, P. Myers, H. Zhang, Hierarchically porous silica monoliths with tuneable morphology, porosity, and mechanical stability, J. Mater. Chem., 21 (2011) 5753–5763. https://doi.org/10.1039/C0JM02664F

[40] G. L. Drisko, A. Zelcer, V. Luca, R. A. Caruso, G. J. D. A. A. Soler-Illia, One-pot synthesis of hierarchically structured ceramic monoliths with adjustable porosity, Chem. Mater. 22 (2010) 4379–4385. https://doi.org/10.1021/cm100764e

[41] B. Platschek, A. Keilbach, T. Bein, Mesoporous structures confined in anodic alumina membranes, Adv. Mater. 23 (2011) 2395–2412. https://doi.org/10.1002/adma.201002828

[42] H. J. Liu, X. M. Wang, W. J. Cui, Y. Q. Dou, D. Y. Zhao, Y. Y. Xia, Highly ordered mesoporous carbon nanofiber arrays from a crab shell biological template and its application in supercapacitors and fuel cells, J. Mater. Chem., 20 (2010) 4223–4230. https://doi.org/10.1039/B925776D

[43] S. G. Hosseini, R. Ahmadi, A. Ghavi, A. Kashi, Synthesis and characterization of α-Fe_2O_3 mesoporous using SBA-15 silica as template and investigation of its catalytic activity for thermal decomposition of ammonium perchlorate particles, Powder Technol. 278 (2015) 316–322. https://doi.org/10.1016/j.powtec.2015.03.032

[44] M. Hu, A. A. Belik, M. Imura, K. Mibu, Y. Tsujimoto, Y. Yamauchi, Synthesis of superparamagnetic nanoporous iron oxide particles with hollow interiors by using Prussian blue coordination polymers, Chem. Mater., 24 (2012) 2698–2707. https://doi.org/10.1021/cm300615s

[45] C. Mijangos, R. Hern´andez, J. Mart´ın, A review on the progress of polymer nanostructures with modulated morphologies and properties, using nanoporous AAO templates, Prog. Polym. Sci. 54–55 (2016) 148–182. https://doi.org/10.1016/j.progpolymsci.2015.10.003

[46] A. H. Lu, F. Sch¨uth, Nanocasting: a versatile strategy for creating nanostructured porous materials, Adv. Mater. 18 (2006) 1793–1805. https://doi.org/10.1002/adma.200600148

[47] Y. Meng, D. Gu, F. Zhang, Y. Shi, H. Yang, Z. Li, C. Yu, B. Tu, D. Zhao, Ordered mesoporous polymers and homologous carbon frameworks: amphiphilic surfactant templating and direct transformation, Angew. Chem., Int. Ed. 44 (2005) 7053–7059. https://doi.org/10.1002/anie.200501561

[48] W. Li, Z. Wu, J. Wang, A. A. Elzatahry, D. Zhao, A Perspective on Mesoporous TiO_2 Materials, Chem. Mater. 26 (2014) 287–298. https://doi.org/10.1021/cm4014859

[49] J. Fan, S. W. Boettcher, C. K. Tsung, Q. Shi, M. Schierhorn, G. D. Stucky, Field-directed and confined molecular assembly of mesostructured materials: basic principles and new opportunities, Chem. Mater. 20 (2008) 909–921. https://doi.org/10.1021/cm702328k

[50] G. J. A. A. Soler-Illia, O. Azzaroni, Multifunctional hybrids by combining ordered mesoporous materials and macromolecular building blocks, Chem. Soc. Rev. 40 (2011) 1107–1150. https://doi.org/10.1039/C0CS00208A

[51] B. P. Bastakoti, Y. Li, S. Guragain, M. Pramanik, S. M. Alshehri, T. Ahamad, Z. Liu, Y. Yamauchi, Synthesis of mesoporous transition-metal phosphates by polymeric micelle Assembly, Chem.– Eur. J. 22 (2016) 7463–7467. https://doi.org/10.1002/chem.201600435

[52] A. B. D. Nandiyanto, T. Ogi, F. Iskandar, K. Okuyama, Highly ordered porous monolayer generation by dual-speed spin-coating with colloidal templates, Chem. Eng. J. 167 (2011) 409–415. https://doi.org/10.1016/j.cej.2010.11.077

[53] B. Liu, M. Louis, L. Jin, G. Li, J. He, Co-template directed synthesis of gold nanoparticles in mesoporous titanium dioxide, Chem. – Eur. J. 24 (2018) 9651–9657. https://doi.org/10.1002/chem.201801223

[54] L. Jin, B. Liu, M. E. Louis, G. Li, J. He, Highly crystalline mesoporous titania loaded with monodispersed gold nanoparticles: Controllable metal–support interaction in porous materials, ACS Appl. Mater. Interfaces, 12 (2020) 9617–9627. https://doi.org/10.1021/acsami.9b20231

[55] P. Wasserscheid, T. Welton, Ionic Liquids in Synthesis, Wiley-VCH, New York, 2003.

[56] Jr J.G. Huddleston, H.D. Willauer, R.P. Swatloski, W.M. Reichert, R. Mayton, S. Sheff, A. Wierzbicki, J.H. Davis, R.D. Rogers, Room temperature ionic liquids as novel media for 'clean' liquid–liquid extraction, Chem. Commun. (1998) 1765–1766. https://doi.org/10.1039/A803999B

[57] M.S. Miran, M. Hoque, T. Yasuda, K. Ueno, M. Watanabe, Key factor governing the physicochemical properties and extent of proton transfer of protic ionic liquids: pK_a or structural chemistry? Phys. Chem. Chem. Phys. 21 (2019) 418–426. https://doi.org/10.1039/C8CP06973E

[58] H. Davy, Researches, Chemical and Philosophical. Biggs and Cottle, Bristol, 1800, 1800.

[59] G. Laus, G. Bentivoglio, H. Schottenberger, V. Kahlenberg, H. Kopacka, Ionic liquids: current developments, potential and drawbacks for industrial applications, Lenzinger Berichte, 84 (2005) 71-85.

[60] W. Ramsay, XXXIV. On picoline and its derivatives, Philos. Mag., Ser. 5(11) (1876) 269-281. https://doi.org/10.1080/14786447608639105

[61] M. M. Islam, S. Ahmed, M. S. Miran, M. A. B. H. Susan, Advances on potential-driven growth of metal crystals from ionic liquids, Progress in Crystal Growth and Characterization of Materials, 68(4) (2022) 100580. https://doi.org/10.1016/j.pcrysgrow.2022.100580

[62] E. Kianfar, S. Mafi, Ionic liquids: properties, application, and synthesis, Fin. Chem. Eng. 2 (2020) 22–31. https://doi.org/10.37256/fce.212021693

[63] H. Tokuda, K. Hayamizu, K. Ishii, M.A.B.H. Susan, M. Watanabe, Physicochemical properties and structures of room temperature ionic liquids. 1. Variation of anionic species, J. Phys. Chem. B 108 (2004) 16593–16600. https://doi.org/10.1021/jp047480r

[64] H. Tokuda, K. Ishii, M.A.B.H. Susan, S. Tsuzuki, K. Hayamizu, M. Watanabe, Physicochemical properties and structures of room-temperature ionic liquids. 3. Variation of cationic structures, J. Phys. Chem. B 110 (2006) 2833–2839. https://doi.org/10.1021/jp053396f

[65] H. Tokuda, K. Hayamizu, K. Ishii, M.A.B.H. Susan, M. Watanabe, Physicochemical properties and structures of room temperature ionic liquids. 2. Variation of alkyl chain length in imidazolium cation, J. Phys. Chem. B 109 (2005) 6103–6110. https://doi.org/10.1021/jp044626d

[66] C.A. Angell, Y. Ansari, Z. Zhao, Ionic Liquids: past, Present and Future, Faraday Discuss 154 (2012) 9–27. https://doi.org/10.1039/C1FD00112D

[67] S. Gabriel, J. Weiner, Ueber einige abkommlinge des propylamins, Berichte der deutschen chemischen gesellschaft 21 (1888) 2669–2679. https://doi.org/10.1002/cber.18880210288

[68] T.L. Greaves, C.J. Drummond, Protic ionic liquids: Properties and applications, Chem. Rev. 108 (2008) 206–237. https://doi.org/10.1021/cr068040u

[69] S. Khazalpour, M. Yarie, E. Kianpour, A. Amani, S. Asadabadi, J.Y. Seyf, M. Rezaeivala, S. Azizian, M.A. Zolfigol, Applications of phosphonium-based ionic liquids in chemical processes, J. Iranian Chem. Soc. 17 (2020) 1775–1917. https://doi.org/10.1007/s13738-020-01901-6

[70] M.S. Miran, H. Kinoshita, T. Yasuda, M.A.B.H. Susan, M. Watanabe, Physicochemical properties determined by ΔpKa for protic ionic liquids based on an organic super-strong base with various Brønsted acids, Phys. Chem. Chem. Phys. 14 (2012) 5178–5186. https://doi.org/10.1039/C2CP00007E

[71] H. Nakamoto, M. Watanabe, Brønsted acid–base ionic liquids for fuel cell electrolytes, Chem. Commun. 24 (2007) 2539–2541. https://doi.org/10.1039/B618953A

[72] A. Noda, M.A.B.H. Susan, S. Mitsushima, K. Hayamizu, M. Watanabe, Brønsted acid base ionic liquids as proton-conducting nonaqueous electrolytes, J. Phys. Chem. B 107 (2003) 4024–4033. https://doi.org/10.1021/jp022347p

[73] T. Tamura, K. Yoshida, T. Hachida, M. Tsuchiya, M. Nakamura, Y. Kazue, N. Tachikawa, K. Dokko, M. Watanabe, Physicochemical properties of glyme-Li salt complexes as a new family of room-temperature ionic liquids, Chem. Lett. 39 (2010) 53–55. http://dx.doi.org/10.1246/cl.2010.753

[74] T. Mandai, K. Yoshida, K. Ueno, K. Dokko, M. Watanabe, Criteria for solvate ionic liquids, Phys. Chem. Chem. Phys. 16 (2014) 8761–8772. https://doi.org/10.1039/C4CP00461B

[75] K. Ueno, K. Yoshida, M. Tsuchiya, N. Tachikawa, K. Dokko, M. Watanabe, Glyme–lithium salt equimolar molten mixtures, concentrated solutions or solvate ionic liquids? J. Phys. Chem. B 116 (2012) 11323–11331. https://doi.org/10.1021/jp307378j

[76] R.F. Rodrigues, A.A. Freitas, J.N. Canongia Lopes, K. Shimizu, Ionic liquids and water: hydrophobicity vs. hydrophilicity, Molecules 26 (2021) 7159–7182. https://doi.org/10.3390/molecules26237159

[77] El Abedin, S. Z.; Endres, F. Ionic Liquids: The Link to High-Temperature Molten Salts? Acc. Chem. Res. 40(11) (2007) 1106–1113. https://doi.org/10.1021/ar700049w

[78] T.-P. Fellinger, A. Thomas, J. Yuan, M. Antonietti, "Cooking carbon with salt": Carbon materials and carbonaceous frameworks from ionic liquids and poly(ionic

Materials Research Forum LLC
https://doi.org/10.21741/9781644902554-2

liquid)s, Adv. Mater. 25(41) (2013) 5838-5855.
https://doi.org/10.1002/adma.201301975

[79] J. Yuan, D. Mecerreyes, M. Antonietti, Poly(ionic liquid)s: An update. Prog. Polym. Sci. 38(7) (2013) 1009-1036. https://doi.org/10.1016/j.progpolymsci.2013.04.002

[80] S. Li, J. L. Banuelos, J. Guo, L. Anovitz, G. Rother, R. W. Shaw, P. C. Hillesheim, S. Dai, G. A. Baker, P. T. Cummings, Alkyl chain length and temperature effects on structural properties of pyrrolidinium-based ionic liquids: A combined atomistic simulation and small-angle X-ray scattering study, J. Phys. Chem. Lett. 3(1) (2012) 125-130. https://doi.org/10.1021/jz2013209

[81] K. Dong, S. J. Zhang, Hydrogen Bonds: A Structural Insight into Ionic Liquids, Chem. Eur. J. 18(10) (2012) 2748-2761. https://doi.org/10.1002/chem.201101645

[82] O. Russina, A. Triolo, L. Gontrani, R. Caminiti, Mesoscopic structural heterogeneities in room-temperature ionic liquids. J. Phys. Chem. Lett. 3(1) (2012) 27-33. https://doi.org/10.1021/jz201349z

[83] A. A. H. Padua, M. F. Gomes, J. N. A. C. Lopes, Molecular solutes in ionic liquids: A structural, perspective, Acc. Chem. Res. 40(11) (2007) 1087-1096. https://doi.org/10.1021/ar700050q

[84] K. Fumino, S. Reimann, R. Ludwig, Probing molecular interaction in ionic liquids by low frequency spectroscopy: Coulomb energy, hydrogen bonding and dispersion forces, Phys. Chem. Chem. Phys. 16(40) (2014) 21903-29. https://doi.org/10.1039/C4CP01476F

[85] T. Ueki, M. Watanabe, Macromolecules in ionic liquids: Progress, challenges, and opportunities, Macromolecules, 41(11) (2008) 3739-3749. https://doi.org/10.1021/ma800171k

[86] E. Husanu, V. Cappello, C. S. Pomelli, J. David, M. Gemmi, C. Chiappe, Chiral ionic liquid assisted synthesis of some metal oxides. RSC Adv. 7(2) (2017) 1154–1160. https://doi.org/10.1039/C6RA25736D

[87] L. Wang, L. Chang, B. Zhao, Z. Yuan, G. Shao, W. Zheng, Systematic investigation on morphologies, forming mechanism, photocatalytic and photoluminescent properties of ZnO nanostructures constructed in ionic liquids. Inorg. Chem., 47(5) (2008) 1443-1452. https://doi.org/10.1021/ic701094a

[88] K. Manjunath, L. S. R. Yadav, T. Jayalakshmi, V. Reddy, H. Rajanaika, G. Nagaraju, Ionic liquid assisted hydrothermal synthesis of TiO_2 nanoparticles: photocatalytic and antibacterial activity, J. Mat. Res. Tech. 7(1) (2018) 7-13. https://doi.org/10.1016/j.jmrt.2017.02.001

[89] Y. Zhou, M. Antonietti, Synthesis of very small TiO_2 nanocrystals in a room-temperature ionic liquid and their self-assembly toward mesoporous spherical

aggregates, J. Am. Chem. Soc. 125 (2003) 14960-14961.
https://doi.org/10.1021/ja0380998

[90] H. Liu, Y. Su, H. Hu, W. Cao, Z. Chen, An ionic liquid route to prepare mesoporous ZrO_2–TiO_2 nanocomposites and study on their photocatalytic activities, Adv. Powder Techol. 24 (2013) 683-688. https://doi.org/10.1016/j.apt.2012.12.007

[91] J. Xia, H. Li, Z. Luo, K. Wang, H. Shu, Y. Yan, Microwave-assisted synthesis of flower-like and leaflike CuO nanostructures via room-temperature ionic liquids, J. Phys. Chem. Solids, 70 (2009) 1461–1464. https://doi.org/10.1016/j.jpcs.2009.08.006

[92] J. Xia, H. Li, Z. Luo, K. Wang, H. Shu, Y. Yan, Ionic liquid-assisted hydrothermal synthesis of three-dimensional hierarchical CuO peachstone-like architectures, Appl. Surface Sci. 256 (2010) 1871-1877. https://doi.org/10.1016/j.apsusc.2009.10.022

[93] V. Taghvaei, A. Habibi-Yangjeh, M. Behboudnia, Preparation and characterization of SnO_2 nanoparticles in aqueous solution of [EMIM][EtSO_4] as a low cost ionic liquid using ultrasonic irradiation, Powder Technol. 195 (2009) 63-67. https://doi.org/10.1016/j.powtec.2009.05.023

[94] S. Z. El Abedin, F. Endres, Electrodeposition of nanocrystalline silver films and nanowires from the ionic liquid 1-ethyl-3-methylimidazolium trifluoromethylsulfonate, Electrochim. Acta, 54 (2009) 5673-5677. https://doi.org/10.1016/j.electacta.2009.05.005

[95] Z. Li, Z. Liu, J. Zhang, B. Han, J. Du, Y. Gao, T. Jiang, Synthesis of single-crystal gold nanosheets of large size in ionic liquids, J. Phys. Chem. B, 109 (2005) 14445-14448. https://doi.org/10.1021/jp0520998

[96] A. Imanishi, M. Tamura, S. Kuwabata, Formation of Au nanoparticles in an ionic liquid by electron beam irradiation, Chem. Commun. 13 (2009) 1775-1777. https://doi.org/10.1039/B821303H

[97] B. Sarmah, R. Srivastava, Highly efficient and recyclable basic ionic liquids supported on SBA-15 for the synthesis of substituted styrenes, carbinolamides, and naphthopyrans, Molecular Catalysis, 427 (2017) 62-72. https://doi.org/10.1016/j.molcata.2016.11.030

[98] S. K. Panja, S. Saha, Recyclable, magnetic ionic liquid bmim [FeCl 4]-catalyzed, multicomponent, solvent-free, green synthesis of quinazolines, RSC Adv., 3(34) (2013) 14495-14500. https://doi.org/10.1039/C3RA42039F

[99] J. Goldberger, D. J. Sirbuly, M. Law, P. Yang, ZnO nanowire transistors, J. Phys. Chem. B, 109 (2004) 9–14. https://doi.org/10.1021/jp0452599

[100] M. S. Arnold, P. Avouris, Z. W. Pan, Z. L. Wang, Field-effect transistors based on single semiconducting oxide nanobelts, J. Phys. Chem. B, 107 (2002) 659–663. https://doi.org/10.1021/jp0271054

[101] H. T. Ng, J. Han, T. Yamada, P. Nguyen, Y. P. Chen, M. Meyyappan, Single crystal nanowire vertical surround-gate field-effect transistor, Nano Lett., 4(7) (2004) 1247-1252. https://doi.org/10.1021/nl049461z

[102] G. Cui, M. Zhang, G. Zou, Resonant tunneling modulation in quasi-2D Cu_2O/SnO_2 p–n horizontal-multi-layer heterostructure for room temperature H_2S sensor application, Sci. Rep. 3 (2013) 1–8.

[103] T. Minami, Y. Nishi, T. Miyata, High-Efficiency Cu_2O-based heterojunction solar cells fabricated using a Ga_2O_3 thin film as n-type layer, Appl. Phys. Express, 6 (2013) 044101. https://doi.org/10.7567/APEX.6.044101

[104] S. Brittman, Y. Yoo, N. P. Dasgupta, S. Kim, B. Kim, P. Yang, Epitaxially aligned cuprous oxide nanowires for all-oxide, single-wire solar cells, Nano Lett., 14(8) (2014) 4665-4670. https://doi.org/10.1021/nl501750h

[105] R. Martins, P. Barquinha, L. Pereira, N. Correia, G. Gonçalves, I. Ferreira, E. Fortunato, Write-erase and read paper memory transistor, Appl. Phys. Lett., 93(20) (2008) 203501. https://doi.org/10.1063/1.3030873

[106] K. Nomura, T. Kamiya, H. Yanagi, E. Ikenaga, K. Yang, K. Kobayashi, M. Hirano, H. Hosono, Subgap states in transparent amorphous oxide semiconductor, In–Ga–Zn–O, observed by bulk sensitive X-ray photoelectron spectroscopy, Appl. Phys. Lett., 92(20) (2008) 202117. https://doi.org/10.1063/1.2927306

[107] M. Law, L. E. Greene, J. C. Johnson, R. Saykally, P. Yang, Nanowire dye-sensitized solar cells, Nature Mater. 4 (2005) 455–459. https://doi.org/10.1038/nmat1387

[108] Y. Sun, J. H. Seo, C. J. Takacs, J. Seifter, A. J. Heeger, Inverted polymer solar cells integrated with a low-temperature-annealed sol-gel-derived ZnO film as an electron transport layer, Adv. Mater. 23 (2011) 1679–1683. https://doi.org/10.1002/adma.201004301

[109] H.-H. Liao, L.-M. Chen, Z. Xu, G. Li, Y. Yang, Highly efficient inverted polymer solar cell by low temperature annealing of Cs_2CO_3 interlayer, Appl. Phys. Lett. 92 (2008) 173303. https://doi.org/10.1063/1.2918983

[110] K. Lee, J. Y. Kim, S. H. Park, S. H. Kim, S. Cho, A. J. Heeger, Air-stable polymer electronic devices, Adv. Mater., 19(18) (2007) 2445-2449. https://doi.org/10.1002/adma.200602653

[111] S. K. Hau, H. Yip, N. S. Baek, J. Zou, K. O'Malley, A. K-Y. Jen, Air-stable inverted flexible polymer solar cells using zinc oxide nanoparticles as an electron selective layer, Appl. Phys. Lett., 92(25) (2008) 225. https://doi.org/10.1063/1.2945281

[112] T. Alammar, A. V. Mudring, Facile ultrasound-assisted synthesis of ZnO nanorods in an ionic liquid, Mater. Lett., 63(9-10) (2009) 732-735. https://doi.org/10.1016/j.matlet.2008.12.035

[113] M. Sabbaghan, A. S. Shahvelayati, K. Madankar, CuO nanostructures: optical properties and morphology control by pyridinium-based ionic liquids, Spectrochim. Acta Part A: Mol. Biomol. Spectrosc., 135 (2015) 662-668. https://doi.org/10.1016/j.saa.2014.07.097

[114] Y. Min, K. Zhang, L. Chen, Y. Chen, Y. Zhang, Ionic liquid assisting synthesis of ZnO/graphene heterostructure photocatalysts with tunable photoresponse properties, Diam. Relat. Mater., 26 (2012) 32-38. https://doi.org/10.1016/j.diamond.2012.04.003

Applications of Emerging Nanomaterials and Nanotechnology   Materials Research Forum LLC
Materials Research Foundations 148 (2023) 63-102   https://doi.org/10.21741/9781644902554-3

Chapter 3

Nanocomposites and their Applications

S. Sreevidya[1], Sushma Yadav[2], Yokraj Katre[1], Ajaya Kumar Singh[2,3]*, Abbas Rahdar[4]

[1]Department of Chemistry, Kalyan PG College, Bhilai Nagar, Durg, 490006, (C.G.), India

[2]Department of Chemistry, Govt. VYT PG Autonomous College, Durg, India

[3] School of Chemistry & Physics, Westville Campus, University of KwaZulu-Natal, Durban 4000, South Africa

[4]Department of Physics, University of Zabol, Zabol, Iran

* ajayaksingh_au@yahoo.co.in

Abstract

The trends in the current developments for utilising the available natural resources as bio-renewable/sustainable supplies have captured the attention of scientists/industrialists in the recent past. The unlimited alarms that are interrelated with the health and environment due to the exhaustive piling up of unwanted and discarded stocks and the damages caused, are indirectly interwound with the foreseeable reduction of fossil/mineral wealth. The bio-renewable/sustainable supplies are hence, opted as platforms for reinforcement in a wide array of applicable functions. Developmental innovations leading to new approaches and perfect tools, as nano-composites through green bio-renewable/sustainable/bio-degradable supplies, submit for note-worthy advantages and are found to be associated with exceptional attributes such as rapid and cost-effective fabrication, with better thermal/mechanical stability, besides easy portability/reusability. It would be worthy mentioning that nano-composites from structured bio-degradable resources find their utility in a broad spectrum of fragments such as food preservation/packaging, agriculture/bio-sensors, bio-medical, optical/energy-storage, water-treatment/marine, automotive/automation, construction, flame-retardant/coating, and many more. Here in this chapter, we comprehensively compare the recent advancement/development of varied applications of nano-composites from bio-renewable/non-bio-renewable reserves. In conclusion, the solutions to the debatable question, and the potential developmental break-ups with new-openings for the naturally agglomerated nano-composites are reflected briefly.

Keywords

Nanocomposites, Anti-Bacterial-Activity, Dye Degradation, Biosensing

Contents

1.   Introduction

The trends in the current developments for utilising the available natural resources as bio-renewable/sustainable supplies have captured the attention of scientists/industrialists in the recent past. The unlimited alarms that are interrelated with the health and environment due to the exhaustive piling up of the unwanted discarded stocks and the damages caused are indirectly interwound with the foreseeable reduction of fossil/mineral wealth [1]. Modulating and manipulating nanotechnology with diverged nano-routes aided by pulsating tools, produces sustainable products that are extremely adaptable for varied applications. The uniquely structured nano-products in the nano-range as nano-materials (NMs) are categorized as nano-particles (NPs)/nano-composites (NCs), where both classes have their applications based on the uniqueness of their properties. NPs are mainly nano-components with a single group of particles in a nano-range with distinctive features [2]. While NCs are those nano-components with matrices having a group of particles (two or more) either in a continuous/dis-continuous phase [3]. Generally, they are got by fortification/reinforcement of NMs/other materials to produce new NMs with advanced characteristics that are extremely adaptable to newer applications [4]. Developmental innovations leading to new approaches with perfect nano-units, as NCs through green bio-renewable/sustainable/bio-degradable supplies submit note-worthy advantages that are

Applications of Emerging Nanomaterials and Nanotechnology   Materials Research Forum LLC
Materials Research Foundations 148 (2023) 63-102   https://doi.org/10.21741/9781644902554-3

associated with exceptional attributes such as rapid and cost-effective fabrication with better thermal/mechanical stability, besides easy portability/reusability [5].

The potential properties (structural/physical/chemical/financial) such as dimensional (size, shape)/mechanical/tensile/thermal / reusability/bio-degradability / cost could be modified depending upon the nature/type of fortification/reinforcement of the matrices that are involved in the formation [6, 7]. Thus, NCs are basically grouped into two major classes: functional (application) and structural (properties). Larger interactions between the reinforced material would generally result in larger ratios between the surface and the volume, leading to increased properties and functionalities [8]. The literature when surveyed indicates that there is an intense, and deep-routed enhanced development while dealing with the fabrication and application of NCs in the recent past. It would be worth mentioning that nanocomposites from structured bio-degradable resources find their utility in a broad spectrum of fragments such as food preservation/packaging, agriculture/bio-sensors, biomedical, optical/energy-storage, water-treatment/marine, automotive/automation, construction, flame retardant/coating, and many more [9 - 11]. However, the complexities that are associated with it are delivering many major issues related to health and the environment. This might be probably due to the acquired high/low-level toxicants in the NCs, as they are formed either directly/indirectly from the raw material supplied as resources (polymers/non-polymers). An unambiguous solution to this debatable question, such that it eliminates the troubleshoots that are associated, would be probably to use NCs that are got from renewable reserves (RR), especially from bio-renewable (BRR) [12, 13]. The potent resourcefulness of BRR is explicitly vivid in the works of Essawy, A.A., Sheer et al., Sharma et al., Parida et al., Yadav et al., Rani et al., Jindal et al., Shafi et al., whose literatures are based on absorptivity/photo-catalytic activity [14], energy-storage devices/application [15], electro-catalysis/energy-storage/waste-water treatment [16], renewable energy-storage [17], food-packaging [18], agricultural bio-mass [19], waste-water remediation [20], and electro-chemical/bio-medical fields [21].

Here in this chapter, we would comprehensively compare the recent advancement/development of varied applications of nano-composites from renewable (RR)/non-renewable (NRR) where the RR would be further addressed as bio-renewable (BRR)/non-bio-renewable reserves (NBRR). In conclusion, the solutions to the debatable question, and the potential developmental break-ups with new-openings for the naturally agglomerated nano-composites would be reflected briefly. Figure – 1 gives us a glimpse of sustainable resources for BRR/NRR NC in a wide array.

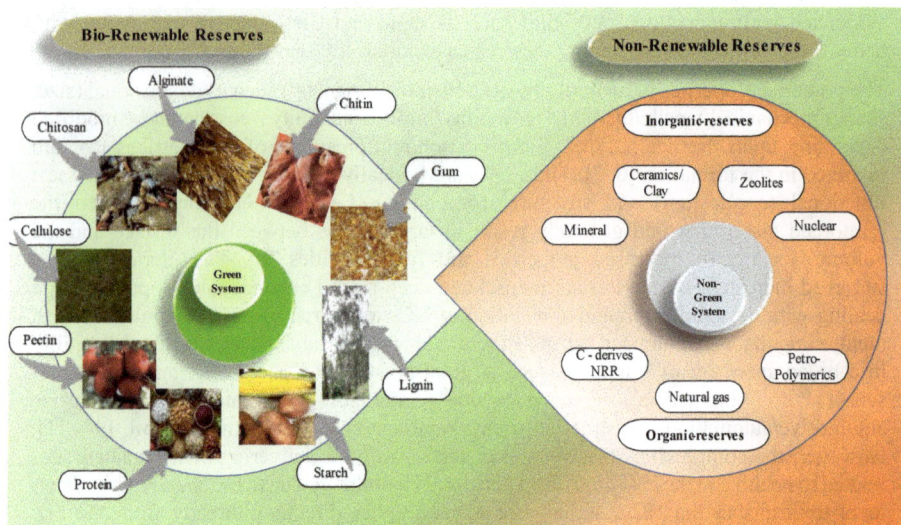

Figure 1: Glimpse of sustainable resources for BRR/NRR NCs in a wide array.

2.   Core reserves for NCS (renewable/non-renewable)

2.1   NCS from non-renewable reserves (NRR)

Depleting materials are the powerful resources of the raw components that are present in the earth's crust and can be categorized here as *non-renewable reserves (NRR)* [22]. With the raising demand in our daily needs, natural minerals/fossil wealth such as petro (polymeric)/mineral-based products are being utilised universally in an enormous way, nearly by all individuals. The most common is being metallics/non-metallics (M/Ns) as metal-metal (MM)/metal-metal-oxides (MMO)/metal-nonmetal (MNM) and metallic-ceramics (MCs)/non-metallic-ceramics (NMCs) from the mineral side and petro-polymerics (PPs) from the fossil side i.e., from the petro-hydrocarbons. However, the essential modifications when done with nanotech and nano-products improve the qualitative/quantitative output [23]. Therefore, interwinding the former (natural minerals) with the latter (PP) as NCs not only improves the value-based characteristics but ensures rapid adaptability and applicability in a wider wing. As a result, metal-based-polymerics (MPs)/ metal-ceramic-based-polymerics (MCPs)/ non-metal-ceramic-based-polymerics (NMCPs)/ ceramic-based-polymerics (CPs) essentially fall under this 3rd group, where the PPs and (M/Ns) as NRR find their source from fossil/mineral wealth. The raw form of the synthetic polymerics is primarily from the natural PP, whereas the synthetic ones are chemically inert/reactive, flexible with resistivity and tough strength [24 - 26]. These PP/metallics/non-metallics as NCs (MPs/MCPs/NCPs/CPs) / (MM/MNO/MCs/NMCs) with high tensile/mechanical/chemical strength/electronic sensibilities seen to have varied functional applicability from road to air to water-ways as tyres/sails/outer-shell. Likewise, these NCs are tapped as components that are remarkably valuable for under-water functions [27]. The degradability rection-time required for NRR-NC is too long a period. Therefore, it gets accumulated on the earth's surface before degradation. Hence, it is a regrettable part with great concerns for the NCs that are produced by using NRRs [28].

2.2   NCS from renewable resources (RR)

The 6th generation of novel materials is focussed in such a way so that the hitches and handicaps that are presently existing while using NRR are completely overcome to a maximum level. Critical minerals/fossils are finite (NRR) and are available in the Earth's crust [29]. NCs with adaptable features would help in resolving present/future global issues. Hence, to narrow and reduce the usage of PP/minerals, the scientific wing is continually probing to fabricate new designs for NCs with enhanced functioning and reduced-cost by indirectly limiting the amount of material used. Contemporary advancements in the fortification/reinforcement philosophies by using new methodologies/materials opted for have eventually resulted in acquiring bio-degradable nano-products. By and large, derives of 'C' as monomers/polymers are used as fillers or plates for reinforcement. Natural and biological derives of 'C' completely satisfy the pitfalls that are commonly faced. Many a time strengthening the properties of NCs leads to non-degradability while using inorganic/PP. Literature survey indicates that biological/natural polymeric materials when reinforced are shown to have positive traits [30 - 34]. This proves that an urge is there for new developments by the technologists, who come up with new formulations using natural/ bio-renewable/bio-degradable reserves. We could notice some deliberate shifts by the researchers/academicians/industrialist towards BRR for fabricating BR NCs. The utility of agricultural supplies as produce/bio-wastes is focussed a lot for BRR and in turn for BR NC [35, 36]. Moreover, the sustainable BRR are to be considered, as the available reserves are depleting faster than the reserve present/replenishing. Rapid fabrication of NCs through quick tools could raise the speed of formation and suffice the need required for application, thus ensuring that the supplies are more than the demand. Moreover, the use of BRR for NCs formation thus would supply very quick nano-product to maintain the ratio between supply and demand. It is expected that the throwaways from the unused and used debris would lead to a harsh environment with their increase to the lowest degree by 3 to 4-folds within the next few years. Despite the negatives seen with the usage of NRR, the positive part is that there is a beneficial and synergistic blend in its structural characteristics such as flexibility/thermal performance/strength [37]. The harmful remains are noxious to the eco-system and to all essential elements of nature (land/water/air) and the living species. This leads to an augmented risk with the raise in illness. Therefore, we would like to mention here that the

BRR and eco-friendly supplies would be the most appropriate solutions to curtail the ill effects despite their limitation [38, 34].

Natural BR bio-polymerics like alginate (Alg), chitosan (Chis), β-cyclodextrin (β_CD), starch (Str), cellulose (Cel), chitin (Chit), pectin (Pec), gelatin (Geln) and lignin (Lig) are widely used as beneficial sources for BR NCs, as they tender for a better advantageous trial, with beneficial outputs such as bio-degradability/bio-compatibility / adaptability/re-usability / non-toxicity/eco-friendliness / low-cost/quick-functionality [39]. The other valuable BRR are hyaluronic acid (HA), terpenes (Ter), vegetable oils (VOs), essential oils (EOs), natural-rubber (plants) and proteins (plants/animals) (keratin/silk/gelatin/gluten/albumin) [40]. On the other hand, lower mechanical/poor water-holding capacity seen as a major dis-advantage can be altered by suitable interwinding substructures, such that it improves the potential applicable properties (tensile/flexural/thermal-impact) [41]. Primarily, the BR NCs fall under the group as NCs with BRR as additives (NC_RRA), NCs with BRR as a matrix (NC_BRRM), and NCs with BRR as additive/matrix (NC_BRRAM). From the articles and reviews seen so far, we can comprehend that BRR functions in an eclectic role with its demand on the raise for new innovations and application of BRR functionalized NCs in almost all fields. Figure 2 gives the needed insights about the core reserves available, to be chosen for safer mode.

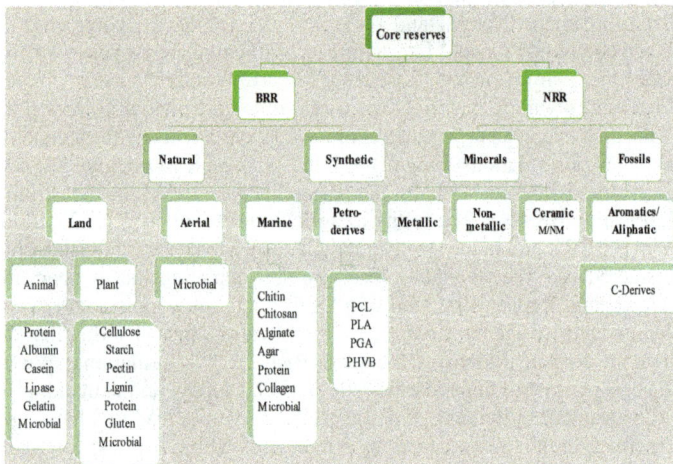

Figure 2. Insights of the core reserves available for safer mode.

Applications of Emerging Nanomaterials and Nanotechnology   Materials Research Forum LLC
Materials Research Foundations 148 (2023) 63-102   https://doi.org/10.21741/9781644902554-3

3. NCS and their application as NRR/BRR

3.1 NCS from non-polymer-based materials (NRR)

3.1.1 Metal-Metal (MM) / Metal-Metal-oxides (MMO) / Metal-Nonmetal (MNM)_based NC

Metal-metal (MM)/metal-metal-oxides (MMO)/metal-nonmetal (MNM) NCs are from the non-polymer-based materials, whose properties (structural/mechanical/ electrical/electrochemical/ optical) basically depend upon the methodology and the base component opted used for the fabrication [42]. Generally, these NCs materials have good mechanical/thermal strength. Thus, they are found to be the most suitable ones with extremely potent capabilities in the area of aerospace/automotive/construction/structural materials formation [43, 44]. However, the complexities that are primarily faced while preparing the NCs are due to the stimulated agglomeration/non-homogeneous dispersals in the medium. Limited time is needed for the growth or nucleation. This enables for suitable matrices before the agglomeration, and the US can aid for a proper wettability in the required phase [45]. Metals/non-metals/salts when inter-fused at suitable condition forms MM/MMO/MNM-based NCs with functional capacities. These NCs find their diversified applicability in environmental-remediation/organic-transformation/hydrogen-evolution/disinfection as photo-catalysts (PC)/adsorbents [46]. High-functioning photo-catalysts could be got by the pairing of semi-conductors (SCs) with metals (M), and they are basically formed due to the increased charge disjunction at the M-SC interface / amplified absorption of light (visible) by the SPR (surface plasmon resonance) of metals. The band-width of the SCs could be adjusted by doping suitable components such that it displays improved properties for light (hv) absorption, but at times might lead to an increased susceptibility of photo-corrosion. Reassimilation of the charges in the doped MSC could take place in an intensified way by the doped group at the interfaces [47]. The structural/functional properties of NC are from the amalgamation of the core and the fillers. The fillers supplement some unique feature such as solubility/bio-compatibility/resistivity/reusability to the NC. Non-polymeric NC with noble/multi-purposeful metals such as Ag, Pd, Au, Pt, Fe, Cu, Ti, Zn, Se, Mo, Ni, Mg, Ca, and many more are notable to be powerful anti-microbials [48 - 50]. Attracting new features like high-electrical/good-thermal conductivity/low-mechanical submissions/needed viscosity have created immense curiosities over the usage of liquid-metal (LM) NC. LM (nano-droplets) are infused in soft polymer-matrix/LM and are mixed with M-NP to form LM NC. Struggles to utilize LM (nano-droplets) in electronics/robotics/biomedical/sensing/energy harvesters are ongoing, with demanding encapsulation methodologies, to tap the properties of the liquid without seeping or smearing [51]. It should be noted that RR supplies the energy as solar/water/wind/tidal that could be effectively used to diminish the discharge of CO_2 into the atmosphere. Photo-catalytic transformation of CO_2 into solar power is a promising note as it addresses increase in energy outputs with no/zero secondary contamination, pertaining to global warming. Cu and its NCs afford suitable solution in this context [52]. Table 1, remarkably signifies the varied application of *MM NCs/ MMO NCs/ MNM NCs.*

Table – 1: The varied application of MM NCs/ MMO NCs/ MNM NCs.

NCs	Application	Remarks	Ref
Bi/Co NCs thin-films	Lubrication-combustion engine/ replacement-Pb/C bearings-engines	High hardness (Co)/ low-coefficient of friction: similar to (soft material-C graphite)/ increases fuel/mechanical-efficiency.	[53]
(GO–Fe_3O_4)/ (GO–Fe_3O_4@ZrO_2): 2D-C framework	Photocatalytic: (UV/hv radiation)/ cationic-dye: Rh B/ anti-bacterial activity: gram-positive/negative bacteria: E. Coli/S. Aurus	Small band-gap/ increased adsorption/photocatalytic action/ shows: absorption/fluorescence/ time: single-photon counting/ nano-second laser-photolysis.	[54]
FLG/Ti, FLG/Al NCs: nano-C (MWCNTs/FLG)	Suitable for structural'-application/aero-space	Light-weight metal-matrix-mechanical-strength, with exceptional functioning.	[55]
$BaTiO_3$/Epoxy-Epoxy Araldite LY1564	Suitable for various application	Active-toughening/ removes fracture-energy/microcracks.	[56]
Co-Fe-Nb-Si-B-MANC Alloy	Suitable for power electronics/magnetics/ transformers/rotating electrical-machinery	High potentials/ soft magnetic-properties.	[57]
$CoFe_2O_4$/Cu(OH)$_2$ NC	Catalytic/magnetic/thermal: activity/ green heterogeneous-catalyst: 1,4-disubstituted β-hydroxy-1,2,3-triazoles	Organic azide-intermediate/ good-regioselectivity/high-product yield/ nontoxic-catalyst with chemical-stability.	[58]
Cu-hemin-MOFs/ GOD-Cu-hemin-MOFs	Electro-chemical glucose (human serum) /biosensing: via catalytic-oxidation	Good-results: detection (glucose)/ electro-chemical sensing: MOFs/GOD.	[59]
Magnetic Cu/$CuFe_2O_4$	Fenton-like catalyst/ pollutant MB removal: waste-water	Radical capture/ coumarin-fluorescent probe system.	[60]
g-C_3N_4/Cu NC	Lubrication/stable additive	Efficient reduction: friction/ lesser friction-coefficient.	[61]
Cu–Ni@SiO_2 alloy NCs	Catalysis: DRM reaction	Good selectivity/improves: H_2 yield/ long lifetime.	[62]
Cu_2O-Cu-CuO: (ternary NC)	Catalytic: 4-nitrophenol reduction	Hydride-transfer/convenient/fast/ cost-effective.	[63]
CuO/Cu-MOF NC	Electro-chemical sensing/detection: NO discharged (live cancer cells)	Linear reaction (broad concentration) interference-negligible (organic molecules ǀ common ions/sensing: excellent stability.	[64]

Cu@Cu$_2$O NC	Photocatalytic/ degradation: MB: waste-water	Microfluid liquid-liquid system/enhanced mass-transfer rates/controlled-oxidation	[65]
Cu-NGr NC.	Bio-sensor: Ampero-metric non-enzymatic glucose/ direct application: detection of glucose-food samples/electro-catalytic	Glucose detection-high/modified glassy C-electrode/high-selectivity/reproducibility/ good-recovery: complex food materials.	[66]
ZnCo$_2$O$_4$/ g-C$_3$N$_4$/Cu	Heterogeneous catalysis: degradation (MNZ)/hv/ suitable for polluted aquatic-system.	Highly effectual photocatalyst/8 runs no loss in efficiency.	[67]

Graphene oxide (GO)/ rhodamine B (RhB)/few-layer graphene (FLG)/ metal amorphous nanocomposite (MANC)/ molecular organic frameworks (MOFs)/glucose-oxidase (GOD)/methylene blue (MB)/methane dry-reforming (DRM) reaction/nitric oxide (NO)/ nitrogen-doped graphene (NGr)/ metronidazole (MNZ)

3.1.2 Metal-Ceramics (MCS) / Non-Metallic-Ceramics (NMCS) based NCS

The modulated significant features of Cer_NC with low-density/wearability renders it to have a better functional applicableness (anti-corrosion/anti-oxidation at increased heat/resistant to wearability/toughness/durability at variable temperature environs) in different fields of industrial/bio-medical [68]. The Cer_NC could be generally categorized a ceramic/carbonaceous-material/metal as nano-phases in a ceramic-matrix. Cer_NC are not restricted only to carbides, but include oxides/nitrides/metallics/non-metallic in them. Nano-phase stability/fracture toughness present in these NCs reveals that metal-oxides of d-block have better potential features than other metallic/organic-oxides at a small nano-range [69, 70]. There is an even distribution of ceramic-matrix with metal/non-metals in Cer_NCs of MCs/NMCs. Usually with the oxides/carbides/nitrides of Al/Si/Ti/Zr/Mg together with W/Mo/Ni/Cu/Zn/Ru/Co/Fe [71]. Cer_NC finds its development/utilisation more progressively with new handling methods such that they enable for quick and easy fabrication of products from the test scale to the marketable ones. The NC, when reinforced as MCs/NMCs improves the product stress and micro-hardness of the newly made-up materials. Additionally perfecting the grain-size can contribute to diminishing the product stress [72]. Fabrication of Cer_NC, by diffusing TiN whiskers in the SiN matrix was seen not to be cost-effective by later researchers due to complexities for the as-prepared NCs by Niihara et al., who were the initiators for designing Cer_NC in 1986 [73]. The NC structured variable ceramic powder coatings are desirable and attractive ones for cutting/wear-resistant. Cer_NCs with NPs as reinforcements delivers outstanding upturns equally at room/high-temperature with exceptional mechanical characteristics when paralleled with pure/raw ceramics, are essentially needed features for variable applications in the aero/engineering/energy harnessing sectors [74]. The affinity of NCs to get agglomerated can be altered/lowered with suitable NPs and procedures, which is the vital step in formulation. Ample reports are available in the literatures with suitable mechanisms that are related to strengthening/toughening/ flaw-size reducing/crack reinforcing/fracture-strength modulating/grain-boundary reinforcement [75]. Similarly, stimulation of ceramic-phase with ionic metal-matrix precedes for a boosted electrical conductivity with improved electrode–electrolyte inter-facial stability, where the properties [(chemistry/size/volume) / (annealing/physical/chemical-matrix/reactivity/temperature)] basically depend upon the ceramic-phase and the metal/non-metal complexes [76]. Another versatile group with its own valuable application, and whose preferred characteristics could be modulated by varying the chemical composition/processing parameters are glass Cer_NC material [77]. This group with oxyhalides/halides (PbF/LaF/LaCl/BaF/BaCl) are glass Cer_NCs, that are found to be incorporated by doping procedures with the rare-earths (RE^{3+}) to improve NC's mechanical-rigidity/chemical-durability with increased up-conversion efficacy [78]. It was noted that the up-conversion efficacy of Cer_NCs improved two-folds when NCs single crystal were doped with RE^{3+} ($La^{3+}/Yb^{3+}/Er^{3+}$) [79]. Similarly, zeolites another group of micro-porous materials with exceptional properties can be finely tuned by altering its composition by fusing with polymeric/mineral composites to form Zeo-NCs. The potential signals like bandgap/luminescence/pore-size/surface-area/ion-exchangeability /chemical-stability can be modulated for potential application as per needs by adopting nano-routes. Literature analysis reveals that Zeo-NCs are potent sources for varied applications in diversified arena with its potency primarily attributed to less/non-corrosive | zero-waste/disposal issues | abundancy/low-cost | high thermal-stability/extreme adaptability/continuous scalability/rapidity. This shows that Zeo-NCs are significantly suitable for water purification/ion-exchange beds/sorbents/catalysts | optically-active resources/micro-electronic chips | polymerization/separation technological tools | photo-electrochemical applications (solar/perovskites cells) | thin film bio-sensors/encapsulation of drugs/pesticides/nano-fertilisers/biomolecules that are mainly targeted for controlled release functions [80]. Table 2 and Table 3 give some potent utilisation of Cer_NC/Zeo-NCs in various adaptable segments.

Table – 2: The potent utilisation of Cer_NCs.

NCs	Application	Remarks	Ref.
TiO$_2$-GO-modified Cer | pristine Cer membrane	Removal of pharmaceuticals: (HA/TA/IBU/ DCF/NAP/CBZ)/bacterial adhesion/anti-fouling.	Gel layer-hydrophobic/high hydrophilicity/ pharmaceutical retention/	[81]
MMT/Ag NC/Sida acuta leaves	Water-treatment: removal MB (99.90%)	Reaction by batch system	[82]
Fe$_2$O$_3$-CeO$_2$ NC-oxide/variable solvents	Electro-chemical H$_2$ storage	Thermal-decomposition/higher H$_2$ storage	[83]
Ce/Y co-doping | BaCe$_{0.16}$Y$_{0.04}$Fe$_{0.8}$O$_{3-\delta}$	Proton-ceramic-fuel cells (PCFCs)	Low-temperature/lesser-AE/greater-IC/highly active/stable cathodes: PCFCs/ORR/ thermo-mechanical compatibility	[84]

$PMo_{12}O_{40}@MnFe_2O_{40}$/Aloe-vera NC	CODS: real/model fuel (DBT/Th/BT) DBT: 98% | real fuel: 96%. (de-sulfuration)	Sol–gel /reusability:96%/5 runs/ effective: industrial-scale/green nano-catalyst	[85]
Hydro-phobic Si/SiOx: (NC-natural palygorskite)	Benzene adsorption (76%).	Volatile/micro/meso/macro-pores	[86]
TiO_2-doped-Ag fibrous NC	Photocatalyst: waste-water treatment MB (94.6%) /anti-microbial: E. coli (99.9%)	Sol–gel | electrospinning | microwave-assisted | green synthesis	[87]
α-Al_2O_3/$CoFe_2O_4$	Removal: MB-model/ibuprofen-real (micro-pollutants)	Highly porous/sponge-like layer/SO4^{--} dominant active radical/ high flux | removal rate/low fouling	[88]
Y_2O_3–MgO Cer NCs	Optical transmittance	GNP | SPS system/high purity/short time | eco-friendly/increase: density/grain growth	[89]
YSZ-TBC- Cer_topcoat	Advanced gas-turbine engines	PS-PVD/highly-phobic (non-wettable)/ lotus-leaf-like-microstructure/protection for molten CMAS deposits/resistant: undesirable environmental-particles/gas-turbine-engines (aircraft propulsion) | electricity generation	[90]
Cu/alumina NC	MEMS/micro-mechanical/ functional micro-nano devices	High interface/volume ratio	[91]
MMNC/CMNC (metallic/Cer NC_ magnetized hybrid-nanofluid) with (Fe_3O_4-TiO_2/H_2O)/(Fe_3O_4-Al_2O_3/H_2O)/ (Al_2O_3-TiO_2/ H_2O)/(TiO_2-Cu/H_2O)	Packaging/bio-medical applications/ suitable for aerospace/automotive/electronics/ biotechnology	High performer /unique design/ extraordinary thermos-physical/ environment friendly	[92]

Ceramic-Cer/humic acid (HA)/tannic acid (TA)/ibuprofen (IBU)/diclofenac (DCF)/naproxen (NAP)/carbamazepine (CBZ)/ montmorillonite/silver nanocomposite (MMT/Ag) /activation-energy (AE)/ionic conductivity(IC)/oxygen reduction reaction (ORR)/catalytic oxidative de-sulfurization (CODS)/(DBT/Th/BT)/ Escherichia coli (E. coli)/ glycine-nitrate process (GNP)/plasma spray physical vapor deposition (PS-PVD)/ metallic/ceramic matrix nanocomposite-magnetized hybrid nanofluid, metallic matrix nanocomposite (MMNC) and ceramic matrix nanocomposites (CMNC) materials.

Table – 3: The potent utilisation of Zeo-NCs.

NCs	Application	Remarks	Ref.
Co-ZSM-5/Co-MOR/Co-BEA	Synthesis: iso-paraffins/ catalytic-activity (light) C5-C12 HC/ isomerization-activity/Fischer-Tropsch synthesis	Meso-porous hierarchical: Zeolites-encapsulated-Co/ ZSM-5: strongest Bronsted acid sites/ higher selectivity to C5-C12 | branched HC	[93]
C-dots@zeolite (CDZ) NCs	Photocatalytic: metronidazole (79%) / Rh B (90%) / real-industrial waste-water	Hydrothermally synthesis/PC:sunlight/UV/ Visible lights/ GC-MS analysis	[94]
RZEO-1: coated meso-porous zeolites RZEO-2: micro–meso-porous NC RZEO-3: meso-porous zeolitic in wall CTAB_Surfactant	Catalytic-activity: alkylation/dehydration/isomerization/ hydro-conversion/cracking/disproportionation/ trans-alkylation reactions.	Desilication/depolymerization/re-assembling micro-porosity/catalytic-activity/selectivity/ hydro-thermal stability	[95]
Ag/zeolite NCs NaBH$_4$-Na$^+$-Y-zeolite powder-SiO$_2$/Al$_2$O$_3$	Anti-bacterial activity: Gram-negative/positive bacteria: E. coli/S. dysentriae/ S. aureus/ methicillin-resistant S. aureus. Suitable for-biological/biomedical applications	Chemical reduction/disk-diffusion	[96]
Magnetic nano/micro-particles clinoptilolite-type of natural zeolite (CZ)	Controlled-drug delivery/ cancer-hyperthermia/ anti-amyloidogenic agents/ carriers: controlled drug delivery/release/ imaging/local heating: biological systems, adsorption: Rh B/ sulfonated aluminum phthalocyanine/hypericin	Unique platform: multifunctional magnetic/ optical probes/ suitable: optical imaging/MRI/ thermos | phototherapy/MCZ: adsorption | release: photodynamic dyes/hyperthermia/ disaggregation/inhibition: amyloid fibrils in vitro/photosensitizers/fluorescence markers	[97]
Zeolite/ZnO NCs	Adsorption: Pb (II) (89%) / As (V) (93%)/ removal: toxic metals	Co-precipitation/low-consumption energy/ facile regeneration	[98]
nZVI–Z | nFe/Cu–Z	Removal: Sr^{2+} (sea-water)/ (sea-water-medium)/(waste-water-streams)	Simple liquid-phase reduction/ water from nuclear accident-Fukushima Daiichi power-station/ Sorption (Sr^{2+}) decreases: in coexisting cations,	[99]

Zeolite-zirconia-copper NCs	Adsorbing asphaltene/upgrading crude-oil/ reduction of CO_2 flooding	Zirconia-copper seen dispersed on zeolite surface/permeability/porosity-reduction	[100]
Zeolite/Fe_2O_3 NCs	Toxicity/cell proliferation/ smart Fe-nanofertilizer/ improves crop yield/soil productivity	Medical applications/in vitro cell studies: zeolites | zeolite/Fe_2O_3 NCs are non-toxic (human-fibroblast cells)/ significantly pernicious to human malignant melanoma-cells/ effective inhibition of malignant/slow-release: iron ions/non-toxic/low-cost raw materials/ low-energy usage/no toxic residues	[101]
TiO_2 – zeolite NC	Photocatalysis: industrial dye waste-water/Reactive Black 5	Sol - gel synthesis/ recovery: spent photocatalysts	[102]
Magnetic Fe_3O_4/zeolite NaA	Adsorption: removal: MB (~96.8%)	Hydrothermal methods/Fe_3O_4: loading increases adsorption/low-cost adsorbent	[103]
η-phase/zeolite |Hombifine N/ zeolite NC/Zeolites-Beta(25)/ZSM-5	Adsorption: P(V) ions (99.48%)/	Amorphous: (η -phase/Beta(25)/ nanocrystalline: (Hombifine N/zeolite)/ microwave/ultrasonic: uniform dispersals of particles (spherical)	[104]

Hydrocarbon (HC)/cetyltrimethyl ammonium bromide (CTAB)/Shigella dysentriae (S. dysentriae)/Staphylococcus aureus (S. Aureus)/ magnetic clinoptilolite zeolite (MCZ)

3.2   NCS from Non-Polymeric-Polymeric (PP/Syn) based materials (NRR)

The science of reinforced poly-NCs reveals that the nature/bonding/size of the fillers have intense influential effect on the applicational properties of the resulting poly-NCs. It might be probably due to inter/intra-facial forces of attraction (matrix-surface/bonding/adhesiveness/dispersion) that are governed by the size of the fillers. This regulates the applicative benches pertaining to reactivity-chemical/catalytic, resistivity-electrical/flow of atomic-molecules, adhesiveness, storage-energy/gas, of the nano-scale polymeric-NCs (fillers/matrix), with at least 1D nano-scales, and finally delivering 3D Poly-NC structures. A broad scale exploration with augmentation of Poly-NCs application are done universally by the scientific wing [105]. Remarkable changes in characteristics (mechanical / thermal-heat resistance / long-term stability / sustainability / rapid fabrication / energy saving / easy handling / cost-efficiency / many more) by modulation, results in a broader segment of functional application (automotive / aerospace / electronic / optical / communication / electrical / photonics / structural / smart-materials / energy-storage / catalytic / environmental-remediation / surface-coatings / flame-retardants / packaging / food-packing / agriculture / bio-medical / anti-microbial /

biodegradable-biomaterials) [106]. The hybridised interfaces between organics and inorganics by weak inter-attractive forces or strong orbital overlap are essential for nano-molecular formation, and primarily depends upon the size of components involved as fillers/matrices for NCs prerequisite application. Inorganics from metals (Au/Ag/Cu/Pt/Pd/Ru/Re/Zn/Hg/Rh/Co/Ni/Li/Fe/Cr) / metal-oxides ($Cu_2O/CuO/CdO/Al_2O_3$/ $MgO/CeO_2/ZrO_2/CeO_2/TiO_2/ZnO/Fe_2O_3/Fe_3O_4/NiO/SnO$ / other particle as $O^{2-}/S^{2-}/Te^{2-}/Se^{2-}$ (Pb/Cd/Si/Al/Ce/Mo/V/Sn) carbon nano-fillers (CNT/G/GO/CF/Flurrenes/nano-diamond) gets embedded or encapsulated or fused as fillers in organic polymerics (Syn/derived) PP/synthetic(polystyrene/epoxies/polyesters/polyamides/polypropylene/low-density-polyethylene (LDPE)/ liquid-crystal-polymers (LCPs)/polyurethane/and others as matrices [107, 108]. However, although NRRs are depleting, and causing alarming signals, it can be noticed that NRR non-polymeric-polymeric NCs are stamped for their unmatched importance in the niche as high-performance advanced nano-materials. However, bounteous profuse work with multi-disciplinary association of material science with physical/chemical/biological/medical/engineering-sciences with nanotechnological tools are foreseeable for an authentic exploration of a new novel group of nano-material [109]. Table 4, delivers versatile recorded application in diversified zones [110-119].

Table – 4: The application in diversified zones.

NCs	Application	Remarks	Ref.
PPy/H-Beta zeolite NC BEA zeolites: SiO_2/Al_2O_3 -host (PPy)	Catalytic applications/ electronic devices electrical conductivity	Chemical oxidation: $FeCl_3$ - polymerization/ superior electronic/magnetic/electrical/ optical properties	[110]
PANI/zeolite NC (Zeolite 13X)	Suitable as anti-ferromagnetic component	Oxidative polymerization/ anti-ferromagnetic component: temperature dependent	[111]
Zeolite-PLA/PEG: stabilizer	Anti-bacterial-activity (E.Coli)	Melt-mixing technique/ good mechanical strength	[112]
Propolis-Embedded Zeolite/PLA/PCL	Anti-bacterial-activity (C. albicans) / dental-implant application	Propolis-embedded-zeolite: PLA/PCL pellets: showed sustained release	[113]
PLA-zeolite/^{60}Co-gamma rays	Radiation-shielding effect	Solution casting/quick degradation/ PLA-increases radiation resistance.	[114]
MANs/poly(MMA-co-BA) NC latex	Heat-insulation applications/ suitable for energy-saving -windows/windshields (cars)	Mini-emulsion polymerization/ enhanced stability/ heat-blocking properties.	[115]
ZnO-PC NC film	Food-Packaging/ anti-bacterial (S.Arus/E.Coli)	Synthesis-blade coating/ enhanced UV-blocking properties/ hydro-phobic nature	[116]
Fe_3O_4@PS/DVB-MNPs	Extraction-atrazine from soil/water /recovery-53%.	Extraction by (MD-SPE)/(DLLME)/ mini-emulsion polymerization.	[117]

APTMS-coated ZrO$_2$ ceramic NCs	3D printing - supportless stereolithography	Improves dispersion stability/ Photo curing-properties/high viscosity	[118]
UiO-AM@POPs (MOFs)@POPs NCs (UMM)	Photo-dynamic therapy (PDT)- bio-medical application/ cancer therapy.	Minimises the damages to nearby tissues, photosensitizer (PS)/ photochemical/ photobiological reactions strong hydro-phobic interactions/ stacking of π-π porphyrins	[119]

Methacryloxypropyltrimethoxysilane (MPS) functionalized antimony-doped tin oxide nanoparticles (MANs)–poly(methyl methacrylate-co-butyl acrylate) (PMMA-co-BA, PMB) nanocomposite/ polypyrrole (PPy)/ polyaniline (PANI)/ poly(lactic acid)(PLA)/ polyethylene glycol (PEG)/ poly(L-lactide) (PLA)/poly(ε"-caprolactone) (PCL)/magnetic polymeric NCs (MNPs)/ poly(styrene/divinylbenzene) (PS/DVB)/ magnetic-dispersive solid-phase extraction (MD-SPE) and dispersive liquid-liquid microextraction (DLLME)/ mini-emulsion polymerization technique/ polycarbonate (PC)/amino based (AM)/porous-organic-polymers (POPs)

3.3   NCS from Polymeric-based materials (NRR)/(BRR)

NCs derived from BRR are to be noted for a sustainable greener advantageous approach rather than from NRR that are from fossil/mineral wealth. Natural bio-organic polymers (Cellu/Lig/Pec/Str/rubber/gum/gluten) plant / (Alg/Chit/Chis) marine / (albumin/gelatin/casein) animal (protein/DNA/EOs) from both and carrageenan from BRR (Figure - 3) gets embedded or encapsulated or fused as fillers/matrices i.e., as polyelectrolytes with cationic/anionic junctional interfaces to deliver versatile properties. The author Ates et al., in his article was of the view that potent structural functionalities (flexibility/thermal-performance / transparency/mechanical-strength / cost-reduction/affordability / rapid preparation/stability) from the derivatives of PP, delivers some constructive, and beneficial synergistic blends despite its undesirable impacts to the eco-system. Likewise, other bio-degradable bio-polymerics as fillers, from natural/synthetic/PP (PLA/PHB/PCL/PGA/PHBV/PBAT) are potent competitors for NCs application especially in bio-medical, food packaging, environmental remediation fields [4, 120, 121]. Appropriate sustainability and adaptability to environmental coordination are achieved when the essentials of bio-degradability are satisfied by the NCs. The pit-falls due to land/water-fills caused by non-degradable reserves have stimulated awakening for a new dawn to overrule the existing conflicts risen by the non-degradable. High performance of the new polymeric NCs is activated by surface modification and physico-chemical intricacies, when NC polymeric matrices are embedded with NMs. The critical parameterises essential in fine-tuning the inter-actions between the host and the fillers are dimensional-geometrical shape/size/surface-ratio of NMs in the output system [107].

Momentous, beneficial, sustainable, and stable NC products, with applicational expansions are practically obtained by infusing bio-polymeric sciences with material-sciences to deliver functionals with their wings spread out in almost all practical needs like (bio-sensing/bio-medicals/catalytics/micro-opto-electronics/magnetics/photonics/energy-storage/food-other   packaging/textile/flame-retardant/   thermal-resistant-surface-coating/smart-materials/bio-materials/automotive-engineering/environmental protectors/high-performing-materials [122]. Basically, bio-polymerics are noticed to possess poor electric conductivity, hence, combining bio-polymeric matrixes with conductive NMs affords for an enhanced conductivity that are suitably applicable for micro/smart electronic counterparts. Low degree of dispersion and minimal surface interactions between the bio-polymer matrixes and NMs, might lead to that decline in thermal resistance and stability of NCs. Adding NMs to bio-polymeric matrix with required dispersion   enhances   the   mechanical/thermal   factors   (tensile strength/modulus/stiffness/flame resistance) essentially needed [108]. However, the challenges that are met with during fabrication/application are modulated as per the situational needs using nano-routes and nano-tools. Alteration thus, delivers signals for improved transparency/lightweight/stability with resistance to chemicals/thermal/climatic variation, that are needed for successful application. Table 5, signifies the recent work done in functional segment with BRR/NRR as resources.

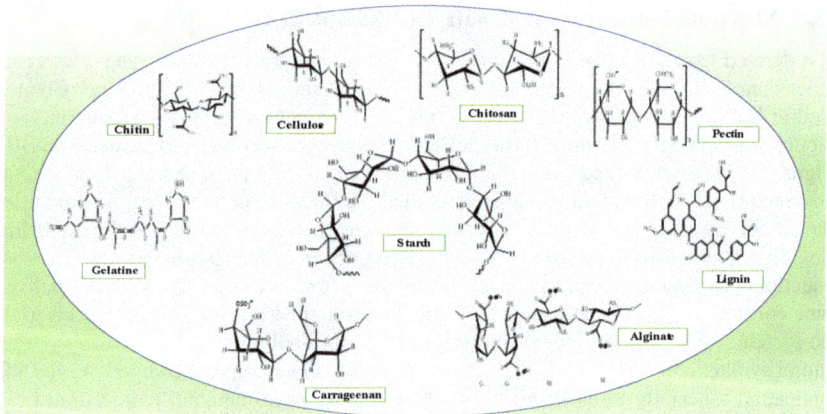

Figure 3: Natural bio-polymerics-BRR

Table – 5: The recent work done in functional segment with BRR/NRR as resources.

Nanocomposites	BRR/NRR	Application	Remarks	Ref.
CNC/CNF/BNC/Chis/Alg/NC	BRR	Food packaging	Thin-flexible film/good-tensile strength/ transparent formulation/ higher hydro-phobicity(non-wetting)	[123]
Alg/nHA/Linum usitatissimum - Extract	BRR	Bone/tissue engineering	Wet-chemical/precipitation/ bioactive/osteo-active/potent anti-oxidant: promotes bone regeneration	[124]
HX-CNT	NRR	Potential-biomaterial cytotoxicity-non-toxic	Good elastic-strength/ decreased: swelling ratios	[125]
MFC/NR film/SA Ncs	BRR	Suitable for: packaging/ bio-degradable wound dressing.	Enhanced tensile-strength/more water adsorption ability/toluene resistance/ bio-degradability	[126]
Alginate/Nano-Cellulose beads (Corncob, nanocellulose, cellulose)	BRR	Protein (bovine serum albumin (BSA) adsorption	Batch method/adsorption increases with increase in NC beads/low density/high mechanical-strength/ high surface-area	[127]
SA/SA-C/SA-P/SA-Ch: beads	BRR/NRR	Large-scale wastewater-treatment: sorption: Cd (II)	Batch scale/highest adsorption: SA-Ch/ maximum break-through time:30 - 48 h	[128]
(CF/T/G): NPs/ (CF/G, T/G) microparticles	NRR	Water-remediation/ removal: $Cu^{2+}/Fe^{3+}/As^{3+}$ ions from water	Removal efficiencies: using: G, CF, T, CF/G and T/G: Cu^{2+}: 91%, 100%, 99.9%, 95%, 98%. Fe^{3+}: 60%, 100%, 100%, 60%, 82%. As^{3+}: 98% - T nano-adsorbent	[129]
Alg-Natural Clay HGs	BRR/NRR	3D printing NC inks-bone-restoration therapy	Freeze-drying fabrication/ bio-compatibility/bio-degradability/ shear-tinning/quick-gelation/ regenerative-medicine.	[130]

Fe_3O_4/CD/AC/SA	BRR/ NRR	Water-remediation: removal: BG/MV/NOX/CPX/Cu^{2+}	Low-cost adsorbents/ desorption/adsorption:4 runs/ environmentally compatible/	[131]
Fe3O4/AC/CD/Alg	BRR/ NRR	Water-remediation/dye-adsorption removal: MB (99.53%/90 min)	Direct mixing (polymer matrix with nanofillers)/ polymer gel/dry-beads	[132]
Met-GO/SA	BRR/ NRR	Water-remediation/antibiotic-adsorption removal: OFX and MOX	Adsorption process:4 runs/easy-handling /eco-friendliness/quick-recovery	[133]
GFMNPECABs	BRR/ NRR	Water-remediation/ adsorption: Hg(II)/leucocrystal violet.	Batch adsorption, reusability:4runs/ low-cost/eco-friendly	[134]
Cellulose-gel beads	BRR	Suitable for transparent-film	Good strength/toughness	[135]
Nanocellulose (80%-cellulose, 17%-hemicellulose, 3%-lignin)/BEK	BRR	Nano-paper - packaging	Reduced energy consumption/good elastic-modulus	[136]
CMC/NFC/CA	BRR	3D-printing/bio-medicinal	No cytotoxicity/high porosity/ Good mechanical-strength/ bio-compatible	[137]
Cellulose/lig HG beads/ Lipase (Candida rugosa)	BRR	Suitable for biocatalytic/biomedical/ bio-electronic fields	Enhanced stability/activity bio-compatibility/bio-degradability.	[138]
Silk Sericin/Lig Blend Beads	BRR	Waste-water remediation-industrial removal: Cr^{6+} (90%)	adsorption-desorption:7 runs/ recycling efficiency (82%)	[139]
Lig-PVA		Anti-plasticizer: polymeric fibers	high crystallinity-lignin gel-spun PVA fibers/high mechanical-strength	[140]
Ligno-cellulosic/cellulosic hydrogel Beads/ Kraft pulp/ functionaliz Magnetization/carboxymethyla	BRR	Adsorption-heavy metals-Cu^{2+}/Pb^{2+}	Functionalized ligno-cellulosic beads-maximum uptake	[141]
Lignin-γ valerolactone (GVL) food additive/ Water (GWBS)	BRR	EO (cinnamaldehyde/eugenol): antimicrobial activity: Penicillium italicum (orange)	Pickering emulsion/ Good-dispersibility (water)/ bio-compatibility/eco-friendliness	[142]

LNPs-SA /kraft lig (LNPs)	BRR	Removal: MB (97.1%)	Potent adsorption/ bio-degradable/bio-compatible	[143]
Soft-wood Kraft lig/PAMAM n Nano-fibers	BRR-NRR	Lignin/PAMAM mats/Suitable for: membranes/filtration/cont rolled release/ drug-delivery	Electrospinning method / bio-degradable/bio-compatible	[144]
Gel-GO NCs	BRR	Kluyveromyces lactis Encapsulation/ probiotics/bioreactor packings/ anti-microbial activity	Probiotic-yeast encapsulation/ enhanced electrical response/solubility/dispe rsibility: water-other solvents/high bio-compatibility	[145]
Lig/polycaprolactone (PCL)	BRR	polymeric-fibers/ soft-tissue engineering	Electrospinning method /bio-degradable/ bio-compatible/appropriate-porosity/ optimum pore-size | fiber diameter/ enhanced tensile-strength	[146]
Chis/Alg/CuO	BRR/ NRR	Nanofertilizers/Micro-nutrient seedling-germination: Fortunella margarita	Slow Cu release/nano-fertilisers/ increased efficacy: seed germination/ bio-degradable coating/eco-friendly	[147]

Hydroxamic derivative of alginate (HX), multiwalled carbon nanotube (CNT)/poly(vinyl alcohol) (PVA)/essential oils: (EO)/γ-valerolactone (GVL)/water binary solvent (GWBS)/sodium alginate (SA)/lignin NPs(LNPs)/methylene blue (MB)/polyamidoamine (PAMAM)/nano-fibrillated cellulose (NFC)/carboxymethyl cellulose (CMC)/citric acid (CA)/hydrogel (HGs)/ gelatin(Gel)/polycaprolactone (PCL)/ natural rubber (NR)/micro-fibrillated cellulose (MFC)/nano-hydroxyapatite (nHA)/methionine functionalized-graphene oxide-sodium alginate polymer (Met-GO/SA)/glycine functionalised magnetic nanoparticles entrapped calcium alginate beads (GFMNPECABs)/SA-Clay (SA-C)/SA-Phosphate (SA-P)/SA activated charcoal (SA-Ch)/cobalt ferrite nanoparticles (CF)/titanate nanotubes (T)/alginate (G)/bleached Eucalytptus-pulp (BEK)

4. Present situation and futuristic outlook

Sustainable equilibrium in/for the eco-system, is a current hot topic withdrawing the attention for utilising bio-renewability and bio-renewable products. With a commitment to utilise the ample BRR available as natures' gift, we are bound to develop new BRR-based products, based on nano-techno-craft innovations [148, 149]. A quick outlook of the diversified wings where, NCs could be effectively used as an operative tool is virtually glimpsed in Figure – 4. It should be noted that ample literatures are seen available for bio-organic-polymerics with minerals/fossils when on comparison with bio-polymeric-bio-polymerics with limited reservations. The bountiful expressions of ample literatures primarily indicate for a gigantic shift towards BRR, however, it is indictive of a novel requirement of only sustainable bio-polymeric NCs for functional application. Table – 6, explicitly delivers information that researchers and scientists as working committees in this 21st century is toiling hard to overcome the barriers caused due to various reason. This fruitful resource principally as an eye-opener, not only gives us information based on functional/structural advances in the NC application using BRRR/NRR but indirectly tries to save the future generation through the knowledge imparted, resourceful potent applications available, and future modulations that are required while using the bounteous supplies of the nature. With the workability as greater advantages of bio-degradation and with lesser dis-advantages about the time involved in degradation (basically due to rapidity), and as the materials disintegrate before the completion of the reaction, now upcoming researches are to be notably focussed on how to re-solve this issue. However, it should be noted that when BRRs are fused with NRRs this problem gets resolved generally but leads to a slow bio-degradation.

Likewise, the next issue is about the durability of the BBRNCs with rich mechanical swings, as they are not at par with NRRNCs/BRR-NRR-NCs. New innovations are to be now focussed on this segment too. Presence of nano-fillers from BRRs in NCs would augment the mechanical/thermal/flame-retardancy superiority of the product. Thus, specifically, nano-fillers from BRRs provides new eye-sights for the best suitable developmental composites, where the surface-interaction between the two module (matrix/fillers) plays a significant role in promoting the applicational activity. Furthermore, the limitations noticed with BRRs-NCs for the industrial scale-ups is another important sector which demands for the required attention. Thus indicates a new scope for the research in this area. Similarly, bio-plastics, water/environment clean-ups, and health sectors too, submit for new challenges with authentic unquestionable BRR-NCs. The major positivity's and negativities of bio-renewable (BRR)/non-bio-renewable reserves (NBRR-PP) that are demarked here in this chapter in Table – 7 would help the research community for augmenting their future work. Probably, with these new insights in bio-renewable/sustainable supplies could be opted as beneficial platforms for reinforcement in a wide array of applicable functions.

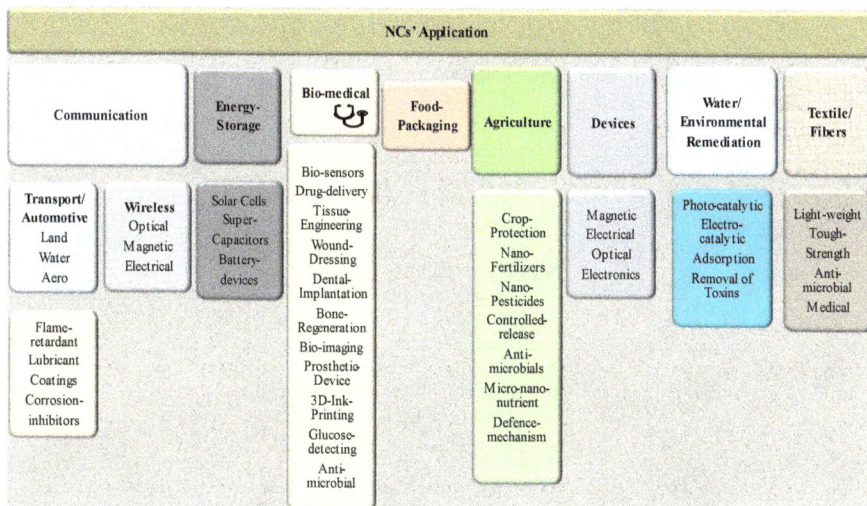

Figure 4: Operative segments of NCs

Table – 6: The researchers and scientists in the working committee who are toiling now to overcome the barriers.

NCs' review based on	Nature	Highlights	Year	Ref.
NCs for Agrochemicals (Variable NC)	BRR/NRR	Synthesis/characterization/ sustained release/ targeted delivery/ application/ commercialization-nano-formulations/ biosafety	2020	[150]
NC' for Environmental / Industrial	Polymerics	Synthesis/applications: water-treatment, electro-magnetic shielding (aerospace)/ sensor devices/ food packaging	2022	[151]
NC' for Food packing	BRR	Types: bio-degradable bio-based matrices/synthetic | natural bio-degradable polymerics (polysaccharide) | protein-based/advantages of nano-technology to improve the functionals	2018	[152]
NCs' for bio-medical applications	BRR/NRR	Properties/ short-comings/ bio-medicine/ drug-cell delivery/ cancer treatment/ tissue-regeneration/ bio-sensing/ bio-imaging/ insulin delivery	2022	[153]
NCs' for supercapacitor applications	BRR/NRR	Merits/demerits of COFs/ recent advancements/ designing features/ types of material/ real applications	2021	[154]

NCs' for C-Capture	NRR	Nano-doped membranes/ carbon capture/ gas separation/effect of physico-chemical properties/ water vapour/ for C-capture/ membrane-module design/ material	2020	[155]
NCs' polymers/applications	BRR/NRR	Areas: sensing/ bio-sensing/ drug-delivery/ actuators/ poly(N-isopropylacrylamide)-based microgels	2017	[156]
NCs' for Sensors (CNTs) Carbon Nanotubes	NRR	CNT devices: sensing bio-molecules/ gas/light/pressure changes/ modification required for good performance	2018	[157]
NCs' for Anti-bacterial	BRR/NRR	Enzyme–metal NC/ synergistic effect/ reduces metallic contents/ increase bio-compatibility/ reduces metallic toxicity/ deactivated enzymatic-functions/ synthesis/bio-application	2022	[158]
NCs' for high performance Li$^+$/ Li metal batteries	BRR/NRR	Increased potential for next-generation LIBs/ synergetic performances by CSSEs/ recent advancements | investigations/ inorganic NC-organic–inorganic NC/ comparative study on types: CSSEs	2020	[159]
NCs' for plants	BRR	Low-cost-Chis NC/ advances: fabrication/application / challenges | future development | opportunities Chit-NC for plants	2021	[160]
NCs' as biocide	BRR/NRR	Antimicrobial activity: Cu-polymeric NC/ application/ physical | chemical features/ influencing factors | variances in susceptibility: gram negative/ positive bacteria	2016	[161]
NCs' for catalytic application	NRR	Synthesis: Ag-NC/ catalytic oxidation/ reduction/ photo-catalysis/ electro-catalysis	2019	[162]
NCs for aero-space	NRR	Synthesis/ Properties/ application/ advances: in mechanical-strength/ durability/ flame-retardancy/ chemical-resistance/ thermal | climatic-stability/ coatings / light-weight structures/ micro-electronic sub-system: aero-space industry	2021	[163]
NCs for CO$_2$ to solar fuels	NRR for BRR	Cu NCs photocatalysts / recent developments/ morphology/ structure/ performance/ reaction mechanisms/ pathways: solar-fuel generation/ enhanced energy conversion-efficiency | production rates.	2017	[164]
NCs' for variables	BRR/NRR	Lig NC: flame retardancy/ food packaging/ plant protection/ electro-active materials/ Energy-storage/ health sciences: application/ critical in-sights: bioeconomy concepts | value-added components.	2021	[165]
NC' for UV protection	BRR/NRR	Lig: agri-biomass/ sustainable approach/ lig-based ZnO | TiO$_2$: UV protective application: coatings/ films/ sunscreen/ economically | environmentally safer alternatives	2021	[166]
NCs' for environmental	BRR/NRR	Alg NC: synthesis/ application/ adsorption mechanisms/ performances/ environmental-remediation: dyes/heavy metals/antibiotics –	2018	[167]

		water/waste-water/ physicochemical/ sorptive-properties		
NCs for electronics	BRR	Bionanomaterials/ source/ design/ performance/ storage/ applications	2022	[168]
NCs' for Future perspectives	BRR	BRR_NCs/ bio-renewable polymers: sources/ types/ synthesis NCs/ applications	2022	[169]
NCs' for bio-medical(nano-heaters)	BRR/NRR	Interfacial-phenomena/ magnetic-heating/ multifunctional applications	2021	[170]

CNTs-carbon nanotubes/ lithium-ion batteries (LIBs)/ composite solid-state electrolytes (CSSEs)

Table – 7: The positivity's and negativities of BRR/ NBRR-PP

Component	Negativity	Positivity
Starch (BRR)	Poor: mechanical-strength/ low-temperature resistance/ high hydro-philicity/ high-biodegradability/ low-decomposition temperature	Low-toxicity/bio-compatibility/ bio-degradability/renewability / low-cost/ easy handling/ easy availability
Cellulose (BRR)	Low-adhesiveness/low-water-retention	Low-toxicity/ bio-compatibility/ bio-degradability/ renewability/ low-cost/ low-density/ good antimicrobial/ easy handling/ quick results/ non-abrasive
β_CD (BRR)	Poor: mechanical-strength	Low-immunogenicity/ Low-toxicity/ high hydro-philicity/ low-viscosity/ high water-solubility
Chitosan (BRR)	Low-porosity/ low-mechanical-strength	Low-toxicity/ bio-compatibility/ bio-degradability/ renewability/ low-cost/ good antimicrobial/ easy handling/ quick results
Chitin (BRR)	Insolubility in neutral/ acidic condition	Low-toxicity/ bio-compatibility/ bio-degradability/ renewability/good-availability/ low-cost/ easy handling/ quick results/ high-mechanical-strength
Alginate (BRR)	Restricted control: mechanical-strength	Low-toxicity/ bio-compatibility/ bio-degradability/ renewability/ good-availability/ low-cost/ easy handling/ quick results/ good anti-microbial/ easily available
Pectin (BRR)	Low-mechanical-ability/highly-hydrophilic/ in-effective for moisture-transfer-in food	Good-solubility/ effective supplemental carrier
Gelatin (BRR)	Hindrances in medicinal drug administration	Good oxygen-barrier property/ good solubility/ high-mechanical activity/ extremely transparent
Lignin (BRR)	Complex-disordered structure	Diversified characteristics with positivity's
NRR-PP	Nondegradable/ detrimental effects: soil/water/air/wildlife/habitat/human health/ highly-hydrophobic	Flexibility-high/ good thermal/ mechanical performance/ transparency/ tough-strength

Conclusion

The new technological developments would get limited by the predictable depletion of NRR from fossils/minerals, however, the utility factor would not be curtailed by the growing global needs. Eventually the combination of BRRs' with material sciences results in NC products that are extremely adaptable and suitable for eco-friendly applications, with perfected mechanical/chemical/physical signals and bio-compatible/bio-degradable elements. Cellulose/Starch, Pectin/Lignin/Guar-gum, Alginate/Gelatine, Chitosan/Chitin, Collagen /Hyaluronan, Albumin/Carrageenan, Soya/Whey/EO/β-CD as natural bio-polymerics are few amongst the BRR, with suitable structural significance for rapid surface interactions, while PGA/PLA/PCL/PAH/PHBV are few among the synthetic-bio-polymerics, that are mainly focused as fillers/matrixes in NC world. NCs from these reserves are effectually applied for biomedical-sciences, environmental-sciences, agri-sciences, power-generation/storage-sciences, communication-sciences (optical/automotive/transport). Hence, for a sustainable balance in the environmental system, initiating the attention for an effective utilisation bio-renewability and bio-renewable products are essential. With a commitment to utilise the ample BRR available as natures' gift, we are bound to develop new BRR-based products, based on nano-techno-craft innovation.

References

[1]   A. Sharif and M.E. Hoque, Renewable resource-based polymers, in: Bio-based polymers and nanocomposites, M. Sanyang, M. Jawaid (Eds.), pp. 1 – 28, Springer, Cham. Switzerland, 2019. https://doi.org/10.1007/978-3-030-05825-8-1

[2]   B.D. Malhotra and Md. A. Ali, Nanocomposite materials: biomolecular devices, in: Nanomaterials for biosensors fundamentals and applications micro and nano technologies, B. D. Malhotra, Md. A. Ali (Eds.), pp 145 – 159, Elsevier, Amsterdam, Netherlands, 2018. https://doi.org/10.1016/B978-0-323-44923-6.00005-4

[3]   E.O. Mikličanin, A. Badnjević, A. Kazlagić and M. Hajlovac, Nanocomposites: a brief review, Health Technol., 10(2020) 51–59, https://doi.org/10.1007/s12553-019-00380-x

[4]   B. Ates, S. Koytepe, A. Ulu, C. Gurses and V.K. Thakur, Chemistry, structures, and advanced applications of nanocomposites from biorenewable resources, Chem. Rev., 120(2020) 9304–9362. https://doi.org/10.1021/acs.chemrev.9b00553

[5]   S. Jafarzadeh, A.M. Nafchi, A. Salehabadi, N.O. Abbasabadi and S.M. Jafari, Application of bio-nanocomposite films and edible coatings for extending the shelf life of fresh fruits and vegetables, Adv. Colloid Interface Sci., 291(2021) 102405. https://doi.org/10.1016/j.cis.2021.102405

[6]   S. Sreevidya, S. Kirtana, Y.R. Katre, A.K. Singh and J. Singh, Functionalized nanomaterial (FNM)–based catalytic materials for water resources, in: Functionalized nanomaterials for catalytic application, C. M. Hussain, S. K. Shukla, B. Mangla (Eds.),

pp 1 – 51, Scrivener Publishing LLC, John Wiley & Sons, Inc., 2021. https://doi.org/ 10.1002/9781119809036.ch1

[7]   S. Sreevidya, S. Kirtana, Y.R. Katre and A.K. Singh, Application of biosurfactant during the process of biostimulation for effective bioremediation of a contaminated environment, in: Green sustainable process for chemical and environmental engineering and science, biosurfactants for the bioremediation of polluted environments, Inamuddin, C.O. Adetunji, (Eds.), pp 291 – 321, Elsevier, Amsterdam, Netherlands, 2021. https://doi.org/10.1016/B978-0-12-822696-4.00003-6

[8]   M.M. Shameem, S.M. Sasikanth, R. Annamalai, R.G. Raman, A brief review on polymer nanocomposites and its applications, Mater. Today: Proc., 45(2021) 2536–2539. https://doi.org/10.1016/j.matpr.2020.11.254

[9]   P.N. Catalano, R.G. Chaudhary, M.F. Desimone, P.L. Santo-Orihuela, A survey on analytical methods for green synthesized nanomaterials, Curr. Pharm. Biotechnol., 22(2021) 813–837. https://doi.org/10.2174/1389201022666210104122349

[10]   R.G. Chaudhary, A.K. Potbhare, P.B. Chouke, A.R. Rai, R.P. Mishra, M. Desimone, A. Abdala, Graphene-based nanomaterials and their nanocomposites with metal oxides: biosynthesis, electrochemical, photocatalytic and antimicrobial applications, Magnetic Oxides and Composites II, Materials Research Forum, 83(2020) 79–116. https://doi.org/10.21741/9781644900970-4.

[11]   S. Wan, J. Peng, L. Jiang, Q. Cheng, Bioinspired graphene-based nanocomposites and their application in flexible energy devices, Adv. Mater., 28(2016) 7862–7898. https://doi.org/10.1002/adma.201601934.

[12]   N. Sumrith, S.M. Rangappa, R. Dangtungee, S. Siengchin, M. Jawaid, C.I. Pruncu, Biopolymers-based nanocomposites: properties and applications. in: Bio-based polymers and nanocomposites, M. Sanyang, M. Jawaid, (Eds.), pp. 255 – 272, Springer, Cham., 2019. https://doi.org/10.1007/978-3-030-05825-8_12

[13]   M. Rajinipriya, F. Gauvin, M. Robert, S. Elkoun, M. Nagalakshmaiah, Structural properties of protein and their role in polymer nanocomposites. in: Bio-based polymers and nanocomposites, M. Sanyang, M. Jawaid, (Eds.), pp. 217 – 232, Springer, Cham., 2019. https://doi.org/10.1007/978-3-030-05825-8

[14]   A.A. Essawy, Biorenewable Nanocomposites as highly adsorptive and potent photocatalyst materials for producing immaculate water, in: Biorenewable nanocomposite materials, vol. 2: desalination and wastewater remediation, D. Pathania, L. Singh, (Eds.), pp 259 – 280, Vol. 1411, ACS Symposium Series, American Chemical Society, 2022. https://doi.org/10.1021/bk-2022-1411.ch010

[15]   F. Sher, M. Ilyas, M. Ilyas, U. Liaqat, E.C. Lima, M. Sillanpää, J.J. Klemeš, Biorenewable nanocomposites as robust materials for energy storage applications, in: Biorenewable nanocomposite materials, vol. 1: electrocatalysts and energy storage, D. Pathania, L. Singh, (Eds.), pp 197 – 224, Vol. 1410, ACS Symposium Series, American Chemical Society, 2022. https://doi.org/10.1021/bk-2022-1410.ch008

[16]   R. Sharma, A. Kumari, Potential applications of biorenewable nanocomposite materials for electrocatalysis, energy storage, and wastewater treatment, in: Biorenewable nanocomposite materials, vol. 1: electrocatalysts and energy storage, D. Pathania, L. Singh, (Eds.), pp 25 – 46, Vol. 1410, ACS Symposium Series, American Chemical Society, 2022. https://doi.org/10.1021/bk-2022-1410.ch002

[17]   S. Parida, D.P. Dutta, Nanostructured materials from biobased precursors for renewable energy storage applications, in: Biorenewable nanocomposite materials, vol. 1: electrocatalysts and energy storage, D. Pathania, L. Singh, (Eds.), pp 307 – 366, Vol. 1410, ACS Symposium Series, American Chemical Society, 2022. https://doi.org/10.1021/bk-2022-1410.ch013

[18]   G. Yadav, M. Ahmaruzzaman, Food Packaging Applications for Biorenewable-Based Nanomaterials, in: Biorenewable nanocomposite materials, vol. 1: electrocatalysts and energy storage, D. Pathania, L. Singh, (Eds.), pp 257 – 267, Vol. 1410, ACS Symposium Series, American Chemical Society, 2022. https://doi.org/10.1021/bk-2022-1410.ch010

[19]   A. Rani, A. Kumari, M. Thakur, K. Mandhan, M. Chandel, A. Sharma, Bionanocomposite synthesized from nanocellulose obtained from agricultural biomass as raw material, in: Biorenewable nanocomposite materials, vol. 1: electrocatalysts and energy storage, D. Pathania, L. Singh, (Eds.), pp 47 – 74, Vol. 1410, ACS Symposium Series, American Chemical Society, 2022. https://doi.org/10.1021/bk-2022-1410.ch003

[20]   R. Jindal, K. Kaur, Khushbu, V. Vaid, Biorenewable nanocomposite: recent advances and its prospects in wastewater remediation, in: Biorenewable nanocomposite materials, vol. 2: desalination and wastewater remediation, D. Pathania, L. Singh, (Eds.), pp 313 – 340, Vol. 1411, ACS Symposium Series, American Chemical Society, 2022. https://doi.org/10.1021/bk-2022-1411.ch012

[21]   A. Shafi, N. Bashar, J. Qadir, S. Sabir, M.Z. Khan, M.M. Rahman, Advanced Biopolymer-based nanocomposites: current perspective and future outlook in electrochemical and biomedical fields, in: Biorenewable nanocomposite materials, vol. 2: desalination and wastewater remediation, D. Pathania, L. Singh, (Eds.), pp 341 – 354, Vol. 1411, ACS Symposium Series, American Chemical Society, 2022. https://doi.org/10.1021/bk-2022-1411.ch013

[22]   Nonrenewable Resources | National Geographic Society https://www.nationalgeographic.org/ encyclopedia/nonrenewable-resources/

[23]   R.M.G. Rajapakse, Depletion of nonrenewable energy resources, entropy crisis and nanotechnology solutions. J. Natal. Sci. Found., 35(2007) 59–61. https://doi.org/https://doi.org/10.4038/jnsfsr.v35i2.3669

[24]   Z. Ayazi, Application of nanocomposite-based sorbents in microextraction techniques: a review, Analyst., 142(2017) 721–739. https://doi.org/10.1039/C6AN02744J

[25] P.H.C. Camargo, K.G. Satyanarayana, F. Wypych, Nanocomposites: synthesis, structure, properties and new application opportunities, Mater. Res., 12(2009) 1–39. https://doi.org/10.1590/S1516-14392009000100002

[26] R. Paul, L. Dai, Interfacial aspects of carbon composites, Compos. Interfaces, 25(2018) 539–605. https://doi.org/10.1080/09276440.2018.1439632

[27] V.T. Rathod, J.S. Kumar, A. Jain, Polymer and ceramic nanocomposites for aerospace applications, Appl. Nanosci.,7(2017) 519–548. https://doi.org/10.1007/s13204-017-0592-9

[28] M.G. Romero, R. Aguado, A. Moral, C. Brindley, M. Ballesteros, From traditional paper to nanocomposite films: analysis of global research into cellulose for food packaging, Food Packag. Shelf Life, 31(2022) 100788. https://doi.org/10.1016/j.fpsl.2021.100788

[29] P. Wang, N. Li, J. Li, W.Q. Chen, Metal-energy nexus in the global energy transition calls for cooperative actions, in: The material basis of energy transitions, A. Bleicher, A. Pehlken, (Eds.), pp 27 – 47, Elsevier, London, 2020. https://doi.org/10.1016/B978-0-12-819534-5.00003-9

[30] N. Saba, M. Jawaid, M. Asim, Nanocomposites with nanofibers and fillers from renewable resources, in: Green composites for automotive applications, part III: nanomaterials and additive manufacturing composites, Woodhead publishing series in composites science and engineering, G. Koronis, A. Silva, (Eds.), pp 145 – 170, Elsevier, UK. 2019. https://doi.org/10.1016/B978-0-08-102177-4.00007-0

[31] S. Sreevidya, S. Kirtana, Y.R. Katre, J. Singh, A.K. Singh, M. Aleksandrova, R. Khenata, Green nanostructures synthesis and spectroscopic characterizations, in: Nanometric spectrometric application, K. Pal, (Eds.), pp 103 – 136, Jenny Stanford Publishing, NY, 2021. https://doi.org/10.1201/9781003160335

[32] N.C. Loureiro, J.L. Esteves, Green composites in automotive interior parts: a solution using cellulosic fibers, in: Green composites for automotive applications, Part II: thermosetting and thermoplastic materials for structural applications, Woodhead publishing series in composites science and engineering, G. Koronis, A. Silva, (Eds.), pp 81 – 97, Elsevier, UK, 2019. https://doi.org/10.1016/B978-0-08-102177-4.00004-5

[33] R. Chakraborty, A. Asthana, A.K. Singh, R. Verma, S. Sreevidya, S. Yadav, S.A.C. Carabineiro, Md.A.B.H. Susan, Chicken feathers derived materials for the removal of chromium from aqueous solutions: kinetics, isotherms, thermodynamics and regeneration studies, J. Dispers. Sci. Technol., 43(2020) 446–460. https://doi.org/10.1080/01932691.2020.1842760

[34] M. Nagalakshmaiah, S. Afrin, R.P. Malladi, S. Elkoun, M. Robert, M.A. Ansari, A. Svedberg, Z. Karim, Biocomposites: present trends and challenges for the future, in: Green composites for automotive applications, part III: nanomaterials and additive manufacturing composites, Woodhead publishing series in composites science and engineering, G. Koronis, A. Silva, (Eds.), pp 81 – 97, Elsevier, UK, 2019. https://doi.org/ s10.1016/B978-0-08-102177-4.00009-4

[35]   H. Abdellaoui, R. Bouhfid, A.K. Qaiss, Preparation of bionanocomposites and bionanomaterials from agricultural wastes, in: Cellulose-reinforced nanofibre composites, production, properties and applications, Woodhead publishing series in composites science and engineering, M. Jawaid, S. Boufi, A. Khalil H.P.S., (Eds.), pp 341 – 371, Elsevier, UK, 2017. https://doi.org/10.1016/B978-0-08-100957-4.00015-2

[36]   J. Jampílek, K. Kráľová, Preparation of nanocomposites from agricultural waste and their versatile applications, in: Multifunctional hybrid nanomaterials for sustainable agri-food and ecosystems, micro and nano technologies, K.A. Abd-Elsalam, (Eds.), pp 51 – 98, Elsevier, Amsterdam, Netherland, 2020. https://doi.org/10.1016/B978-0-12-821354-4.00004-2

[37]   P.S. Kumar, E. Gunasundari, Nanocomposites: recent trends and engineering applications, NHC, 20(2018) 65–80. https://doi.org/10.4028/www.scientific.net/NHC.20.65

[38]   S.P. Cestari, D.F.S. Freitas, D.C. Rodrigues, L.C. Mendes, Recycling processes and issues in natural fiber-reinforced polymer composites, in: Green composites for automotive applications, Part IV: life cycle assessment and risk analysis, Woodhead publishing series in composites science and engineering, G. Koronis, A. Silva, (Eds.), pp 285 – 299, Elsevier, UK, 2019. https://doi.org/10.1016/B978-0-08-102177-4.00012-4

[39]   B. Sharma, P. Malik, P. Jain, Biopolymer reinforced nanocomposites: a comprehensive review, Mater. Today Commun., 16(2018) 353–363. https://doi.org/10.1016/j.mtcomm.2018.07.004

[40]   N.A.S. Abdullah, Z. Mohamad, Z.I. Khan, M. Jusoh, Z.Y. Zakaria, N. Ngadi, Alginate based sustainable films and composites for packaging: a review, Chem. Eng. Trans., 83(2021) 271–276. https://doi.org/10.3303/CET2183046

[41]   N. Basavegowda, K.H. Baek, Advances in functional biopolymer-based nanocomposites for active food packaging applications, Polymers, 13(2021) 4198. https://doi.org/10.3390/polym13234198

[42]   P. Gupta, D. Kumar, M.A. Quraishi, O. Parkash, Metal matrix nanocomposites and their application in corrosion control, in: Advances in nanomaterials, advanced structured materials, M. Husain and Z.H. Khan (Eds.), 79, pp 231 – 246, Springer India, 2016. https://doi.org/10.1007/978-81-322-2668-0_6

[43]   P.S.S.R. Kumar, P.M. Mashinini, S.J. Alexis, Metal matrix nanocomposites in: Nanotechnology in the automotive industry, micro and nano technologies, H. Song, T.A. Nguyen, G. Yasin, N.B. Singh, R.K. Gupta, (Eds.), pp 199 – 213, Elseveir, Amsterdam, Netherland, 2022. https://doi.org/10.1016/B978-0-323-90524-4.00010-4

[44]   S. Sreevidya, S. Kirtana, Y.R. Katre, A.K. Singh, Nanomaterials for environmental hazard: analysis, monitoring, and removal in: Nanomaterials: in bionanotechnology: fundamentals and applications, R. P. Singh, K.R.B. Singh (Eds.) pp 159 – 188, First Edition, Boca Raton, CRC-Press, 2021. https://doi.org/10.1201/9781003139744-7

[45] P.H.C. Camargo, K.G. Satyanarayana, F. Wypych, Nanocomposites: synthesis, structure, properties and new application opportunities, Mat. Res., 12(2009) 1–39. https://doi.org/10.1590/S1516-14392009000100002

[46] R.K. Mishra, P. Mishra, K. Verma, A. Mondal, R.G. Chaudhary, M.M. Abolhasani, Loganathan, S., Electrospinning production of nanofibrous membranes, Envirn. Chem. Letter, 17(2019) 767–800. https://doi.org/10.1007/s10311-018-00838-w

[47] Y.S. Fu, J. Li, J. Li, Metal/Semiconductor nanocomposites for photocatalysis: fundamentals, structures, applications and properties, Nanomaterials, 9(2019) 359. https://doi.org/10.3390/nano9030359

[48] I.A. Shurygina, M.G. Shurygin, B.G. Sukhov, Nanobiocomposites of metals as antimicrobial agents, in: Antibiotic resistance, mechanisms and new antimicrobial approaches, K. Kon, M. Rai. (Eds.), pp 167 – 186, Elseveir, UK, 2016. https://doi.org/10.1016/B978-0-12-803642-6.00008-3

[49] R. Pachaiappan, S. Rajendran, P.L. Show, K. Manavalan, M. Naushad, Metal/metal oxide nanocomposites for bactericidal effect: a review, Chemosphere, 272(2021) 128607. https://doi.org/10.1016/j.chemosphere.2020.128607

[50] S.J. Owonubi, N.M. Malima, N. Revaprasadu, Metal oxide–based nanocomposites as antimicrobial and biomedical agents, in: Antibiotic materials in healthcare, V. Kokkarachedu, V. Kanikireddy, R. Sadiku, (Eds.), pp 287 – 323, Elsevier, UK, 2020. https://doi.org/10.1016/B978-0-12-820054-4.00016-1

[51] M.H. Malakooti, M.R. Bockstaller, K. Matyjaszewski, C. Majidi, Liquid metal nanocomposites, Nanoscale Adv., 2(2020) 2668–2677. https://doi.org/10.1039/d0na00148a

[52] H. Xie, J. Wang, K. Ithisuphalap, G. Wu, Q., Li, Recent advances in Cu-based nanocomposite photocatalysts for CO_2 conversion to solar fuels, J. Energy Chem., 26(2017) 1039–1049. https://doi.org/10.1016/j.jechem.2017.10.025

[53] N.A. Luechinger, R.N. Grass, E.K. Athanassiou, W.J. Stark, Bottom-up fabrication of metal/metal nanocomposites from nanoparticles of immiscible metals, Chem. Mater., 22(2010) 155–160. https://doi.org/10.1021/cm902527n

[54] N.M.E. Shafai, M.E.E. Khouly, M.E. Kemary, M.S. Ramadana, M.S. Masoud, Graphene oxide–metal oxide nanocomposites: fabrication, characterization and removal of cationic rhodamine B dye, RSC Adv., 8(2018) 13323. https://doi.org/10.1039/c8ra00977e

[55] S.E. Shin, H.J. Choi, J.Y. Hwang, D.H. Bae, Strengthening behavior of carbon/metal nanocomposites, Sci. Rep., 5(2015) 16114. https://doi.org/10.1038/srep16114

[56] D. Li, J. Barrington, S. James, D. Ayre, M. Słoma, M.F. Lin, H.Y. Nezhad, Electromagnetic field controlled domain wall displacement for induced strain tailoring in $BaTiO_3$-epoxy, nanocomposite, Sci. Rep., 12(2022) 7504. https://doi.org/10.1038/s41598-022-11380-9

[57] K. Byerly, P.R. Ohodnicki, S.R. Moon, A.M. Leary, V. Keylin, M.E. Mchenry, S. Simizu, R. Beddingfield, Y. Yu, G. Feichter, R. Noebe, R. Bowman, S. Bhattacharya, Metal amorphous nanocomposite (MANC) alloy cores with spatially tuned permeability for advanced power magnetics applications, JOM, 70(2018) 879–891. https://doi.org/10.1007/s11837-018-2857-5

[58] R. Eisavi, A. Karimi, $CoFe_2O_4/Cu(OH)_2$ magnetic nanocomposite: an efficient and reusable heterogeneous catalyst for one-pot synthesis of β-hydroxy-1,4-disubstituted-1,2,3-triazoles from epoxides, RSC Adv., 9(2019) 29873. https://doi.org/10.1039/c9ra06038c

[59] J. He, H. Yang, Y. Zhang, J. Yu, L. Miao, Y. Song, L. Wang, Smart nanocomposites of Cu-hemin metal-organic frameworks for electrochemical glucose biosensing, Sci. Rep., 6(2016) 36637. https://doi.org/10.1038/srep36637

[60] Z. Li, J. Lyu, M. Ge, Synthesis of magnetic $Cu/CuFe_2O_4$ nanocomposite as a highly efficient Fenton-like catalyst for methylene blue degradation, J. Mater. Sci., 53(2018) 15081–15095. https://doi.org/10.1007/s10853-018-2699-0

[61] J. Yang, H. Zhang, B. Chen, H. Tang, C. Li, Z. Zhang, Fabrication of the g-C_3N_4/Cu nanocomposite and its potential for lubrication applications, RSC Adv., 5(2015) 64254–64260. https://doi.org/10.1039/C5RA13683K

[62] T. Wu, W. Cai, P. Zhang, X. Song, L. Gao, Cu–Ni@SiO_2 alloy nanocomposites for methane dry reforming catalysis, RSC Adv., 3(2013) 23976. https://doi.org/10.1039/c3ra43203c

[63] A.K. Sasmal, S. Dutta, T. Pal, A ternary nanocomposite Cu_2O-Cu-CuO: a catalyst for intriguing activity. Dalton Trans., 45(2016) 3139–3150. https://doi.org/10.1039/C5DT03859F

[64] N. Alizadeh, A. Salimi, T.K. Sham, CuO/Cu-MOF nanocomposite for highly sensitive detection of nitricoxide released from living cells using an electrochemical microfluidic device, Microchim. Acta., 188(2021) 240. https://doi.org/10.1007/s00604-021-04891-1

[65] L. Xu, C. Srinivasakannan, J. Peng, L. Zhang, D. Zhang, Synthesis of Cu-CuO nanocomposite in microreactor and its application to photocatalytic degradation, J. Alloys Compd., 695(2017) 263–269. https://doi.org/ 10.1016/j.jallcom.2016.10.195

[66] L. Shabnam, S.N. Faisal, A.K. Roy, E. Haque, A.I. Minett, V.G. Gomes, Doped graphene/Cu nanocomposite: a high sensitivity non-enzymatic glucose sensor for food, Food Chem., 221(2017) 751–759. https://doi.org/10.1016/j.foodchem.2016.11.107

[67] R. Jahanshahi, A. Mohammadi, M. Doosti, S. Sobhani. J.M. Sansano, $ZnCo_2O_4$/g-C_3N_4/Cu nanocomposite as a new efficient and recyclable heterogeneous photocatalyst with enhanced photocatalytic activity towards the metronidazole degradation under the solar light irradiation, Environ. Sci. Pollut. Res., 9(2022) 65043–65060. https://doi.org/10.1007/s11356-022-19969-3

[68] Vidya, L. Mandal, B. Verma, P.K. Patel, Review on polymer nanocomposite for ballistic & aerospace applications, Mat. Today: Proc., 26(2020) 3161–3166. https://doi.org/10.1016/j.matpr.2020.02.652

[69] S.K. Hubadillah, Z.S. Tai, M.H.D. Othman, Z. Harun, M.R. Jamalludin, M.A. Rahman, J.J. Ahmad, F. Ismail, Hydrophobic ceramic membrane for membrane distillation: a mini review on preparation, characterization, and applications, Sep. Purif. Technol., 217(2019) 71–84. https://doi.org/10.1016/j.seppur.2019.02.014

[70] A.M.K. Kirubaharan, P. Kuppusami, Corrosion behavior of ceramic nanocomposite coatings at nanoscale, in: Corrosion protection at the nanoscale micro and nano technologies, S. Rajendran, T.ANH. Nguyen, S. Kakooie, M. Yeganeh, Y. Li, (Eds.), pp 295 – 314, Elseveir, Amsterdam, Netherlands, 2020. https://doi.org/10.1016/B978-0-12-819359-4.00016-7

[71] M.S. Hasnain, S.A. Ahmad, M.A. Minhaj, T.J. Ara, A.K. Nayak, Nanocomposite materials for prosthetic devices, in: Applications of nanocomposite materials in orthopedics, Woodhead publishing series in biomaterials, Inamuddin, A.M. Asiri, A. Mohammad, (Eds.), pp 127 – 144. Elsevier, UK, 2019. https://doi.org/10.1016/B978-0-12-813740-6.00007-7

[72] J.D.R. Selvam, I. Dinaharan, R.S. Rai, Matrix and reinforcement materials for metal matrix composites, in: Encyclopedia of materials: composites, vol 2, pp 615 – 639, 2021. https://doi.org/ 10.1016/B978-0-12-803581-8.11890-9

[73] K. Niihara, A. Nakahira, T. Sekino, New nanocomposite structural ceramics, Mat. Res. Soc. Symp. Proc., 286(1992) 405–412. https://doi.org/10.1557/PROC-286-405.

[74] P. Palmero, Structural ceramic nanocomposites: a review of properties and powders' synthesis methods, Nanomaterials, 5(2015) 656–696. https://doi.org/10.3390/nano5020656

[75] H. Porwal, R. Saggar, Ceramic matrix nanocomposites, in: Reference module in materials science and materials engineering, comprehensive composite materials II, P.W.R. Beaumont, C.H. Zweben, (Eds.), volume 6, pp 138 – 161, Elseveir, Amsterdam, Netherlands, 2018. https://doi.org/10.1016/B978-0-12-803581-8.10029-3

[76] B. Kumar, Ceramic nanocomposites for energy storage and power generation, in: Ceramic nanocomposites, Woodhead publishing series in composites science and engineering, R. Banerjee, I. Manna, (Eds.), pp 509 – 529, Elsevier, UK, 2013. https://doi.org/10.1533/9780857093493.4.509

[77] A. Tarafder, A.R. Molla, B. Karmakar, Advanced glass-ceramic nanocomposites for structural, photonic, and optoelectronic applications, in: Glass nanocomposites, synthesis, properties and applications, B. Karmakar, K. Rademann, A.L. Stepanov, (Eds.), pp 299 – 338, Elsevier, UK, 2016. https://doi.org/10.1016/B978-0-323-39309-6.00013-4

[78] N.R. Bose, Thermal shock resistant and flame retardant ceramic nanocomposites, in: Ceramic nanocomposites, Woodhead publishing series in composites science and

engineering, R. Banerjee, I. Manna, (Eds.), pp 3 – 50, Elsevier, UK, 2013.
https://doi.org/10.1533/9780857093493.1.3

[79]   B. Karmakar, Fundamentals of glass and glass nanocomposites, in: Glass
nanocomposites, synthesis, properties and applications, B. Karmakar, K. Rademann,
A.L. Stepanov, (Eds.), pp 3 – 53, Elsevier, UK, 2016. https://doi.org/10.1016/B978-0-
323-39309-6.00001-8

[80]   G.G. Kumar, Zeolites and composites, in: Nanomaterials and nanocomposites:
zero- to three-dimensional materials and their composites, P.M. Visakh, M. José, M.
Morlanes, (Eds.), pp 187 – 221, Wiley-VCH Verlag GmbH & Co. KGaA, 2016.
https://doi.org/10.1002/9783527683772.ch6

[81]   C. Li, W. Sun, Z. Lu, X. Ao, C. Yang, S. Li, Systematic evaluation of TiO_2-GO
modified ceramic membranes for water treatment: retention properties and fouling
mechanisms, Chem. Eng. J., 378(2019) 122138.
https://doi.org/10.1016/j.cej.2019.122138

[82]   N. Choudhary, V.K. Yadav, K.K. Yadav, A.I. Almohana, S.F. Almojil, G.
Gnanamoorthy, D.H. Kim, S. Islam, P. Kumar, B.H. Jeon, Application of green
synthesized MMT/Ag nanocomposite for removal of methylene blue from aqueous
solution, Water, 13(2021) 3206. https://doi.org/10.3390/w13223206

[83]   F.S. Sangsefidi, M.S. Niasari, Fe_2O_3–CeO_2 ceramic nanocomposite oxide:
characterization and investigation of the effect of morphology on its electrochemical
hydrogen storage capacity, ACS Appl. Energy Mater., 1(2018) 4840–4848. doi.
10.1021/acsaem.8b00907

[84]   D. Zou, Y. Yi, Y. Song, D. Guan, M. Xu, R. Ran, W. Wang, W. Zhou, Z. Shao,
$BaCe_{0.16}Y_{0.04}Fe_{0.8}O_{3-\delta}$ nanocomposite: a new high-performance cobalt-free triple-
conducting cathode for protonic ceramic fuel cells operating at reduced temperatures,
J. Mater. Chem. A, 10(2022) 5381–5390. https://doi.org/10.1039/D1TA10652J

[85]   M.A. Rezvani, M. Hadi, H. Rezvani, Synthesis of new nanocomposite based on
ceramic and heteropolymolybdate using leaf extract of aloe vera as a high-
performance nanocatalyst to desulfurization of real fuel, Appl. Organomet. Chem.,
35(2021) 1-15. https://doi.org/10.1002/aoc.6176

[86]   Q. Chen, R. Zhu, L. Deng, L. Ma, Q. He, J. Du, H. Fu, J. Zhang, A. Wang, One-
pot synthesis of novel hierarchically porous and hydrophobic Si/SiOx composite from
natural palygorskite for benzene adsorption, Chem. Eng. J., 378(2019) 122131.
https://doi.org/10.1016/j.cej.2019.122131

[87]   N.Y. Soylu, A. Akturk, O. Kabak, M.E. Taygun, F.K. Guler, S. Küçükbayrak,
TiO_2 nanocomposite ceramics doped with silver nanoparticles for the photocatalytic
degradation of methylene blue and antibacterial activity against Escherichia coli, Eng.
Sci. Technol., (2022) 101175. In Press, Corrected Proof.
https://doi.org/10.1016/j.jestch.2022.101175

[88]   X. Wang, Y. Li, H. Yu, F. Yang, C.Y. Tang, X. Quan, Y. Dong, High-flux robust
ceramic membranes functionally decorated with nano-catalyst for emerging micro-

pollutant removal from water, J. Membr. Sci., 611(2020) 118281.
https://doi.org/10.1016/j.memsci.2020.118281

[89]   N.A. Safronova, O.S. Kryzhanovska, M.V. Dobrotvorska, A.E. Balabanov, A.V. Tolmachev, R.P. Yavetskiy, S.V. Parkhomenko, R. Brodskii, V.N. Baumer, D.Y. Kosyanov, O.O. Shichalin, E.K. Papynov, J. Li, Influence of sintering temperature on structural and optical properties of Y_2O_3–MgO composite SPS ceramics, Ceram. Int., 46(2020) 6537–6543, https://doi.org/10.1016/j.ceramint.2019.11.137.

[90]   B. Zhang, W. Song, L. Wei, Y. Xiu, H. Xu, D.B. Dingwell, H. Guo, Novel thermal barrier coatings repel and resist molten silicate deposits, Scr. Mater., 163(2019) 71–76. https://doi.org/10.1016/j.scriptamat.2018.12.028

[91]   X.F. Zhang, G. Harley, L.C.D. Jonghe, Co-continuous metal–ceramic nanocomposites, Nano Lett., 5(2005) 1035–1037. https://doi.org/10.1021/nl050379t

[92]   M.Z.A. Qureshi, S. Bilal, M.Y. Malik, Q. Raza, E.S.M. Sherif, Y.M. Li, Dispersion of metallic/ceramic matrix nanocomposite material through porous surfaces in magnetized hybrid nanofluids flow with shape and size effects, Sci. Rep., 11(2021) 12271. https://doi.org/10.1038/s41598-021-91152-z

[93]   A. Carvalho, M. Marinova, N. Batalha, N.R. Marcilio, A.Y. Khodakov, V.V. Ordomsky, Design of nanocomposites with cobalt encapsulated in the zeolite micropores for selective synthesis of isoparaffins in Fischer–Tropsch reaction, Catal. Sci. Technol., 7(2017) 5019–5027. https://doi.org/10.1039/C7CY01945A

[94]   A. Rathi, S. Basu, S. Barman, Efficient eradication of antibiotic and dye by C-dots@zeolite nanocomposites: performance evaluation, and degraded products analysis, Chemosphere, 298(2022) 134260. https://doi.org/10.1016/j.chemosphere.2022.134260

[95]   I.I. Ivanova, E.E. Knyazevaa, Micro–mesoporous materials obtained by zeolite recrystallization: synthesis, characterization and catalytic applications, Chem. Soc. Rev., 42(2013) 3671–3688. https://doi.org/10.1039/c2cs35341e

[96]   K. Shameli, M.B. Ahmad, M. Zargar, W.M.Z.W. Yunus, N.A. Ibrahim, Fabrication of silver nanoparticles doped in the zeolite framework and antibacterial activity, Int. J. Nanomedicine, 6(2011) 331–341. https://doi.org/10.2147/IJN.S16964.

[97]   V. Hovhannisyan, K. Siposova, A. Musatov, S.J. Chen, Development of multifunctional nanocomposites for controlled drug delivery and hyperthermia, Sci. Rep., 11(2021) 5528. https://doi.org/10.1038/s41598-021-84927-x

[98]   A.A. Alswata, M.B. Ahmad, N.M.A. Hada, H.M. Kamari, M.Z.B. Hussein, N.A. Ibrahim, Preparation of zeolite/zinc oxide nanocomposites for toxic metals removal from water, Results Phys., 7(2017) 723–731. https://doi.org/10.1016/j.rinp.2017.01.036

[99]   T. Shubair, O. Eljamal, A. Tahara, Y. Sugihara, N. Matsunaga, Preparation of new magnetic zeolite nanocomposites for removal of strontium from polluted waters, J. Mol. Liq., 288(2019) 111026. https://doi.org/10.1016/j.molliq.2019.111026

[100] M. Mansouri, Y. Ahmadi, Applications of zeolite-zirconia-copper nanocomposites as a new asphaltene inhibitor for improving permeability reduction during CO_2 flooding, Sci. Rep., 12(2022) 6209. https://doi.org/10.1038/s41598-022-09940-0

[101] H. Jahangirian, R.R. Moghaddam, N., Jahangirian, B., Nikpey, S. Jahangirian, N. Bassous, B. Saleh, K. Kalantari, T.J. Webster, Green synthesis of zeolite/Fe_2O_3 nanocomposites: toxicity & cell proliferation assays and application as a smart iron nanofertilizer, Int. J. Nanomedicine, 15(2020) 1005–1020. https://doi.org/10.2147/IJN.S231679

[102] M.N. Chong, Z.Y. Tneu, P.E. Poh, B. Jin, R. Aryal, Synthesis, characterisation and application of TiO_2–zeolite nanocomposites for the advanced treatment of industrial dye wastewater, J. Taiwan Inst. Chem. Eng., 000(2014) 1–9. https://doi.org/10.1016/j.jtice.2014.12.013

[103] N.B.T. Tran, N.B. Duong, N.L. Le, Synthesis and characterization of magnetic Fe_3O_4/Zeolite NaA nanocomposite for the adsorption removal of methylene blue potential in wastewater treatment, J. Chem., (2021) 1–14. https://doi.org/10.1155/2021/6678588

[104] G.V. Kravchenko, E.N. Domoroshchina, G.M. Kuz'micheva, A.A. Gaynanova, S.V. Amarantov, L.V. Pirutko, A.M. Tsybinsky, N.V. Sadovskaya, E.V. Kopylova, Zeolite–titanium dioxide nanocomposites: preparation, characterization, and adsorption properties, Nanotechnologies in Russia, 11(2016) 579–592. https://doi.org/10.1134/S1995078016050098

[105] E.I. Akpan, X. Shen, B. Wetzel, K. Friedrich, Design and synthesis of polymer nanocomposites, in: Polymer composites with functionalized nanoparticles, synthesis, properties, and applications, micro and nano technologies, K. Pielichowski, T.M. Majka, (Eds.) pp 47 – 83, Elsevier, Amsterdam, Netherlands, 2019. https://doi.org/10.1016/B978-0-12-814064-2.00002-0

[106] V. LazićJovan, M. Nedeljković, Organic–inorganic hybrid nanomaterials: synthesis, characterization, and application, in: Nanomaterials synthesis, design, fabrication and applications, micro and nano technologies, Y.B. Pottathara, N. Kalarikkal, Y. Grohens, S. Thomas, V. Kokol, (Eds.), pp 419 – 449, Elsevier, Amsterdam, Netherlands, 2019. https://doi.org/10.1016/B978-0-12-815751-0.00012-2

[107] N. Karak, Fundamentals of nanomaterials and polymer nanocomposites, in: Nanomaterials and polymer nanocomposites, raw materials to applications, N. Karak (Eds.), pp 1 – 45, Elsevier, Amsterdam, Netherlands, 2019. https://doi.org/10.1016/B978-0-12-814615-6.00001-1

[108] S.H. Zaferani, Introduction of polymer-based nanocomposites, in: Polymer-based nanocomposites for energy and environmental applications, Woodhead publishing series in composites science and engineering, M. Jawaid, M.M. Khan, (Eds.), pp 1 – 25, Elsevier, UK, 2018. https://doi.org/10.1016/B978-0-08-102262-7.00001-5

[109] A. Sheikhi, Emerging cellulose-based nanomaterials and nanocomposites, in: Nanomaterials and polymer nanocomposites, raw materials to applications, N. Karak

(Eds.), pp 307 – 351, Elsevier, Amsterdam, Netherlands, 2019.
https://doi.org/10.1016/B978-0-12-814615-6.00009-6

[110] K. Yu, N. Kumar, J. Roine, M. Pesonen, A. Ivaska, Synthesis and characterization of polypyrrole/H-Beta zeolite nanocomposites, RSC Adv., 4(2014) 33120–33126. https://doi.org/10.1039/C4RA03864A

[111] H.L. Frisch, H. Song, J. Ma, M. Rafailovich, S. Zhu, N.L. Yang, X. Yan, Antiferromagnetic pairing in polyaniline salt−zeolite nanocomposites, J. Phys. Chem. B, 105(2001) 11901–11905. https://doi.org/10.1021/jp012278z

[112] D. Bendahou, A. Bendahou, Y. Grohens, H. Kaddami, New nanocomposite design from zeolite and poly(lactic acid), Ind. Crops Prod., 72(2015) 107–118. https://doi.org/10.1016/j.indcrop.2014.12.055

[113] J.S. Son, E.J. Hwang, L.S. Kwon, Y.G. Ahn, B.K. Moon, J. Kim, D.H. Kim, S.G. Kim, S.Y. Lee, Antibacterial activity of propolis-embedded zeolite nanocomposites for implant application, Materials, 14(2021) 1193. https://doi.org/10.3390/ma14051193

[114] H. Faghihian, M. Moayed, A. Firooz, M. Iravani, Synthesis of a novel magnetic zeolite nanocomposite for removal of Cs^+ and Sr^{2+} from aqueous solution: kinetic, equilibrium, and thermodynamic studies, J. Colloid Inter. Sci., 393(2017) 445–451. https://doi.org/10.1016/j.jcis.2012.11.010

[115] S. Gong, H. Chen, X. Zhou, S. Gunasekaran, Synthesis and applications of MANs/poly(MMA-co-BA) nanocomposite latex by miniemulsion polymerization, R. Soc. Open Sci. 4(2017) 170844. https://doi.org/10.1098/rsos.170844

[116] V. Dhapte, N. Gaikwad, P.V. More, S. Banerjee, V.V., Dhapte, S. Kadam, P.K. Khanna, Transparent ZnO/polycarbonate nanocomposite for food packaging application, Nanocomposites, 1(2015) 106–112. https://doi.org/10.1179/2055033215Y.0000000004

[117] M. Kermani, H. Sereshti, N. Nikfarjam, Application of a magnetic nanocomposite of cross-linked poly(styrene/divinylbenzene) as an adsorbent for the magnetic dispersive solid phase extraction-dispersive liquid–liquid microextraction of atrazine in soil and aqueous samples, Anal. Methods, 12(2020) 1834–1844. https://doi.org/10.1039/D0AY00374C

[118] S.Y. Song, M.S. Park, D. Lee, J.W. Lee, J.S. Yun, Optimization and characterization of high-viscosity ZrO_2 ceramic nanocomposite resins for supportless stereolithography, Mater. Design., 180(2019) 107960. https://doi.org/10.1016/j.matdes.2019.107960

[119] X. Zheng, L. Wang, Q. Pei, S. He, S. Liu, Z. Xie, Metal-Organic frameworks@ porous organic polymers nanocomposite for photodynamic therapy, Chem. Mater., 29(2017) 2374 – 2381. https://doi.org/10.1021/acs.chemmater.7b00228

[120] M. Ramesh, M. Muthukrishnan, Biodegradable polymer blends and composites for food-packaging applications, in: Biodegradable polymers, blends and composites, Woodhead publishing series in composites science and engineering, S. Mavinkere,

R.J. Parameswaranpillai, S. Siengchin, M. Ramesh, (Eds.), pp 693 – 716, Elseveir, UK, 2022. https://doi.org/10.1016/B978-0-12-823791-5.00004-1

[121] Z. Zhang, O. Ortiz, R. Goyal, J. Kohn, Biodegradable polymers, in: Handbook of polymer applications in medicine and medical devices, plastics design library, K. Modjarrad, S. Ebnesajjad, (Eds.), pp 303 – 335, Elsevier, Oxford, USA, 2014. https://doi.org/10.1016/B978-0-323-22805-3.00013-X

[122] P. Bhagabati, Biopolymers and biocomposites-mediated sustainable high-performance materials for automobile applications, in: Sustainable nanocellulose and nanohydrogels from natural sources, micro and nano technologies, F. Mohammad, H.A. Al-Lohedan, M. Jawaid, (Eds.), pp 197 – 216, Elsevier, Amsterdam, Netherlands, 2020. https://doi.org/10.1016/B978-0-12-816789-2.00009-2

[123] G. Lavri˘c, A. Oberlintner, I. Filipova, U. Novak, B. Likozar, U.V. Brodnjak, Functional nanocellulose, alginate and chitosan nanocomposites designed as active film packaging materials, Polymers, 13(2021) 2523. https://doi.org/10.3390/polym13152523

[124] M. Mohammadpour, H. Samadian, N. Moradi, Z. Izadi, M. Eftekhari, M. Hamidi, A. Shavandi, A. Quéro, E. Petit, C. Delattre, R. Elboutachfaiti, Fabrication and characterization of nanocomposite hydrogel based on alginate/nano-hydroxyapatite loaded with linum usitatissimum extract as a bone tissue engineering scaffold, Mar. Drugs, 20(2020) 20. https://doi.org/10.3390/md20010020

[125] A.M.F. Lima, M.D.F. Lima, O.B.G. Assis, A. Raabe, H.C. Amoroso, V.A.D.O. Tiera, M.B.D. Andrade, M.J. Tiera, Synthesis and physicochemical characterization of multiwalled carbon nanotubes/hydroxamic alginate nanocomposite scaffolds, J. Nanomater., (2018) 1–12. https://doi.org/10.1155/2018/4218270

[126] G. Supanakorn, N. Varatkowpairote, S. Taokaew, M. Phisalaphong, Alginate as dispersing agent for compounding natural rubber with high loading microfibrillated cellulose, Polymers, 13(2021) 468. https://doi.org/10.3390/polym13030468

[127] B. Piluharto, U. Salamah, D. Indarti, Preparation of alginate/nanocellulose nanocomposite for protein adsorption, Macromol. Symp., 391(2020) 1900141. https://doi.org/10.1002/masy.201900141

[128] F. Aziz, M.E. Achaby, A. Lissaneddine, K. Aziz, N. Ouazzani, R. Mamouni, L. Mandi, Composites with alginate beads: a novel design of nano-adsorbents impregnation for large-scale continuous flow wastewater treatment pilots, Saudi J. Biol. Sci., 27(2020) 2499–2508. https://doi.org/10.1016/j.sjbs.2019.11.019

[129] M. Esmat, A.A. Farghali, M.H. Khedr, I.M.E. Sherbiny, Alginate-based nanocomposites for efficient removal of heavy metal ions, Int. J. Biol. Macromol., 102(2017) 272–283. https://doi.org/10.1016/j.ijbiomac.2017.04.021

[130] R.L. Alexa, R. Ianchis, D. Savu, M. Temelie, B. Trica, A. Serafim, G.M. Vlasceanu, E. Alexandrescu, S. Preda, H. Iovu, 3D Printing of alginate-natural clay hydrogel-based nanocomposites, Gels, 7(2021) 211. https://doi.org/10.3390/gels7040211

[131] S. Yadav, A. Asthana, A.K. Singh, R. Chakraborty, S. Sreevidya, M.A.B.H. Susan, S.A.C. Carabineiro, Adsorption of cationic dyes, drugs and metal from aqueous solutions using a polymer composite of magnetic/β-cyclodextrin/activated charcoal/Na alginate: isotherm, kinetics and regeneration studies, J. Hazard. Mater., 409(2021) 124840. https://doi.org/10.1016/j.jhazmat.2020.124840

[132] S. Yadav, A. Asthana, R. Chakraborty, B. Jain, A.K. Singh, S.A.C. Carabineiro, M.A.B.H. Susan, Cationic dye removal using novel magnetic/activated charcoal/β-cyclodextrin/alginate polymer, nanocomposite, Nanomaterials, 10(2020) 170. https://doi.org/10.3390/nano10010170

[133] Yadav, A. Asthana, A.K. Singh, R. Chakraborty, S.S Vidya, A. Singh, S.A.C. Carabineiro, Methionine-functionalized graphene oxide/sodium alginate bio-polymer nanocomposite hydrogel beads: synthesis, isotherm and kinetic studies for an adsorptive removal of fluoroquinolone antibiotics, Nanomaterials, 11(2021) 568. https://doi.org/10.3390/nano11030568

[134] S. Lilhare, S.B. Mathew, A.K. Singh, S. Sreevidya, A simple spectrophotometric study of adsorption of Hg(II) on glycine functionalised magnetic nanoparticle entrapped alginate beads, Int. J. Environ. Anal. Chem., 2021. https://doi.org/10.1080/03067319.2021.1884242

[135] H. Li, M. Kruteva, M. Dulle, Z. Wang, K. Mystek, W. Ji, T. Pettersson, L. Wågberg, Understanding the drying behavior of regenerated cellulose gel beads: the effects of concentration and nonsolvents, ACS Nano, 16(2022) 2608−2620. https://doi.org/10.1021/acsnano.1c09338

[136] W. Kargupta, R. Seifert, M. Martinez, J. Olson, J. Tanner, W. Batchelor, Preparation and benchmarking of novel cellulose nanopaper, Cellulose, 29(2022) 4393–4411. https://doi.org/10.1007/s10570-022-04563-0,

[137] A.D. Štiglic, F. Gürer, F. Lackner, D. Bračič, A. Winter, L. Gradišnik, D. Makuc, R. Kargl, I. Duarte, J. Plavec, U. Maver, M. Beaumont, K.S. Kleinschek, T. Mohan, Organic acid cross-linked 3D printed cellulose nanocomposite bioscaffolds with controlled porosity, mechanical strength, and biocompatibility, iScience, 25(2022) 104263. https://doi.org/10.1016/j.isci.2022.104263

[138] S. Park, S.H. Kim, J.H. Kim, H. Yu, H.J. Kim, Y.H. Yang, H. Kim, Y.H. Kim, S.H. Ha, S.H. Lee, Application of cellulose/lignin hydrogel beads as novel supports for immobilizing lipase, J. Mol. Catal. B: Enzym., 119(2015) 33–39. https://doi.org/10.1016/j.molcatb.2015.05.014

[139] H.W. Kwak, M. Shin, H. Yun, K.H. Lee, Preparation of silk sericin/lignin blend beads for the removal of hexavalent chromium ions, Int. J. Mol. Sci., 17(2016) 1466. https://doi.org/10.3390/ijms17091466

[140] J. Lin, Y. Cheng, Z. Li, Y. Zheng, B. Xu, C. Lu, Synthesis of modified lignin as an antiplasticizer for strengthening poly(vinyl alcohol)–lignin interactions toward quality gel-spun fibers, ACS Appl. Polym., Mater., 4(2022) 1595–1607. https://doi.org/10.1021/acsapm.1c01384

[141] X. Shen, Y. Xie, Q. Wang, X. Yi, J.L. Shamshina, R.D. Rogers, Enhanced heavy metal adsorption ability of lignocellulosic hydrogel adsorbents by the structural support effect of lignin, Cellulose, 26(2019) 4005–4019. https://doi.org/10.1007/s10570-019-02328-w

[142] L. Chen, Y. Shi, B. Gao, Y. Zhao, Y. Jiang, Z., Zha, W. Xue, L. Gong, Lignin nanoparticles: green synthesis in a γ-valerolactone/water binary solvent and application to enhance antimicrobial activity of essential oils, ACS Sustainable Chem. Eng., 12(2019) 17. https://doi.org/10.1021/acssuschemeng.9b06716

[143] T. Luo, Y. Hao, C. Wang, W. Jiang, X. Ji, G. Yang, J. Chen, S. Janaswamy, G. Lyu, Lignin nanoparticles and alginate gel beads: preparation, characterization and removal of methylene blue, Nanomaterials, 12(2022) 176. https://doi.org/10.3390/nano12010176

[144] S. Akbari, A. Bahi, A. Farahani, A.S. Milani, F. Ko, Fabrication and characterization of lignin/dendrimer electrospun blended fiber mats, Molecules, 26(2021) 518. https://doi.org/10.3390/molecules26030518

[145] J.L. Patarroyo, E. Fonseca, J. Cifuentes, F. Salcedo, J.C. Cruz, L.H. Reyes, Gelatin-graphene oxide nanocomposite hydrogels for kluyveromyces lactis encapsulation: potential applications in probiotics and bioreactor packings, Biomolecules, 11(2021) 922. https://doi.org/10.3390/biom11070922

[146] M.A. Salami, F. Kaveian, M. Rafienia, S. Saber-Samandari, A. Khandan, M. Naeimi, Electrospun polycaprolactone/lignin based nanocomposite as a novel tissue scaffold for biomedical applications, J. Med. Sign. Sens., 7(2017) 228–238.

[147] M. Leonardi, G.M. Caruso, S.C. Carroccio, S. Boninelli, G. Curcuruto, M. Zimbone, M. Allegra, B. Torrisi, F. Ferlito, M. Miritello, Smart nanocomposites of chitosan/alginate nanoparticles loaded with copper oxide as alternative nanofertilizers, Environ. Sci. Nano, 8(2021) 174. https://doi.org/10.1039/d0en00797h

[148] S. Sreevidya, S. Kirtana, Y.R. Katre, A. Kumar, A.K. Singh, Plant extract: isolation, purification, and applications of green nanomaterials stabilization, in: Green nanomaterials sustainable technologies and applications, K. Pal, (Eds.), New York, 1st Edition, Apple Academic Press, USA, pp 189 – 218, 2022. https://doi.org/10.1201/9781003130314-8.

[149] R. Chakraborty, A. Asthana, A.K. Singh, S. Yadav, M.A.B.H. Susan, S.A.C. Carabineiro, Intensified elimination of aqueous heavy metal ions using chicken feathers chemically modified by a batch method, J. Mol. Liq., 312(2020) 113475. https://doi.org/10.1016/j.molliq.2020.113475

[150] T. Guha, G. Gopal, R. Kundu, A. Mukherjee, Nanocomposites for delivering agrochemicals: a comprehensive review, J. Agric. Food Chem., 68(2020), 3691–3702. https://doi.org/10.1021/acs.jafc.9b06982

[151] Darwish, M.S.A., Mostafa, M.H., Al-Harbi, L.M., Polymeric nanocomposites for environmental and industrial applications, Int. J. Mol. Sci., 23(2023) 1023. https://doi.org/10.3390/ijms23031023

[152] E. Fortunati, F. Luzi, W. Yang, J.M. Kenny, L. Torre, P. Puglia, Bio-based nanocomposites in food packaging, in: Nanomaterials for food packaging materials, processing technologies, and safety issues, micro and nano technologies, M.Â.P.R. Cerqueira, J.M. Lagaron, L.M.P. Castro A.A.M.de O.S. Vicente, (Eds.), pp 71 – 110, Elseveir, Amsterdam, Netherlands, 2018. https://doi.org/10.1016/B978-0-323-51271-8.00004-8

[153] S. Huang, X, Hong, M. Zhao, N. Liu, H. Liu, J. Zhao, L. Shao, W. Xue, H. Zhang, P. Zhud, R. Guo, Nanocomposite hydrogels for biomedical applications, Bioeng. Transl. Med., 7(2022) e10315. https://doi.org/10.1002/btm2.10315

[154] M. Sajjad, W. Lu, Covalent organic frameworks based nanomaterials: design, synthesis, and current status for supercapacitor applications: a review, J. Energy Storage, 39(2021) 102618. https://doi.org/10.1016/j.est.2021.102618

[155] E.E. Okoro, R. Josephs, S.E. Sanni, Y. Nchila, Advances in the use of nanocomposite membranes for carbon capture operations, Int. J. Chem. Eng., (2021)22. https://doi.org/10.1155/2021/6666242

[156] M. Wei, Y. Gao, X. Li, M.J. Serpe, Stimuli-responsive polymers and their applications, Polym. Chem., 8(2017) 127. https://doi.org/10.1039/c6py01585a

[157] L. Camilli, M. Passacantando, Advances on sensors based on carbon nanotubes, Chemosensors, 6(2018) 62. https://doi.org/10.3390/chemosensors6040062

[158] J. Xiong, X. Cai, J. Ge, Enzyme–metal nanocomposites for antibacterial applications, Particuology, 64(2022) 134–139. https://doi.org//10.1016/j.partic.2021.02.003

[159] T. Zhang, W. He, W. Zhang, T. Wang, P. Li, Z.M., Sun, X. Yu, Designing composite solid-state electrolytes for high performance lithium ion or lithium metal batteries, Chem. Sci., 11(2020) 8686–8707. https://doi.org/10.1039/d0sc03121f

[160] J. Yu, D., Wan, N., Geetha, K.M., Khawar, S. Jogaiah, M. Mujtaba, Current trends and challenges in the synthesis and applications of chitosan-based nanocomposites for plants: a review, Carbohydr. Polym., 261(2021) 117904. https://doi.org/10.1016/j.carbpol.2021.117904

[161] L. Tamayo, M. Azócar, M. Kogan, A. Riveros, M. Páez, Copper-polymer nanocomposites: an excellent and cost-effective biocide for use on antibacterial surfaces, Mater. Sci. Eng. C, 69(2016) 1391–1409. https://doi.org/10.1016/j.msec.2016.08.041

[162] G. Liao, J. Fang, Q. Li, S. Li, Z. Xu, B. Fang, Ag-Based nanocomposites: synthesis and applications in catalysis, Nanoscale, 11(2019) 7062–7096. https://doi.org/10.1039/C9NR01408J

[163] A. Bhat, S. Budholiya, S.A. Raj, M.T.H. Sultan, D. Hui, D., A.U.M. Shah, S.N.A. Safri, Review on nanocomposites based on aerospace applications, Nanotechnol. Rev., 10(2021) 237–253. https://doi.org/ntrev-2021-/10.15150018

[164] H. Xie, J. Wang, K. Ithisuphalap, G. Wu, Q. Li, Recent advances in Cu-based nanocomposite photocatalysts for CO_2 conversion to solar fuels, J. Energy Chem., 26(2017) 1039–1049. https://doi.org/10.1016/j.jechem.2017.10.025

[165] E. Lizundia, M.H. Sipponen, L.G. Greca, M. Balakshin, B.L. Tardy, O.J. Rojas, D. Puglia, Multifunctional lignin-based nanocomposites and nanohybrids, Green Chem., 23(2021) 6698. https://doi.org/10.1039/d1gc01684a

[166] R. Kaur, S.K. Bhardwaj, S. Chandna, K.Y. Kim, J. Bhaumik, Lignin-based metal oxide nanocomposites for UV protection applications: a review, J. Cleaner Prod., 317(2021) 128300. https://doi.org/10.1016/j.jclepro.2021.128300

[167] B. Wang, Y. Wan, Y. Zheng, X. Lee, T. Liu, Z. Yu, J. Huang, Y.S. Ok, J. Chen, B. Gao, Alginate-based composites for environmental applications: a critical review, Crit. Rev. Environ. Sci. Technol., 49(2018) 318–356. https://doi.org/10.1080/10643389.2018.1547621

[168] O. Faruk, D. Hosen, A. Ahmed, M.M. Rahman, Functional bionanomaterials—embedded devices for sustainable energy storage, in: Biorenewable nanocomposite materials, vol. 1: electrocatalysts and energy storage, D. Pathania, L. Singh, (Eds.), pp 1 – 23, Vol. 1410, ACS Symposium Series, American Chemical Society, 2022. https://doi.org/10.1021/bk-2022-1410.ch001

[169] M. Thakur, M. Chandel, A. Rani, A. Sharma, Introduction to biorenewable nanocomposite materials: methods of preparation, current developments, and future perspectives, in: Biorenewable nanocomposite materials, vol. 2: desalination and wastewater remediation, D. Pathania, L. Singh, (Eds.), pp 1 – 24, ACS Symposium Series Vol. 1411, American Chemical Society, 2022. https://doi.org/10.1021/bk-2022-1411.ch001

[170] G.C. Lavorato, R. Das, J.A. Masa, M.H. Phan, H. Srikanth, Hybrid magnetic nanoparticles as efficient nanoheaters in biomedical applications, Nanoscale Adv., 3(2021) 867. https://doi.org/10.1039/d0na00828a

Applications of Emerging Nanomaterials and Nanotechnology   Materials Research Forum LLC
Materials Research Foundations 148 (2023) 103-126   https://doi.org/10.21741/9781644902554-4

Chapter 4

Engineered Nanomaterials for Energy Conversion Cells

Mohammad Harun-Ur-Rashid[1] and Abu Bin Imran[2*]

[1]Department of Chemistry, International University of Business Agriculture and Technology, Dhaka 1230, Bangladesh

[2]Department of Chemistry, Bangladesh University of Engineering and Technology, Dhaka 1000, Bangladesh

*abimran@chem.buet.ac.bd

Abstract

Day by day, the energy demand is exceeding due to consumption by increasing world population and fast-growing industrialization. As a result, the biggest problems of the 21st century are energy demand and how it affects the environment. The disquiet is caused by the excessive reliance on fossil fuels as raw materials for the production of energy, such as coal, oil, and natural gas. Around 13 terawatts of energy are needed every day by more than 6.5 billion people around the world. However, the scarcity of currently used fossil fuels and the environmental deterioration corresponding to fuel rectification processes have triggered the compulsion to produce renewable, non-polluting, and eco-friendly energy generation and conversion technologies. Suitable technologies for the conversion and storage of energy will play a vital role in addressing the current challenges associated with the increasing demand for clean, renewable, sustainable, transferable, benign, eco-friendly, nominal, and ceaseless power supplies for users. The substitution of fossil fuels could be clean energies, for example, solar, hydroelectric, wind, geothermal, biogas, and tidal energies. Generally, alternative renewable energy conversion requires various complicated physical and chemical processes on the surface and interfaces of cell components and transporting electrons, positive holes, ions, and molecules through the entire system. The harnessing of energy requires new and novel nanomaterials and evolution of nanocomposite and multifunctional nanostructured materials, including metal, ceramic, polymer matrix, and amalgamation. Various essential advantages of using engineered nanomaterials, such as high surface area, unique physicochemical properties, mechanical strength, and favorable transport properties, are crucial to energy harnessing applications. Electrocatalysis-based energy conversion devices are widely studied to get high yield and optimum performance of energy conversion services. The structural engineering of nanomaterials is associated with the fabrication of size, spatial array, hetero architecture, and shape of nanostructures, thereby producing a well-defined novel nanomaterial, which could be used for high-performance energy conversion system applications. The development and the innovations introduced in nanotechnology and material chemistry are making key breakthroughs for amplifying these devices' performance for perceiving the objective of renewable and sustainable clean energy technologies. The engineered nanomaterials such as nanoparticles, nanorods, nanospheres, nanosheets, nanotubes, and nanowires have drawn the attention of many nanotechnologists because of their attractive physical and chemical properties attributed to their significantly smaller size. The applications of zero (0-), one (1-), two (2-), and three (3-) dimensional nanostructures in the construction of high performance and cost-effective systems for harnessing energy by using renewable and sustainable technologies have been reported in many works of literature. This chapter will focus on the basic characteristics and idea of engineered nanomaterials for energy conversion cells with well-built prominence on the connection between structural features and resultant performances. In addition to emphasizing the applications of various nanomaterials in energy conversion cells, the apparent advantages, disadvantages, limitations, and challenges will be addressed. Finally, the outlook regarding the prospective futures of engineered nanomaterials for energy conversion will be discussed.

Keywords

Nanomaterials, Energy Conversion, Solar Cell, Nanotechnology, Supercapacitor, Aerogel, Graphene Quantum Dots, Polymeric Nanoparticles, Fuel Cell, Rechargeable Batteries

Contents

Applications of Emerging Nanomaterials and Nanotechnology   Materials Research Forum LLC
Materials Research Foundations 148 (2023) 103-126   https://doi.org/10.21741/9781644902554-4

1. Introduction

Nanomaterials play a very significant role in developing many necessary technologies. They are quite different from bulk and micro-sized materials for their usual proportions and, size and their unique physical properties. Gradually, the conventional bulk materials have been replaced by engineered nanomaterials such as nanocomposite gels [1-3], nanocomposite films [4-6], structural colored nanomaterials [7-9], molecular machines [10], nanomaterial based biosensors [11], etc. in different industrial sectors [12-15] to overcome the limitations that hinder the process of successful commercialization. The available properties of nano-fabricated materials rely predominantly on their characteristic structures at different levels, such as microscale, mesoscale, and particle levels. The first microscopic level can be described as the arrangement of atoms in the molecules and the electronic structure and distributions in atoms. The crystal surface, lattice orientation, crystal dislocations, and defects can influence the functional properties of nanomaterials. In addition, the particle size and surface morphology remarkably impact the unique properties of nanomaterials. Substantial research has already been done on the structural design of nanomaterials belonging to the first and third levels, such as microscopic and particle size [16, 17]. Some examples, like how quantum confinement leads to a higher bandgap when the characteristic dimension of a semiconductor is reduced to a certain size [18, 19], can help explain why nanomaterials have such unique properties. The bandgap can be precisely controlled by changing the size of the nanomaterials. By adjusting the bandgap, it is possible to fine-tune the absorption and emission spectra to fulfill the requirements of the desired applications [20, 21]. Because of the effect of surface plasmon resonance absorption, even the golden color of gold nanoparticles (AuNPs) swaps to pink color when the gold particle size is decreased to several tens of nm [22-24]. More interestingly, the AuNPs can show outstanding catalytic activity when the particle size is further reduced to less than 3 nm because of the comparatively smaller shrinking of the d-orbitals than the s- and p-orbitals [25-27].

The versatile and multidimensional utilizations of fabricated nanomaterials are not only for their excellent physical properties but also for the expanded specific surface area and surface energy related to their nanostructure. The enhanced surface energy associated with the surface area of nanomaterials has opened several critical applications in divergent sectors. The vapor pressure and solubility of materials vary according on their surface morphology, particularly the curvature of nanostructured materials [28]. The remarkable lowering of melting point and the superparamagnetic behavior of magnetic materials are attributed to the materials having reduced particle size at the nanometer scale [29-32]. The mechanical properties of nanomaterials are superior to that of bulk materials because of having less probability of structural imperfection and construction faults [33]. Engineered or fabricated nanomaterials put forward a considerable number of advantages towards developing current technologies, providing a suitable platform to create exciting and new technologies. Nanomaterials have already found potential applications in the drug delivery system for carrying the drug to a specific target site [34]. For energy conversion purposes, nanomaterials play an essential role and provide plenty of advantages. Energy conversion requires physical interaction and chemical changes occurring at the surface level and face-to-face involving materials. As a result, specific surface area, energy, and surface chemistry are essential parameters for energy conversion applications. Nanomaterials are the best candidates for providing such surface properties and mechanical strengths that significantly persuade at the interface of materials involved in the process and facilitating mass, energy, and charge transfer for energy conversion cells [35].

2. Nanostructured materials for solar cell applications

The harnessing of renewable and clean solar energy by photovoltaic (PV) conversion is the most promising technology. The first-generation (1G) solar cells were constructed using a single p-n junction diode composed of amorphous silicon, monocrystalline silicon, and polycrystalline silicon because of their optimal bandgap, light absorption, and high conversion efficiency [36, 37]. In order to keep costs down, thin-film cells made of materials including polycrystalline silicon, amorphous silicon, copper indium gallium selenide, copper zinc tin sulfur, and cadmium telluride are commonly used to build second-generation (2G) solar cells [38, 39]. The third-generation (3G) solar cells are fabricated using cheap and straightforward techniques and are fabricated from low-cost, non-polluting, and available materials. The characteristic features of 3G solar cells are large surface area with lightweight. Its application is in various new materials, such as quantum dots, nanotubes, nanocrystals, nanowires, solar ink, organic pigments, and conductive polymers besides silicon. The 3G solar cells are quite different from 1G and 2G solar cells as they do not have traditional p–n junctions dependent on setting apart the photogenerated charge carriers [40, 41]. Fourth-generation (4G) solar cells are constructed using hybrid nanocrystal cells composed of nanoparticles and polymer composite to facilitate the easy electron-proton transfer and generate superior voltage and current. Instead of the traditional p-n junction used in earlier generations, 4G solar cells are constructed following multi-junction-oriented technologies. Multi-junction technology makes the 4G solar cells cheaper but more efficient than their ancestors. Though the 4G solar cells can transform a higher percentage of solar energy, the conversion rate is not yet satisfactory. That's why researchers, scientists, and technologists are still working hard and accumulating their knowledge and experience to increase the energy conversion rate in solar cells by utilizing engineered nanomaterials to make solar energy a more pragmatic alternative energy source [42]. Semonin et al. reported solar cells, composed of organic-inorganic hybrid materials, convert the total amount of solar energy into electric energy [43]. Nanostructured solar cells, which are classified as either 3G or 4G solar cells, show considerable promise for advancing novel methods of converting solar energy into electrical or chemical energy [44]. For example, 4-tricyanovinyl-N,N-diethylaniline (TCVA) is an acceptor-donor disubstituted benzene-based organic molecule used as a dye (the molecular structure is illustrated in Fig. 1a). It is common practice to use p-type materials, such as TCVA, in optoelectronic devices. Thermal evaporation was used by Al Garni and colleagues to create ITO/InSe/TCVA/Au-configured hybrid organic-inorganic solar cells [45]. They estimated the capacitance-voltage and current density-voltage relationships to learn more about the solar cell's electrical transport mechanisms. With a rectification ratio of 157 at ± 1 V , the reported cell showed exemplary improvement in conversion efficiency. The experimental data suggested that the energy conversion rate was directly proportional to the system's operating temperature. Engineered nanomaterials are the potential for operating reversible reactions and are promising for making energy conversion cells for harnessing solar energy through a sustainable procedure [46].

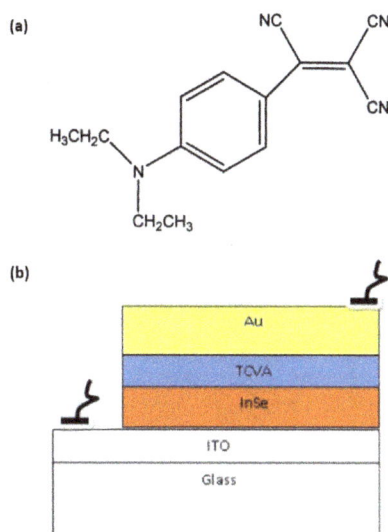

Figure 1. (a) The molecular structure of TCVA, and (b) a schematic representation of the ITO/InSe/TCVA/Au hybrid solar cell. The figure has been modified and reproduced with the permission from [45].

Upconversion luminescence (UCL) of rare earth (RE) nanomaterials that facilitate anti-stokes emission is crucial for many energy conversion cells [47]. Anti-stokes shifts, durability, and sharp emissions have multifaceted applications. RE ions emit very energetic UV and visible photons with the help of low-energy near-infrared (NIR) photons in the exciting UC process. Plasmon-enhanced UCL in RE doped NaYF4 has been used in solar cell production for years. Fig. 2 shows the simplified mechanism of UCL.

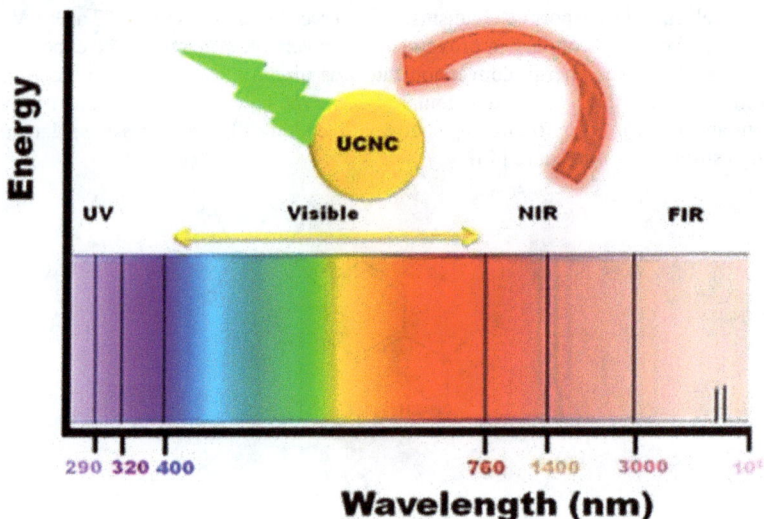

Figure 2. The simplified version of the upconversion luminescence (UCL) operating principle. The figure has been reproduced with the permission from [47].

Integrating the PV cell with a UC component that harvests infrared (IR) photons and emits visible light can improve PV performance. This method was introduced in the 1990s, but recently substantial breakthrough has been made. The intriguing UC nanomaterial family for PV technology is lanthanide (Ln^{3+}) ions embedded in host lattices. Ln^{3+} upconversion has several mechanisms (illustrated in Fig. 3). In the excited state absorption (ESA) mechanism, one Ln^{3+} uninterruptedly absorbs two photons to form a metastable intermediary state that absorbs a second photon to release a single high-energy photon [48]. One Ln^{3+} acts as an activator and the other as a sensitizer in energy transfer upconversion (ETU). The sensitizer Ln^{3+} absorbs a photon to gain an excited metastable intermediate state and transfers the energy non-radiatively to the activator Ln^{3+}. The sensitizer ion absorbs another photon and transfers it to the activator, which is stimulated to a higher energy state. Fig. 4 shows a typical schematic of the UC layer location in several PV cell types.

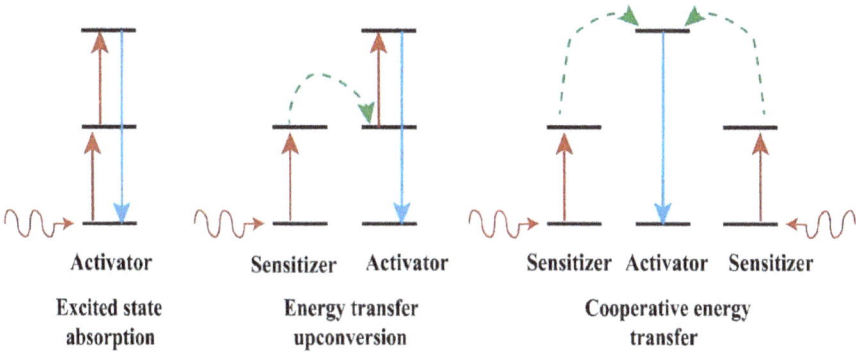

Figure 3. Important UC operating concepts that are applicable to ions containing
Ln^{3+}. The figure has been reproduced with the permission from [48].

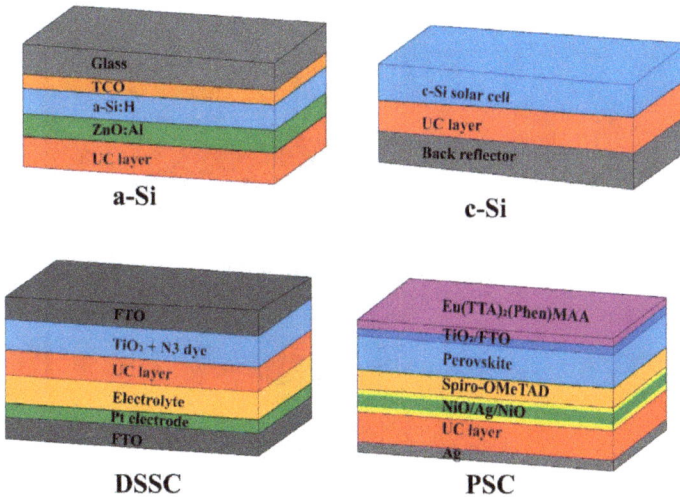

Figure 4. Integration of UC layers into various types of solar cells, including amorphous
silicon (a-Si), bifacial crystalline silicon (c-Si), dye-synthesized solar cells (DSSC), and
perovskite solar cell (PSC). The figure has been reproduced with the permission from
[48].

Dye-sensitized solar cells (DSSC) are composed of Titanium dioxide (TiO_2), the most frequently used photo-anode because of its unique features [49]. A larger band edge, dyes filling, exterior area, and the arrangement of lepton make TiO_2 suitable for an electrode. TiO_2 based on DSSC shows an energy conversion rate of around 13%. For the construction of DSSC, several types of engineered nanostructured oxide of metal, such as Fe_2O_3, Al_2O_3, ZrO_2, SnO_2, CeO_2, and V_2O_5, are used as photoanodes in DSSCs [50]. Kabir et al. fabricated DSSC using a photoanode composed of TiO_2 fused with AuNPs [51]. The plasmonic effect of AuNPs enhanced DSSC performance. In order to assess the efficiency of solar cells, photocurrent and photovoltage measurements are carried out. The solar-to-electric power efficiency of the DSSC by AuNPs was about 50% greater than those without AuNPs. The basic mechanism of the operation of the DSSC composed of TiO_2 fused with AuNPs is shown in Fig. 5. The transmission electron microscopy (TEM) imaging of the synthesized AuNPs and AuNPs with dye sensitized TiO_2, was conducted to explore the morphology nanoparticles. The TEM images of the AuNPs and AuNPs with dye sensitized TiO_2 are shown in Fig. 6. The AuNPs were spherical shaped and smaller than the TiO_2 nanoparticles.

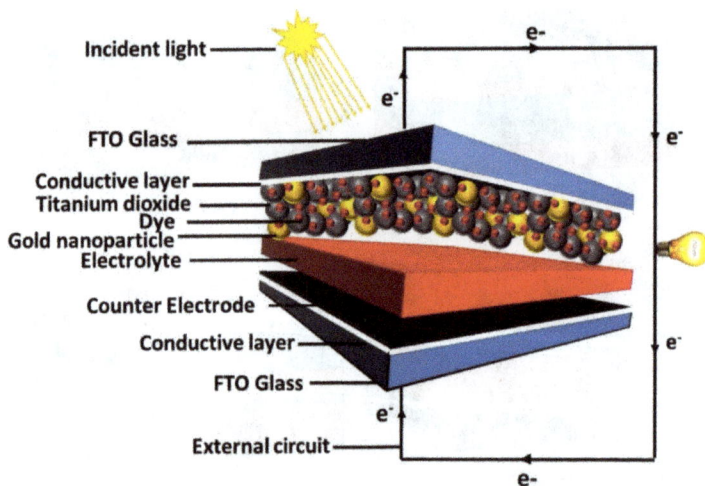

Figure 5. Dye-sensitized solar cells (DSSC) are schematically depicted, which explains how their basic operating principle works. The figure has been reproduced with permission from [51].

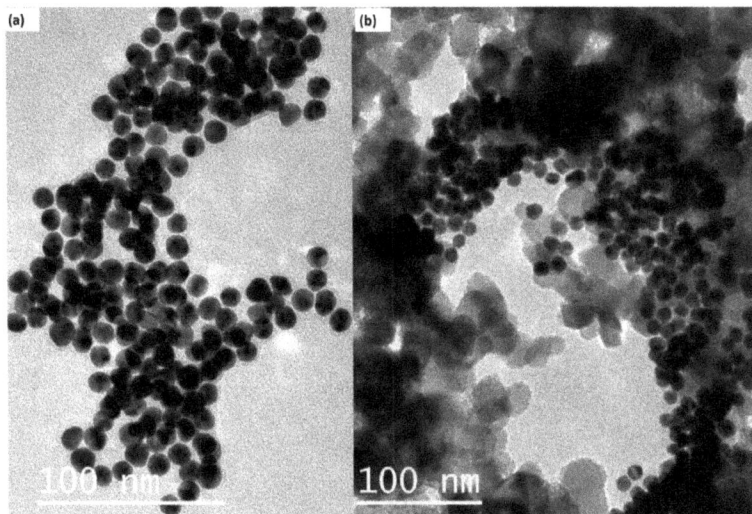

Figure 6. Transmission Electron Microscopy (TEM) images of gold nanoparticles (a) and gold nanoparticles with TiO₂ (b). The figure has been reproduced with permission from [51].

Carbonaceous nanomaterials are extensively used in energy conversion applications because of their unique structures, dimensional versatility, physical properties, and chemical features. Different allotropes of carbon, such as graphene quantum dots, carbine, and graphdiyne of various hybridizations, can construct energy conversion devices. Graphene quantum dots (GQDs) can absorb a wide range of light, making them very promising for building more productive solar cells [52]. Solar cell energy conversion efficiency can be increased by lowering hot-carrier relaxation rates. GQDs cool slowly because the energy of phonons is lower than the energy of their distinct electronic levels. Mueller and his team found that carrier cooling in GQDs is two orders of magnitude slower than in bulk materials, allowing hot charge carriers to be harvested to increase solar energy conversion efficiency. Novak et al. [53] found that low-molecular-weight polyethylene glycol (PEG) GQDs significantly improved their interaction with the poly 3-hexylthiophene: 6,6-phenyl-C61-butyric acid methyl ester (P3HT:PCBM) active layer. Oxygen plasma treatment of GQDs on a blocking TiO₂ layer improved planar-type perovskite solar cells (PSCs) [54]. They found that the GQDs can influence the fast extraction of electrons and induce the formation of upgraded perovskite quality. The experimental data suggested that the cells with perfectly sized-GQDs demonstrated about 10% higher conversion efficiency than those without GQDs. Fig. 7a and 7b show the GQD-inserted planar-type PSC and cell energy band diagram of PSC cells. The thickness of

GQDs/TiO$_2$, perovskite, hole transport material (HTM), and Au layers were 50 nm, 300 nm, 220 nm, and 60 nm, respectively, as illustrated in Fig. 7c.

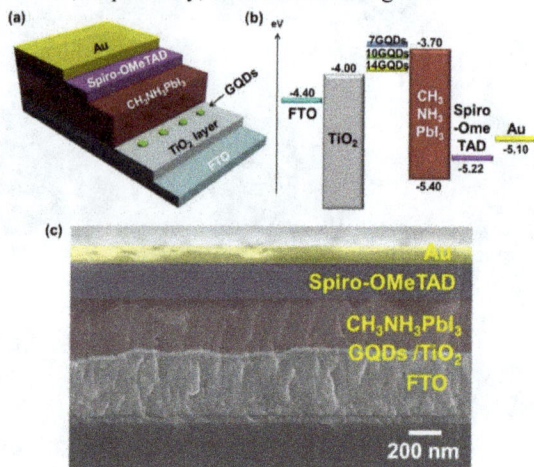

Figure 7. (a) A graphical representation of the prepared planar structure of the PSCs (b) An energy diagram of the materials that were exploited in the work. (c) A cross-sectional image taken with a scanning electron microscope of certain GQD-applied planar PSC. The figure has been reproduced with the permission from [54].

Aerogels are a special type of three-dimensional non-fluidic, interlinked, colloidal, and porous nanostructured materials where the particles are imprecisely packed with the ability to expand throughout the volume of the system [55]. They have extremely low density and large specific surface area and are prepared from wet gels by exchanging the interstitial liquid by air component and keeping the interlinked network structure stable without any shrinkage. Thus, the aerogels are composed of air, which is 95% of its total volume. They are classified as a new state of matter, interlinking between the liquid and gaseous phases. Since aerogels have very low density, they are being used in space technologies and gradually becoming popular in other applications. TiO$_2$ based metal-oxide type aerogel is an important substitution for silica-based aerogel, which is suitable for being used as photoanode to fabricate DSSC because of their large specific surface area, including the wide range of bandgap. The efficiency of DSSC is dependent on photon absorption, which is interrelated to the amount of dye adsorption. Initially, Baia et al. reported the potentiality of TiO$_2$ based aerogel for PV applications [56], and Rolison et al. demonstrated the application of TiO$_2$ based aerogel as photoanode material in DSSC [57]. Both showed that the performance of TiO$_2$ based aerogel is better in the case of longer wavelengths because of greater light scattering features. Chiang et al. demonstrated that the performance of TiO$_2$

aerogel photoanode is higher than that of P25 photoanode in DSSCs [58]. Recent investigations have suggested that the TiO$_2$ aerogel-based photoanodes synthesized by the sol-gel method can show interesting features used for DSSCs construction [59].

Mu and Li reported innovative forms-stable composite phase change materials (FS-CPCM) for solar energy storage and conversion [60]. Vacuum impregnation yielded FS-CPCM (illustrated in Fig. 8). Lauric acid (LA) was used as the phase change material (PCM) and grafted onto graphene aerogel (GA) via a reduced esterification reaction to provide supporting material and boost thermal conductivity. LA/LA-GA-1 FS-CPCMs displayed 352.1% and 32.6% higher thermal conductivities than LA and LA/GA FS-CPCMs. DSC showed that the LA/LA-GA-1 FS-CPCM has excellent phase change properties, excellent thermal cycling, and low undercooling stability. They have higher thermal stability and a melting-freezing enthalpy of 207.3 and 205.8 J/g. The LA/LA-GA FS-CPCM has exceptional photon absorption and 80.6% light-to-heat efficiency. Fig. 9 depicts a system built with FS-CPCM that converts light energy into thermal energy. It was found that both the LA and FS-CPCMs have naturally occurring shape-stabilizing features, which were studied by heating the materials in an oven. Glass-surfaced jars were used to maintain an 80 °C temperature for both the LA and FS-CPCMs (see Fig. 10). Within 30 minutes, the intrinsic LA melted, but the FSCPCMs remained solid.

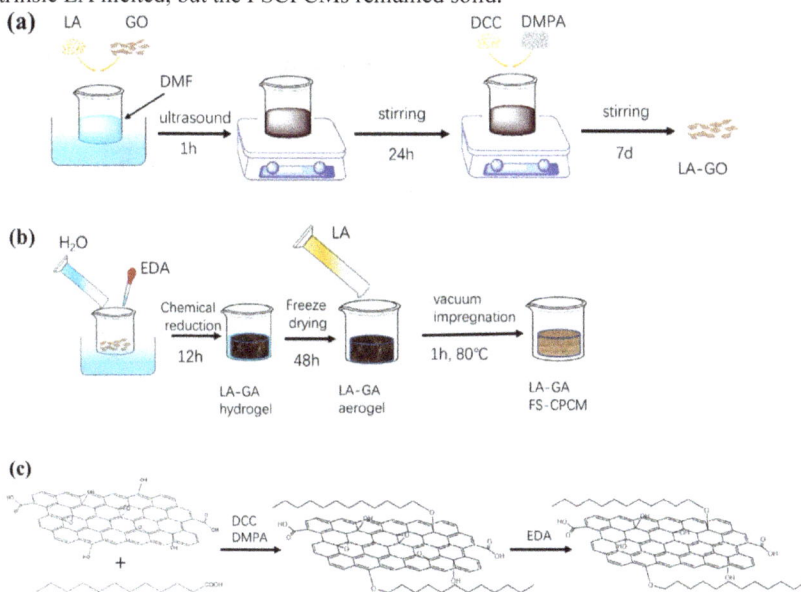

Figure 8. (a) A schematic illustration of the process that led to the formation of LA-GA, (b) the LA-GA FS-CPCM, and (c) the process that leads up to the formation of LA-GA. The figure has been reproduced with the permission from [60].

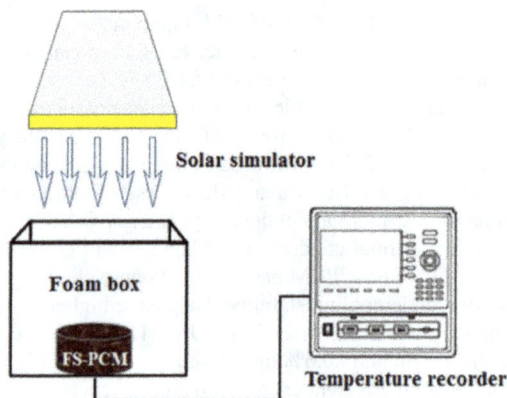

Figure 9. A system for the conversion of light into thermal energy that was built using FS-CPCM. The figure has been reproduced with the permission from [60].

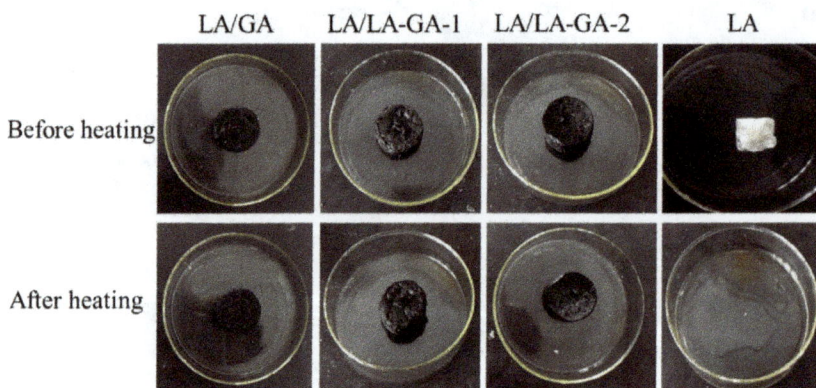

Figure 10. Photographs before and after heating of the LA, LA/GA, and LA/LA-GA FS-CPCMs. The figure has been reproduced with the permission from [60].

3. Nanostructured materials for supercapacitor applications

Mesoporous metal oxides with greater specific surface area and porosity are very good competitors for supercapacitor applications. Due to the large surface area and well-distributed porous nature, the specific capacitance of the mesoporous metal oxides-based electrode is high [61]. Copper oxide nanostructures with voids provide a stable platform with significantly better electrochemical characteristics compared to solid analogues of the

Applications of Emerging Nanomaterials and Nanotechnology   Materials Research Forum LLC
Materials Research Foundations 148 (2023) 103-126   https://doi.org/10.21741/9781644902554-4

same surface structures. The results of the cyclic voltammetry (CV) to determine the specific capacitance and charge-discharge (CD) are highly promising. Because of the proposed low-cost synthesis process, these materials are now ready for widespread use in alternative energy storage and conversion devices. The electrochemical performance of transition metal oxides (TMOs) is significantly influenced by the morphologies of the particles. SEM and TEM micrographs in Fig. 11a–d reveal two hollow and solid Cu_2O structures with a diameter of ~200–300 nm and a cavity size of ~150 nm (Fig. 11a,b). Fig. 11e,f shows Cu_2O particle size distributions. Fig. 12 shows artificial Cu_2O crystal morphologies and growth mechanisms of Cu(HN) and Cu(OCT). Fig. 13 shows ion-electrode interaction for Cu(HN) and Cu(OCT) electrodes. Gholivand et al. developed CuO/PANI nanocomposite via in situ polymerization method [62]. The composite exhibited 75% capacitance retention and 185 F/g specific capacitance, while pure CuO nanoparticles exhibited 30% and 76 F/g, respectively. Ates's group reported the highest specific capacitance of 286.35 F/g for CuO/PANI nanocomposite [63]. Carbon-based well-defined nanomaterials also have the potential for making supercapacitors because of their low price, availability, superior mechanical properties, excellent stability, large and porous surface area, high conductivity, and electrolyte accessibility [64]. There are other reports on using well-defined carbon-based nanomaterials for fabricating supercapacitors, which are used for energy conversion cells [65].

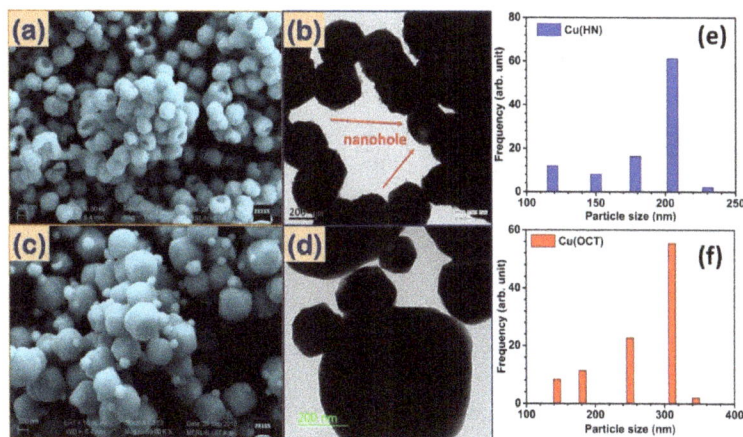

Figure 11. (a,c) SEM images, (b,d) TEM images, and (e,f) particle size distribution of Cu(HN) and Cu(OCT), respectively. The figure has been reproduced with the permission from [61].

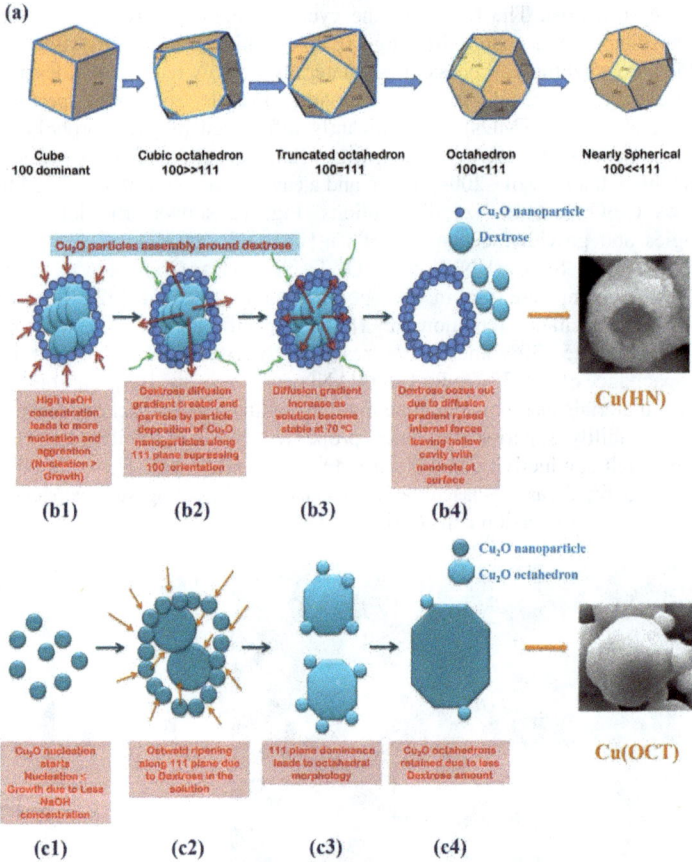

*Figure 12. (a) Artificial crystal morphologies of the Cu₂O crystal to explain the growth of Cu(HN) and Cu (OCT), (**b1-b4**) Growth mechanism Cu(HN) and (**c1-c4**) Cu(OCT). The figure has been reproduced with the permission from [61].*

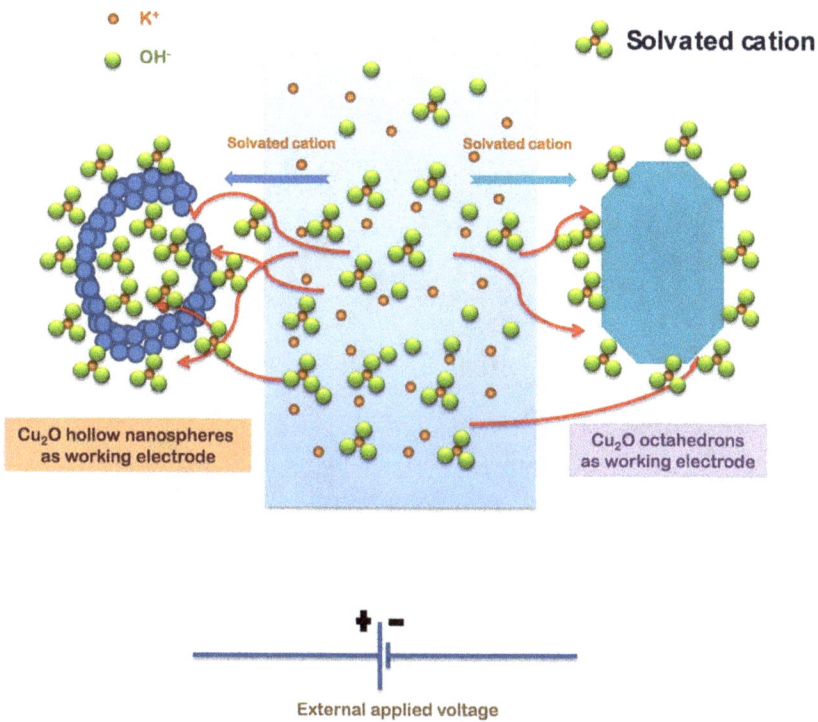

Figure 13. Ion-electrode interaction for Cu(HN) and Cu(OCT) electrodes. The figure has been reproduced with the permission from [61].

4. Nanomaterials for battery applications

Moretti et al. produced layered V_2O_5 to intercalate Na^+ ions for sodium battery anodes [66]. Li-ion batteries with layered V_2O_5 anodes exhibited 300 mAh/g discharge capacity. In the V_2O_5 based anode, the average cell voltage became 2.5 V, greater than vanadium redox flow and nickel-metal hydride batteries. Maloney et al. used lithium titanate-based anode for Li-ion battery and found a specific capacity of 137 mAh/g, similar to commercially available lithium titanate anode material [67]. Gradually and steadily, well-defined carbon-based nanomaterials are taking over the necessary position to enhance the efficiency of energy conversion devices and solve rechargeable batteries. Nanoscale materials have a substantially greater impact on the performance of rechargeable batteries due to their architectural uniqueness, increased surface area, pore size, and distribution pattern. Different nanomaterials such as nanoalloys, TMOs, or silicon with carbon nanomaterial-based complexes can potentially be utilized in batteries for energy conversion and storage [68]. Applications of various nanomaterials for lithium ion batteries are illustrated in Fig. 14.

Figure 14. Schematic diagram of the recent anode materials for lithium ion batteries. The figure has been reproduce with permission from [68].

5.   Nanostructured materials for fuel cell applications

Production of hydrogen by splitting water is an excellent technique for creating renewable and clean energy. The water-splitting process involves oxygen evolution at anode and hydrogen evolution at cathode. Several investigations have already been reported regarding water-splitting techniques with divergent elementary formulations and nanostructured materials. The development of catalysts having a ternary and quaternary structure is essential to enhance the water-splitting process for making fuel. A recent study regarding the applications of inorganic aerogels in energy conversion technologies shows that SiO_2 aerogels can be used as a catalyst in a fuel cells [69]. They block the Pt nanoparticle conglomeration, but the higher concentration results in the loss in mass transport at high current density. Polymeric nanomaterials are used in microbial fuel cells as coating or modifier for anode, hydrogen fuel cells as supporting components of cathode catalyst, in fabricating the anode catalyst for direct methanol fuel cells. The fuel cell is a prospective energy source due to its huge advantages and applications. The effective and productive implementation of nanostructured materials has brought new innovations and finding to overcome the existing limitations that hamper successful commercialization. The utilization of nanomaterials in various components of fuel cell, such as electrodes, catalyst, electrolyte/membrane, etc. can beat many restrictions. The novel properties of nanomaterials can significantly enhance the entire work efficiency and the overall cell performance [70]. The successful applications of various nanomaterials in fuel cells have been illustrated in Fig. 15.

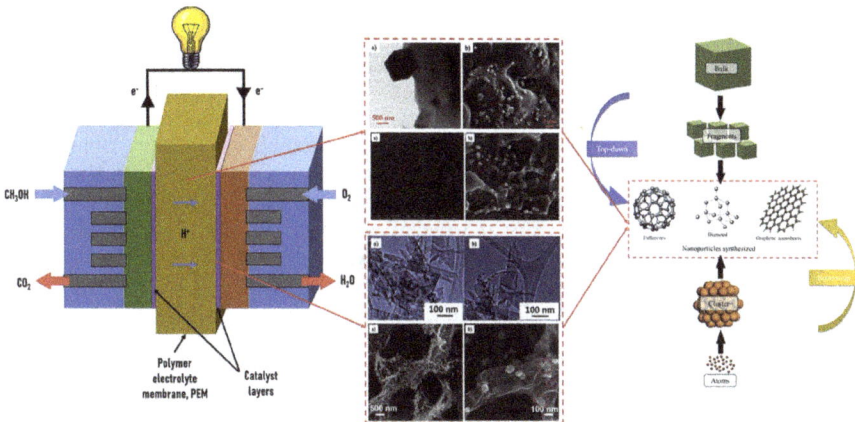

Figure 15. The successful applications of various nanomaterials in fuel cells. The figure has been reproduced with the permission from [70].

6. Challenges and limitations of engineered nanomaterials for energy conversion

Engineered nanomaterials are highly promising for fabricating energy conversion devices because of their excellent and exceptional physical and chemical features. Nanostructured materials have many advantages for manufacturing and improving the efficiency of energy conversion cells. However, some limitations still correspond to the construction materials of such cells. Usually, two fundamental limitations for all solar cells are the absence of sunlight at night and the supply of large amounts of light energy. Scientists and technologists are trying to overcome the drawbacks and limitations of nanostructured materials used in energy conversion cells. Nanostructured materials-based solar cells have excellent features especially when those are evaluated from profit-oriented perspectives; however, such cells accompany some limitations from a mechanical point of view. The inappropriate passiveness of internal surfaces obstructs the steadiness of solar energy conversion cells for a long time. However, the stability of devices is undeniably one of the significant criteria for solar energy conversion technologies because the overall cost will be minimized if the system runs for a long time without or with minimum maintenance. The stability issues present in the conversion devices are either due to the chemical process or the system's configuration. The commercial feasibility of the nanostructured solar cell might be increased by encapsulation technologies, which could be a possible solution to the stability problem. The initial maintenance and operational cost and the efficiency of cells might be kept optimum to increase the use of such devices by potential consumers. The commercialization of energy conversion devices continues to face some difficulties even though the uses of nanostructured materials are thoroughly examined on a lab scale. Besides, to gain high electrical conductivity, the synthesis of composite will be highly important in the future.

The most critical problem of fuel cells is accompanying the rate of reactions [71]. Other issues are long initiation time, low conversion efficiency, reaction by-product, catalyst poisoning, durability, high cost, and the scarcity of construction materials [72]. The factors mentioned above seriously limit the commercial prospects of nanostructured fuel cells. So, it is crucial to carry out more research to create such materials which could eliminate or minimize the factors associated with poor performance and the price of devices. The application of polymer nanomaterials is still experimental-based. Many studies have already been performed to describe the use of polymeric nanomaterials in fuel cells; however, most of them are based on different synthetic routes and physicochemical analyses. The availability of such applications in the commercial sector is minimal due to a lack of further research and development.

7. Summary and future perspectives

Though there are some vital drawbacks, nanostructured solar cells hold brilliant prospects to attain a higher energy conversion rate for all types of applications, including commercial sectors. Latest technology-based solar cells like DSSC and perovskite solar cells have a bright future due to their higher efficiency, close to 27.3%. The next-generation solar cells expect to reduce the demand and supply chain gap by increasing the energy conversion rate and lowering the overall cost. The quick and steady success in the development of nanotechnology makes us confident in achieving the goal. The limitations and challenges associated with the properties and synthesis might be overcome through extensive research. Especially the synthesis of composite with the combination of graphene, carbon nanotube, and conducting polymers will provide the required advantages for energy conversion applications. The adjustment and alternation of aerogel's surface through physical and chemical treatment will enhance the adsorption rate to find potential applications in the future. The introduction of single metallic or manifold metallic networks to nanostructures will increase the deposition on the surface. Composite aerogels are expected to have exceptionally high electrocatalytic properties and stability for energy conversion in fuel cell fabrication. The rate of catalytic transformation and the energy conversion efficiency in fuel cells might be improved by employing nanostructured materials, especially polymeric nanoparticles, to support the catalyst to enhance the efficiency of expensive Pt and Pt-based catalysts. The comprehensive exhibition of fuel cells constructed by polymeric nanomaterials is very rare, but the published results seem very promising to overcome all sorts of drawbacks limiting the commercialization process by increasing efficiency and lowering the manufacturing cost. The polymeric nanostructured particles have excellent carbon monoxide tolerance, good interaction with the deposited metal catalyst, high electrocatalytic efficiency, and high electronic and proton conductivities. They also have a high holding and retention capability of the expensive catalyst, and uniform and homogeneous dispersion and distribution of the catalyst nanoparticles. These features are the major criteria of polymeric nanomaterials to overcome drawbacks related to poor performance and overall installation and maintenance cost of energy conversion cells.

The above-mentioned examples, applications, discussions, and prospects of engineered nanomaterials for energy conversion cells depict clearly that the future scientific research and developments of engineered nanomaterials are greatly expected and needed; however, to achieve such a goal is still challenging. In such circumstances, effective and fruitful initiatives should be taken to make national and international participation in implementing sustainable energy conversion technologies. The chapter has been composed to provide worthy particulars regarding the applications of engineered nanomaterials for energy conversion technologies. Hopefully, the information cited in the chapter will supply relevant information for the researchers and technologists who intend to deal with the structural design and performance control of energy conversion cells composed of engineered nanomaterials.

Acknowledgements

A.B. Imran cordially acknowledges the Ministry of Education, People's Republic of Bangladesh, and the Committee for Advanced Studies and Research (CASR) in BUET for the funding.

References

[1] M. Harun-Ur-Rashid, T. Seki, Y. Takeoka, Structural colored gels for tunable soft photonic crystals, Chem. Rec. 9 (2009) 87-105. https://doi.org/10.1002/tcr.20169

[2] M. Harun-Ur-Rashid, A.B. Imran, Superabsorbent hydrogels from carboxymethyl cellulose", in Ibrahim H. Mondal (ed.) Carboxymethyl Cellulose. Volume I: Synthesis and Characterization, Nova Science Publishers, New York, 2019, pp. 159-182.

[3] M.R. Karim, Harun-Ur-Rashid, A.B. Imran, Highly stretchable hydrogel using vinyl modified narrow dispersed silica particles as cross-linker, ChemistrySelect, 5 (2020) 10556-10561. https://doi.org/10.1002/slct.202003044

[4] M. Harun-Ur-Rashid, T. Foyez, A.B. Imran, Fabrication of stretchable composite thin film for superconductor applications. In Sensors for Stretchable Electronics in Nanotechnology, CRC Press, 2021, pp. 63-78. https://doi.org/10.1201/9781003123781-5

[5] M. N. Huda, T. Seki, H. Suzuki, A. N. M. H. Kabir, M. Harun-Ur-Rashid, Y. Takeoka, Characteristics of high-density poly (N-isopropylacrylamide)(pNIPA)

brushes on silicon surface by atom transfer radical polymerization. Trans. Mater. Res. Soc. Japan 35(4) (2010) 845-848.

[6] A-N. Chowdhury, J. Shapter, and A. B. Imran, Innovations in Nanomaterials, Nova Science Publishers, Inc., NY, USA, 2015

[7] M. Harun-Ur-Rashid, A. B. Imran, T. Seki, Y. Takeoka, M. Ishii, H. Nakamura, Template synthesis for stimuli-responsive angle independent structural colored smart materials, Trans. Mater. Res. Soc. 34 (2009) 333-337. https://doi.org/10.14723/tmrsj.34.333

[8] M. Harun-Ur-Rashid, A. B. Imran, T. Seki, M. Ishii, H. Nakamura, Y. Takeoka, Angle-independent structural color in colloidal amorphous arrays, ChemPhysChem, 11 (2010) 579-583. https://doi.org/10.1002/cphc.200900869

[9] Y. Takeoka, S. Yoshioka, M. Teshima, A. Takano, M. Harun-Ur-Rashid, M., T. Seki, Structurally coloured secondary particles composed of black and white colloidal particles, Sci. Rep. 3 (2013) 1-7. https://doi.org/10.1038/srep02371

[10]   A. B. Imran, M. Harun-Ur-Rashid, Y. Takeoka, Polyrotaxane Actuators. In Soft Actuators, Springer, Singapore, 2019, pp. 81-147. https://doi.org/10.1007/978-981-13-6850-9_6

[11]   M. Harun-Ur-Rashid, T. Foyez, I. Jahan, K. Pal, A. B. Imran, Rapid diagnosis of COVID-19 via nano-biosensor-implemented biomedical utilization: a systematic review, RSC Advances, 12 (2022) 9445-9465. https://doi.org/10.1039/D2RA01293F

[12]   M. Harun-Ur-Rashid, A. B. Imran, M. A. B. H. Susan, Green Polymer Nanocomposites in Automotive and Packaging Industries, Curr. Pharm. Biotechnol. 24(1) (2023). https://doi.org/10.2174/1389201023666220506111027.

[13]   M. Harun-Ur-Rashid, A. B. Imran, Nanomaterials in the Automobile Sector. In Emerging Applications of Nanomaterials, Materials Research Forum LLC, 141 (2023) 124-150. https://doi.org/10.21741/9781644902295-6

[14]   M. Harun-Ur-Rashid, A. B. Imran, M. A. B. H. Susan, Prospective Nanomaterials for Food Packaging and Safety. In Emerging Applications of Nanomaterials, Materials Research Forum LLC, 141 (2023) 327-352. https://doi.org/10.21741/9781644902295-13

[15]   A. B. Imran, M. A. B. H Susan, Natural fiber-reinforced nanocomposites in automotive industry. In Nanotechnology in the Automotive Industry, Elsevier, 2022, pp. 85-103. https://doi.org/10.1016/B978-0-323-90524-4.00005-0

[16]   Y. Chen, S. Ji, C. Chen, Q. Peng, D. Wang, Y. Li, Single-atom catalysts: Synthetic strategies and electrochemical applications. Joule 2 (2018) 1242-1264.

[17]   X. Zhu, Q. Zhang, C. Huang, Y. Wang, C. Yang F. Wei, Validation of surface coating with nanoparticles to improve the flowability of fine cohesive powders. Particuology 30 (2017) 53-61.

[18] A. P. Alivisatos, Semiconductor Clusters, Nanocrystals, and Quantum Dots. Science 271 (1996) 933-937.

[19] Q. Zhang, E. Uchaker, S. Candelaria G. Cao, Nanomaterials for energy conversion and storage. Chem. Soc. Rev. 42 (2013) 3127-3171.

[20] Y. Wang, N. Herron, Nanometer-sized semiconductor clusters: materials synthesis, quantum size effects, and photophysical properties. J. Phys. Chem. 95 (1991) 525-532.

[21] V. I. Klimov, Semiconductor and metal nanocrystals: synthesis and electronic and optical properties, CRC, November 2003.

[22] S. K. Ghosh, T. Pal, Interparticle Coupling Effect on the Surface Plasmon Resonance of Gold Nanoparticles: From Theory to Applications. Chem. Rev. 107 (2007) 4797-4862.

[23] N. Felidj, J. Aubard, G. Levi, J. Krenn, A. Hohenau, G. Schider, A. Leitner, F. Aussenegg, Optimized surface-enhanced Raman scattering on gold nanoparticle arrays. Appl. Phys. Lett. 82 (2003) 3095-3097.

[24] K. Kneipp, M. Moskovits, H. Kneipp, Surface-enhanced Raman scattering: physics and applications. Springer, vol.103, pp. 1-17, 2006.

[25] R. Sardar, A. M. Funston, P. Mulvaney, R. W. Murray, Gold Nanoparticles: Past, Present, and Future. Langmuir 25 (2009) 13840-13851.

[26] M. Haruta, M. Date, Advances in the catalysis of Au nanoparticles. Appl. Catal. A 222 (2001) 427-437.

[27] M. Cortie, E. Van der Lingen, Catalytic gold nanoparticles. Mater. Forum 26 (2002) 1-4.

[28] E. Roduner, Size matters: why nanomaterials are different, Chem. Soc. Rev. 35 (2006) 583-592.

[29] G. Cao, Nanostructures & nanomaterials: synthesis, properties & applications. World Scientific Publishing Company, pp.448, April 2004.

[30] K. Koga, T. Ikeshoji, K. Sugawara, Size- and Temperature-Dependent Structural Transitions in Gold Nanoparticles. Phys. Rev. Lett. 92 (2004) 115507.

[31] H. Petrova, M. Hu, G. V. Hartland, Photothermal Properties of Gold Nanoparticles. Phys. Z. Chem. 221 (2007) 361-376.

[32] D. Leslie-Pelecky, R. Rieke, Magnetic Properties of Nanostructured Materials Chemistry of Materials. 8 (1996) 1770-1783.

[33] M. Terrones, Science and Technology of the Twenty-First Century: Synthesis, Properties, and Applications of Carbon Nanotubes. Annu. Rev. Mater. Res. 33 (2003) 419-501.

[34] R. Singh, J. W. Lillard Jr, Nanoparticle-based targeted drug delivery. Exp. Mol. Pathol. 86 (2009) 215-223.

[35]　G. Wang, L. Zhang, J. Zhang, A review of electrode materials for electrochemical supercapacitors. Chem. Soc. Rev. 41 (2012) 797-828.

[36]　T. Shiyani, T. Bagch, Hybrid nanostructures for solar-energy-conversion applications. Nanomater. Energy 9 (2020) 39-46.

[37]　S. Chander, A. Purohit, A. Sharma, Arvind, S. P. Nehra, M. S. Dhaka, A study on photovoltaic parameters of mono-crystalline silicon solar cell with cell temperature. Energy Rep. 1 (2015) 104-109.

[38]　M. A. Green, E. D. Dunlop, D. H. Levi, J. Hohl-Ebinger, M. Yoshita, Ho-Baillie, W. Y. Anita, Solar cell efficiency tables (version 54). Prog. Photovolt.: Res. Appl. 27 (2019) 565-575.

[39]　J. Zheng, H. Mehrvarz, F. J. Ma, C. F. J. Lau, M. A. Green, S. Huang, A. W. Y. Ho-Baillie, 21.8% Efficient Monolithic Perovskite/Homo-Junction-Silicon Tandem Solar Cell on 16 cm^2. ACS Energy Lett. 3 (2018) 2299-2300.

[40]　O. Yehezkeli, R. Tel-Vered, J. Wasserman, A. Trifonov, D. Michaeli, R. Nechushtai, I. Willner, Integrated photosystem II-based photo-bioelectrochemical cells. Nat. Commun. 3:742 (2012) 1-7.

[41]　J. Li, X. Feng, Y. Jia, Y. Yang, P. Cai, J. Huang, J. Li, Co-assembly of photosystem II in nanotubular indium–tin oxide multilayer films templated by cellulose substance for photocurrent generation. J. Mater. Chem. A 5 (2017) 19826-19835.

[42]　D. Lan, M. A. Green, The potential and design principle for next-generation spectrum-splitting photovoltaics: Targeting 50% efficiency through built-in filters and generalization of concept. Prog. Photovolt.: Res. Appl. 27 (2019) 899-904.

[43]　O. E. Semonin, J. M. Luther, S. Choi, H. Y. Chen, J. Gao, A. J. Nozik, M. C. Beard, Peak external photocurrent quantum efficiency exceeding 100% via MEG in a quantum dot solar cell. Science 334:6062 (2011) 1530-1533.

[44]　X. Yan, J. Zhang, Copper(II) complexes based on 4-R-terpyridine: Synthesis, structures, and photocatalytic properties. Chem. Res. Chin. Univ. 33:1 (2017) 1-6.

[45]　S. E. Al Garni, A. A. A. Darwish, Photovoltaic performance of TCVA-InSe hybrid solar cells based on nanostructure films. Sol. Energy Mater. Sol. Cells 160 (2017) 335-339.

[46]　C. Masquelier, L. Croguennec, Polyanionic (Phosphates, Silicates, Sulfates) frameworks as electrode materials for rechargeable Li (or Na) batteries. Chem. Rev. 113 (2013) 6552-6591.

[47]　D. Kumar, S. K. Sharma, S. Verma, V. Sharma, V. Kumar, A short review on rare earth doped NaYF4 upconverted nanomaterials for solar cell applications, Mater. Today: Proc. 21 (2020) 1868-1874. https://doi.org/10.1016/j.matpr.2020.01.243

[48]   A. Ghazy, M. Safdar, M. Lastusaari, H. Savin, M. Karppinen, Advances in upconversion enhanced solar cell performance. Sol. Energy Mater. Sol. Cells 230 (2021) 111234.

[49]   R. Jose, V. Thavasi, S. Ramakrishna, Metal oxides for dye-sensitized solar cells. J. Am. Ceram. Soc. 92 (2009) 289-301.

[50]   N. G. Park, J. van de Lagemaat, A. J. Frank, Comparison of dye-sensitized rutile- and anatase-based TiO2 solar cells. J. Phys. Chem. B 104:38 (2000) 8989-8994.

[51]   D. Kabir, T. Forhad, W. Ghann, B. Richards, M. M. Rahman, M. N. Uddin, M. R. J. Rakib, M. H. Shariare, F. I. Chowdhury, M. M. Rabbani, N. M. Bahadur, Dye-sensitized solar cell with plasmonic gold nanoparticles modified photoanode. Nano-Struct. Nano-Objects 26 (2021) 100698.

[52]   Y. Wang, P. Yang, L. Zheng, X. Shi, H. Zheng, Carbon nanomaterials with sp or/and sp hybridization in energy conversion and storage applications: A review. Energy Storage Mater. 26 (2020) 349-370.

[53]   T. G. Novak, J. Kim, S. H. Song, G. H. Jun, H. Kim, M. S. Jeong, S. Jeon, Fast P3HT Exciton Dissociation and Absorption Enhancement of Organic Solar Cells by PEG-Functionalized Graphene Quantum Dots. Small, 12 (2016) 994-999. doi:10.1002/smll.201503108

[54]   J. Ryu, J.W. Lee, H. Yu, J. Yun, K. Lee, J. Lee, D. Hwang, J. Kang, S.K. Kim, and J. Jang, "Size effects of a graphene quantum dot modified-blocking TiO2 layer for efficient planar perovskite solar cells," J. Mater. Chem. A, vol. 5, pp. 16834-16842, 2017.

[55]   S. Alwin, S. Shajan, Aerogels: promising nanostructured materials for energy conversion and storage applications. Mater. Renew. Sustain. Energy 9:2 (2020) 1-27. doi:10.1007/s40243-020-00168-4

[56]   L. Baia, A. Peter, V. Cosoveanu, E. Indrea, M. Baia, J. Popp, and V. Danciu, Synthesis and nanostructural characterization of TiO₂ aerogels for photovoltaic devices. Thin Solid Films 511 (2006) 512-516.

[57]   J. J. Pietron, A. M. Stux, R. S. Compton, D. R. Rolison, Dyesensitized titania aerogels as photovoltaic electrodes for electrochemical solar cells. Sol. Energy Mater. Sol. Cells 91:12 (2007) 1066-1074.

[58]   Y.-C. Chiang, W.-Y. Cheng, and S.-Y. Lu, Titania aerogels as a superior mesoporous structure for photoanodes of dye-sensitized solar cells. Int. J. Electrochem. Sci. 7 (2012) 6910-6919.

[59]   S. Alwin, X. S. Shajan, R. Menon, P. Nabhiraj, K. Warrier, G. M. Rao, Surface modification of titania aerogel films by oxygen plasma treatment for enhanced dye adsorption. Thin Solid Films 595 (2015) 164-170.

[60]   B. Mu, M. Li, Synthesis of novel form-stable composite phase change materials with modified graphene aerogel for solar energy conversion and storage. Sol. Energy Mater. Sol. Cells 191 (2019) 466-475.

[61] V. Sharma, I. Singh, A. Chandra, Hollow nanostructures of metal oxides as next generation electrode materials for supercapacitors. Sci. Rep. 8:1 (2018) 1-12.

[62] M. B. Gholivand, H. Heydari, A. Abdolmaleki, H. Hosseini, Nanostructured CuO/PANI composite as supercapacitor electrode material. Mater. Sci. Semicond. Proc. 30 (2015) 157-161.

[63] M. Ates, M. A. Serin, I. Ekmen, and Y. N. Ertas, Supercapacitor behaviors of polyaniline/CuO, polypyrrole/CuO and PEDOT/CuO nanocomposites. Polym. Bull. 72 (2015) 2573-2589.

[64] Y. Zhang, Z. Shang, M. Shen, S. P. Chowdhury, A. Ignaszak, S. Sun, Y. Ni, Cellulose Nanofibers/Reduced Graphene Oxide/Polypyrrole Aerogel Electrodes for High-Capacitance Flexible All-Solid-State Supercapacitors. ACS Sustainable Chem. Eng. 7:13 (2019) 11175-11185.

[65] A. Rahman, Solaiman, T. Foyez, M. A. B. H. Susan, A. B. Imran, Self-Healable and Conductive Double-Network Hydrogels with Bioactive Properties. Macromol. Chem. Phys. 221:7 (2020) 2000207.

[66] A. Moretti, M. Secchiaroli, D. Buchholz, G. Giuli, R. Marassi, S. Passerini, Exploring the low voltage behavior of V_2O_5 aerogel as intercalation host for sodium ion battery. J. Electrochem. Soc. 162:14 (2015) A2723-A2728.

[67] R. P. Maloney, H. J. Kim, J. S. Sakamoto, Lithium titanate aerogel for advanced lithium-ion batteries. ACS Appl. Mater. Interfaces 4:5 (2012) 2318-232.

[68] W. Qi, J. G. Shapter, Q. Wu, T. Yin, G. Gao, D. Cui, Nanostructured anode materials for lithium-ion batteries: principle, recent progress and future perspectives. J. Mater. Chem. A 5(37) (2017) 19521-19540.

[69] J. B. Gerken, J. Y. C. Chen, R. C. Massé, A. B. Powell, S. S. Stahl, Development of an O_2-Sensitive Fluorescence-Quenching Assay for the Combinatorial Discovery of Electrocatalysts for Water Oxidation. Angew. Chem. Int. Ed. 51 (2012) 6676-6680.

[70] N. F. Raduwan, N. Shaari, S. K. Kamarudin, M. S. Masdar, R. M. Yunus, An overview of nanomaterials in fuel cells: Synthesis method and application. Int. J. Hydrog. Energy 47 (2022) 18468-18495.

[71] P. Kumar, K. Dutta, S. Das, P. P. Kundu, An overview of unsolved deficiencies of direct methanol fuel cell technology: factors and parameters affecting its widespread use. Int. J. Energy Res. 38 (2014) 1367e1390. https://doi.org/10.1002/er.3163.

[72] A.S. Aricò, S. Srinivasan, V. Antonucci, DMFCs: From fundamental aspects to technology development. Fuel Cells 1 (2001) 133e161. https://doi.org/10.1002/1615-6854(200107)1:2<133::AID-FUCE133>3.0.CO;2-5.

Applications of Emerging Nanomaterials and Nanotechnology   Materials Research Forum LLC
Materials Research Foundations 148 (2023) 127-169   https://doi.org/10.21741/9781644902554-5

Chapter 5

Emerging Trends of Nanotechnology in Cosmetics

Garima Nagpal[1*], Rashi Chaudhary[2], Ratiram G Chaudhary[3] and N.B. Singh[4,5]

[1]Department of Environmental Sciences, Sharda University, Greater Noida, India

[2]Department of Life Sciences, Sharda University, Greater Noida, India

[3]Post Graduate Department of Chemistry, S. K. Porwal College, Kamptee-441001, India

[4]Department of Chemistry and Biochemistry, Sharda University, Greater Noida, India

[5]Research Development Cell, Sharda University, Greater Noida, India

*garima.nagpal@sharda.ac.in

Abstract

Cosmetics attract most age groups from teenagers to old age. Cosmetic goods now contain a variety of nanoparticle and nanomaterial types. Nano cosmeceuticals have changed the era of cosmetics as they have advanced delivery mechanisms with task specifications. They are used in nail, hair, lip, and skin care products by cosmetic giants including Estee Lauder, L'oreal, Nivea, Zelens, and Derma Swiss, etc., and have patented the use of dozens of "nanosome particles." The global market for cosmetics using nanotechnology is worth millions of dollars and increasing at 7.14% annually. Liposomes, niosomes, nanostructured lipid carriers, solid lipid nanoparticles, gold nanoparticles, nanoemulsions, and nanosomes are novel nanocarriers that are now used in a variety of cosmeceuticals for drug delivery to achieve site specification, improved stability, biocompatibility, extended action, and increased drug-loading capacity. In this chapter use of various nanocarriers in cosmetic applications with their safety concerns will be discussed.

Keyword

Nanomaterial, Nanotechnology, Nano Cosmeceuticals, Cosmetics

Contents

1.   Introduction

Nanotechnology seek people's attention all over the world in recent years. It is a wide range of technology used to design, develop or manipulate materials at the nanoscale (size range of 1 – 100nm). At the nanoscale, there is a change in the characteristics of materials, which results in the surface-to-volume ratio rising, the number of particles present in unit weight, and the quantum confinement effect [1]. These exclusive properties of nanoparticles make them fit for various industries including household appliances, agriculture, pollutant removal, foods, communication, medical, transportation, military, cosmetics, and cosmeceuticals [2]. The substantial increase in the customer's interest and need for better

Applications of Emerging Nanomaterials and Nanotechnology   Materials Research Forum LLC
Materials Research Foundations 148 (2023) 127-169   https://doi.org/10.21741/9781644902554-5

performance, alluring appearance, retention, and safety has resulted in the expansion of formulations in cosmetics [3].

In 1961, Raymond Reed, a founding member of the US Society of Cosmetic Chemists, invented the term "cosmetics." Cosmetics are the substances or mixtures which come in direct interaction with the human body's external components. They are mainly used for changing the look, eliminating body odors, longer persistence of color, etc. [4]. Civilization has been using cosmetics for centuries. Western women began using cosmetics in secret using domestic goods in the late 19th century, and by the early 20th century, there was no longer any need for concealment. By the twenty-first century, cosmetics were widely used, and as technology advanced, creative cosmetic compositions were created by using the newest techniques [5]. Cosmetics with physiologically active ingredients that have therapeutic effects on the surface applied are known as cosmeceuticals. These are used in cosmetics because they purport to improve appearance. Between medications and personal care items, there is a gulf called cosmetics. Cosmeceutical products are used to treat a variety of disorders, such as hair loss, wrinkles, photoaging, skin dryness, dark spots, uneven skin tone, hyperpigmentation, and others. They, therefore, have an observable therapeutic effect on the skin [6].

The cosmetics business started utilizing nanotechnology to innovate and improve the quality of active components. There are a large number of factors including drug composition, polymers, additives, component interactions (physical or chemical), and manufacturing process, which regulate the drug release from carriers. Nano cosmeceuticals allow the regulated release of active compounds. They prolong perfumes, make them last longer, and boost the effectiveness of sunscreens by enhancing UV protection. Their use in hair care preparations includes hair fall treatment and avoiding hair from going grey. The surface area is boosted by having very small particle sizes, which enables the bitter of these active substances into the skin [7].

Nevertheless, some of these nanomaterials have the potential to harm the environment or human tissues and organs. To counter the adverse effect, natural or biodegradable substances were used for the framing of nanomaterials without compromising their performance. In this chapter advantages and disadvantages of nanoparticles in cosmetics and their utilization and patent by cosmetic giants will be discussed. The chapter gives a brief outline of the use of Nano cosmeceuticals in the formulation of cosmetics and their possible benefits followed by the mode of action of different carriers in Nano cosmetic systems. Later the chapter highlights the green nanomaterials in cosmetics and the mechanism of action of Nano cosmeceuticals. It also covers the various facets of how nanoparticles affect cosmeceuticals and human health.

2.   Nanotechnology in cosmetic industry

It would be an uphill battle to set a timeframe for the origin and use of cosmetics because they wish to stay and appear decent as such has been as old as man itself and it is the key driving force for cosmetic use. Egyptians were credited with cosmetic use around 4000

BC. Later Americans, Greeks, Chinese, Romans, and Japanese were also reported to use cosmetics. Close to the turn of the 20th-century household items were used as cosmetics. By the 20th century, cosmetics were used without seclusion. In the 21st century technological integration has developed inventive cosmetic formulations that are used immensely nowadays [8]. When these cosmetic products are integrated with biologically vital therapeutic additives to escalate appearance are called Cosmeceuticals [9]. They are used in the treatment of hyperpigmentation, wrinkles, uneven complexion, photoaging, skin dryness, dark spots, hair damage, etc. Several surveys have exhibited that nanosized materials are now employed by all leading cosmetic fabricators in their various products. Many streams of chemical sciences have been changed to the pharmaceuticals as well as get linked with nomenclature, Nano cosmetics. A large number of metals (AgNPs, AuNPs), metal oxide nanoparticles (TiO_2NPs, ZnO NPs, Fe_2O_3NPs), and carbon-based NPs are used in the formulation of cosmetics [10–12]. All the cosmetic giants including Estee Lauder, L'oreal, Nivea, Zelens, Derma Swiss, etc. are using nanoparticles in the formulation of cosmetics (Table 1-5). The industry for nanotechnology-based cosmetics is worth millions of dollars worldwide. The biggest cosmetics corporation in the world, L'Oreal, is investing a significant portion of its earnings in nano patents and has already earned patents for numerous "nanosome particles" [13]. TiO_2 has properties of UV filtration (UVA and UVB filter) and is thus mainly used in the fabrication of products like sunscreens and moisturizers. It is also utilized in the production of lip balm, foundation, and daycare cream since it shields the skin from the damaging effects of UV radiation. Proctor and Gamble, Dermatone, Colore science, and Boots are some giants using TiO_2 and ZnO as nanoparticles. Nanoparticles made of silver and gold have anti-bacterial and anti-fungal characteristics [14].These nanoparticles are used in manufacturing deodorants, sanitizers, and anti-aging creams. Lancôme is one of the leading cosmetic manufacturers using nanocapsules of vitamin E in moisturizer with the name Hydra Flash Bronzer Daily Face moisturizer and it declares for natural, healthy glowing skin. Similarly, one of the hand sanitizer manufacturing company, Evolut claim high effectiveness for skin protection and disinfectant as they use silver nanoparticles for antibacterial properties. Nail polishes containing nanomaterials have been patented as they are not damaged easily, are long-lasting, and are resistant to scratches [15]. One of the nail polish manufacturers Nano lab corp has used lacquer containing nanoparticles which make its application easy and protect it from shock, scratch, or crack [16]. Similar to this, a variety of products with distinct qualities of nanoparticles (nanogold and nanosilver) integrated into nail cleaning items are accessible in the market [17].

The fabrication of beauty treatments using nanotechnology has greatly increased in recent years as it results in the long-lasting fragrance of perfumes, sunscreens with UV protection, antiaging creams with prolonged dermal hydration, and more persistence of hair colors, etc. [6].

3. Classification of nanocosmeceuticals

The personal care industry's fastest-expanding sector is thought to be cosmetics. Products for caring for the skin, hair, lips, and nails comprise nano cosmeceuticals. Some of the classes in nano cosmeceuticals are shown in figure 1.

3.1 Skin care

Skincare products helps skin function and texture by promoting collagen formation and fending off the damaging effects of free radicals. By preserving the keratin's healthy structural integrity, they get better well-being of the skin. ZnO and TiO_2 nanoparticles are the most effective minerals for skin protection and are used in sunscreen lotions. They allow the product to become less greasy, odorous, and transparent by penetrating the deeper layers of the skin [18]. SLNs, Nanoemulsions, Liposomes, and Nanosomes are widely used in the creation of moisturisers due to their capacity to build thin films of humectants and hold onto moisture for protracted periods of time. Products marketed as anti-aging Nanocosmeceuticals that incorporate nanocarriers show properties of collagen regrowth, skin rejuvenation, and forming and preserving the skin [19].

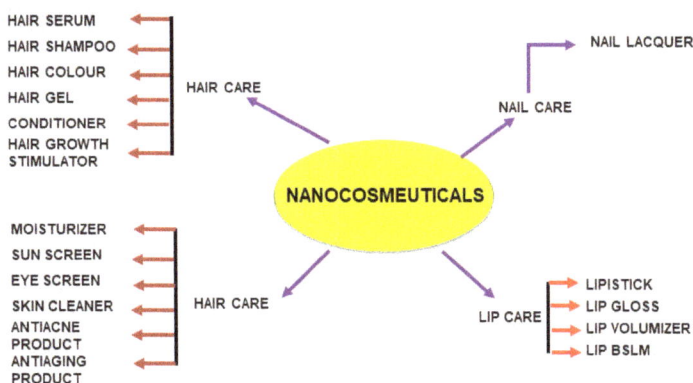

Figure 1. Nanocosmeceuticals Classes

3.2 Hair care

Shampoos, conditioners, hair growth accelerators, coloring agents, and styling treatments all fall under the category of hair Nano cosmeceuticals. The intrinsic qualities and distinct size of nanoparticles enable them to target the hair follicle, and shaft, and boost the amount of active substance.

The inclusion of nanoparticles in shampoos optimizes residing contact time with the scalp and hair follicles by generating a protective layer, which locks moisture within the cuticles [20]. Nano cosmeceutical conditioners are designed to add softness, sheen, silkiness, and gloss while improving hair detangling. Novel carriers such as liposomes, nanospheres, microemulsions, and niosomes have the primary purpose of texture and gloss restoration, healing damaged cuticles and making hair less brittle, glossy, and oily [21].

3.3   Lip care

The lip care line includes the use of lipstick, lip balm, lip gloss, and lip volumizer. Lip gloss and lipstick can contain a variety of nanoparticles combined to minimize trans epidermal water loss, soften lips, stop lip pigmentation, and keep lips colored for longer. Liposomes present in lip volumizer fill in lip wrinkles, moisturizes and define the lips, and improve lip volume [22]

3.4   Nail care

Products available for caring for nails that use nano cosmeceuticals are far superior to those that do not. The advantages of nail paints based on nanotechnology include greater toughness, quick-drying, more durability, crack resistance, and ease of application because of their elastic properties [23]. Amalgamating silver and metal oxide nanoparticles are used for the treatment of toenails as they have antifungal characteristics [24].

4.   Carriers in nano cosmetic systems

Carrier technology, which offers an acute strategy for the distribution of active substances, is used for the delivery of nano cosmeceuticals. Different cutting-edge nanocarriers for the distribution of cosmeceutical products are seen in Figure 2.

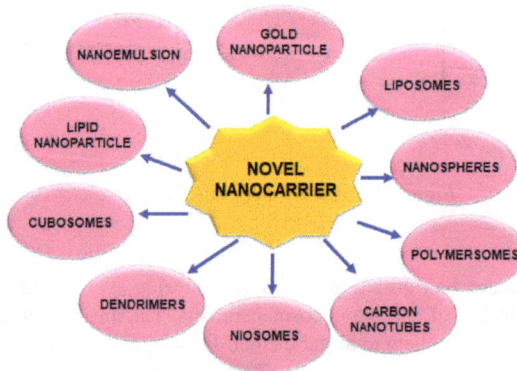

Figure 2. Nano cosmeceutical carriers

4.1   Liposomes

The most common use for liposomes is in cosmeceutical products. They are membranous structures that are surrounded by a hydrophobic lipid bilayer and possess an aqueous core [25]. Since phospholipids make up the majority of the lipid bilayer in liposomes and are typically considered to be nontoxic substances, the likelihood of adverse consequences is decreased [26]. Liposomes encapsulate drugs and release their active elements gradually to prevent metabolic breakdown [27].

The transport of both hydrophilic and hydrophobic substances is possible using liposomes. They can be either multilamellar or unilamellar in structure, and they range in size from 20 nanometers to several micrometers [28]. Due to their distinct benefits, liposomes, which are versatile functional carriers, have found use in the cosmetics business, including an increase in cosmetic chemicals solubility, compatibility with human skin, decreased toxicity, and improved drug accumulation at the targeted site. The fact is that they frequently display regulated release kinetics while also guarding the medication against external degradation [29,30]. The cosmetics sector can use liposome-based Nano formulations to create antiperspirants, lotions, lipsticks, deodorants, moisturizers, and other beauty goods. Capture, an anti-aging cream introduced by Dior in 1986, was the first beauty product made with liposomes. Additionally, liposomes can be employed to transport active biomolecules like vitamins A, E, and K, as well as antioxidants like Coenzyme Q10, lycopene, and carotenoids. Liposomes aid in the skin's moisture as well.

Lipid molecules like cholesterol and ceramides can be easily integrated into liposomes, which also aid in rebuilding the skin's epidermal layers [31,32]. Phosphatidylcholine, the primary component of liposomes, has been used in a variety of skin care formulations, including moisturizing creams and other products, as well as hair care formulations, including shampoo and conditioner, due to its softening and conditioning properties. Due to their compatibility with living tissues, biodegradability, and nontoxic properties, liposomes are utilized in numerous cosmeceuticals as they contain active components [33]. Vegetable phospholipids are utilized frequently for transdermal treatment and cosmetics as they contain large amounts of essential fatty acid esters. These phospholipids facilitate the entry of linoleic acid into the epidermis. It results in the improvement of skin within a short period as it improves the skin's barrier function and reduces water loss [34,35]. The distribution of flavors, phytonutrients, and vitamins from dehydrated compositions like antiperspirants, body sprays, deodorants, and lipsticks is now being worked on using liposomes. They are also utilized in sunscreen, anti-aging, deep moisturizing, cosmetics, and hair loss treatments [36]. Figure 3 [37] discusses several liposome advantages and disadvantages. The different commercial formulations are listed in Table 1 [31,38–40].

Applications of Emerging Nanomaterials and Nanotechnology   Materials Research Forum LLC
Materials Research Foundations 148 (2023) 127-169   https://doi.org/10.21741/9781644902554-5

Table 1. Formulations of liposomes

S.No.	Brand Name	Promoted by	Uses
1.	Rehydrating Liposome Day Crème	Kerstin Florian	Moisturizing
2.	Liposome Face & Neck Lotion	Clinicians Complex	Skin is nourished, and photoaging is avoided.
3.	Capture Totale	Dior	when paired with sunscreen, eliminates wrinkles, and dark spots, and provides a radiant look.
4.	Dermosome	Microfluidics	Demulcent
5.	Liposome face cream	Decorte	Demulcent
6.	Liposome eye cream	Decorte	moisturizes while firming and brightening the sensitive skin around the eyes.
7.	Natural progesterone liposomal skin cream	NOW Solution	keeping a healthy female balance
8	Advanced night repair protective recovery complex	Estee Lauder	Skin regrowth
9.	Fillderma lips lip volumizer	Sesderma	Given more volume, wrinkles are filled in, the skin is moisturized, and the lips are outlined.
10.	Luminescence eye cream	Aubrey Organics	Solidification and anti-aging
11.	Russell organics liposome concentrate	Russell organics	Skin becomes more hydrated and rejuvenated, firmer, softer, and smoother.

POSITIVE ASPECTS

- Increased Stability and ease of penetration in derma layer
- Biocompatible and biodegradable
- Increased efficacy and reduced toxicity

NEGATIVE ASPECTS

- High production cost and low solubility
- Leakage of drug and inadequate stability
- occasionaly oxidation and hydrolysis reaction and osmotically sensitive

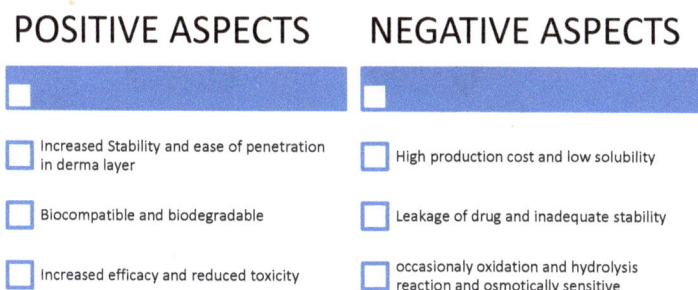

Figure 3. Positive and Negative aspects of liposomes [37]

4.2   Niosomes

When combined with or without cholesterol or similar lipids, hydrated nonionic surfactants, it results in the spontaneous formation of niosomes [41]. Niosomes are membrane-enclosed vesicles that are multilamellar or unilamellar, bilayer-organized surfactant macromolecules that can surround a mixture of solute and lipophilic substances in water [42]. Small unilamellar vesicles, multilamellar vesicles, and large unilamellar vesicles have relative sizes of 0.025–0.05 m, =>0.05 m, and 0.10 m [43]. The manufacture of Niosomes uses nonionic surfactants, polyoxymethylene alkyl ether, surfactants linked with steroids, as well as cholesterol as major constituents [44]. Niosomes are effective for delivering both hydrophilic and hydrophobic substances. Niosomes can be employed as a unique medication delivery mechanism for pharmaceuticals that are not readily absorbed [45]. It gives the drug encapsulation, which prolongs its time in systemic circulation and increases its penetration of the target tissue. Niosomes are an improvement over liposomes, which have drawbacks such as instability issues, high cost, and oxidation susceptibility [46]. Niosomes are utilized in cosmetic and skin care products because they can reversibly lower the horny layer's barrier resistance, which allows chemicals to penetrate the skin more quickly and reach live tissues. The nature of the encapsulated drug, the content of the membrane, and the hydration temperature are only a few of the many parameters that affect the creation of niosomes [47]. L'Oreal first created niosomes in 1970 through the study and creation of artificial liposomes.

Niosomes were created under the trade name Lancome and were given a patent by L'Oreal in 1987. There are many niosome cosmeceuticals products in the market, including anti-aging creams, creams for whitening and hydrating the skin, and shampoos and conditioners for restoring damaged hair [48]. Figure 4 [49–50] discusses several niosome benefits and drawbacks. Table 2 [10,51–52] discusses the usage of various commercially available goods.

Advantages

1. Physically unstable
2. Aggregation and expensive
3. Insufficient drug loading capacity
4. Specialized equipments required for maintenance
5. Leakage of entrapped drug
6. time consuming techniques required for formulation

Disadvantages

1. controlled and targeted drug delivery and osmotically active
2. Increased dermal penetration and bioavailability
3. no special conditions required for handling and storage of surfactants.
4. Improved therapeutic performance of drug
5. Used for parental and oral as well as topical routes

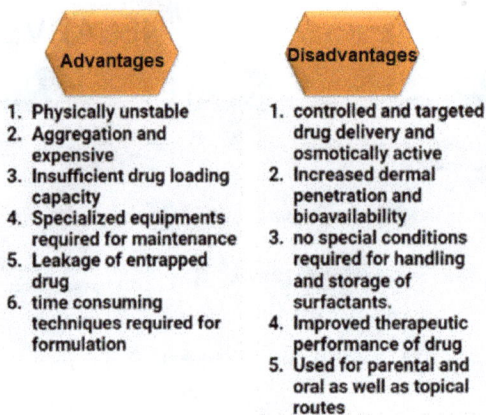

Figure 4 Advantages and Disadvantages of Niosomes [49-50]

Table 2. Marketed formulation of Niosome

S.No.	Product Name	Marketed By	Uses
1.	Niosome + Perfect age treatment	Lancome	eliminates crease
2.	Niosome +	Lancome	wrinkle-reducing foundation cream, clean skin tone
3.	Anti-Age Response cream	simply man match	Crease reducing
4.	Identik Shampooing Floral repair	Identik	Shampoo for hair patch
5.	Identik Masque Floral repair	Identik	Masque for hair patch
6.	Eusu Niosome Maka Pom Whitening Facial Cream	Eusu	whitened skin
7.	Mayu Niosome Base cream	Laon cosmetics	Hydrating and lightening

There are many noisome-based cosmetic formulations available for use in hair and skin care. In comparison to traditional liposomes and niosomes, Novasomes are a new development in liposomal technology. Novasomes have a central core surrounded by many bi-lipid layers (2-7) with a very high drug-loading capability. Drugs that are both

Applications of Emerging Nanomaterials and Nanotechnology   Materials Research Forum LLC
Materials Research Foundations 148 (2023) 127-169   https://doi.org/10.21741/9781644902554-5

hydrophobic and hydrophilic, as well as those that can interact with others, can all be delivered at the same time in various layers of novasomes [53]. They are made from a combination of monoesters of polyoxyethylene fatty acids, free fatty acids, and cholesterol (non-phospholipid surfactant). IGI laboratories, NOVAVAX, created the nova some technology, which includes this ground-breaking encapsulation method [54]. In comparison to liposomes and niosomes, novasomes have a substantially cheaper production cost and a high drug entrapment efficiency. These nanosystems are also more stable, remaining stable in a pH range of 2 to 13 and a temperature range of 0 to 100^0 C. Novasome offer tremendous potential in a variety of sectors, including the pharmaceutical, cosmetic, and agricultural industries. There are many nova some-based products in the merchandise, one of them is AcneWorx by Dermworx, which claim to treat acne with novasomes having salicylic acid [55]. Terconazone (TCZ)-containing novasomes was recently developed for the cure of newborn napkin candida albicans, and their curative efficacy was compared to that of traditional TCZ solutions. The nova some-based formulation was discovered to be more therapeutically efficacious than the suspension and to have a somewhat higher volume of skin deposition than the noisome-based formulation [54].

4.3 Lipid nanoparticles

4.3.1 Solid lipid nanoparticles (SLNs)

The first-generation lipoidal carriers, known as Solid Lipid Nanoparticles (SLNs), were created in the early 1990s. It consists of a solid lipid core in the centre that is disseminated in a water-based media with the help of surfactants. The medication and cosmetic ingredient can be loaded in the lipid matrix if it is hydrophilic, lipophilic, or poorly soluble in water. Due to the use of lipids that are both biocompatible and physiologically stable, SLNs avoid toxicity issues. Precipitation and high-pressure homogenization are the two main methods used to synthesize SLNs. SLNs carrying drugs within the shell result in blast release whereas SLNs carrying drugs within the core show sustained discharge [29,63]. SLNs are useful in cosmeceuticals and medicines. Because of their diminutive size and direct contact with the stratum corneum, active substances can penetrate the skin more easily [64].

Combining SLNs with sunscreen can improve photoprotection while minimising negative effects because they have UV-resistant qualities and function as physical sunscreens on their own [26]. SLNs have occlusive qualities that can be employed to raise the skin's water content and thus its hydration [65]. Since they spread out the release of scent over a longer length of time, SLNs are also used in perfume compositions and work best in day creams [66–67]. Because they are solid in nature and the mobility of the active molecules is limited, they have greater stability coalescence when compared to liposomes, which prevents leaking from the carrier [68–69]. Figure 5 [70-72] illustrates the benefits and downsides of SLNs. The usage of various commercially available products is listed in Table 3.

Table:3 Marketed formulations of Solid Lipid Nanoparticle

S.No.	Product Name	Uses	Promoted By
1.	Phyto NLC Active Cell Repair	Skin firming, hydration, and reduction of tan	Sireh Emas
2.	Allure Eau Parfum Spray	Scent	Chanel
3.	Allure parfum bottle	Scent	Chanel
4.	Allure Body Cream	Lotion for Body	Chanel
5.	Sossion Facial Lifting Cream SLN Technology	Antiaging Cream	Soosion

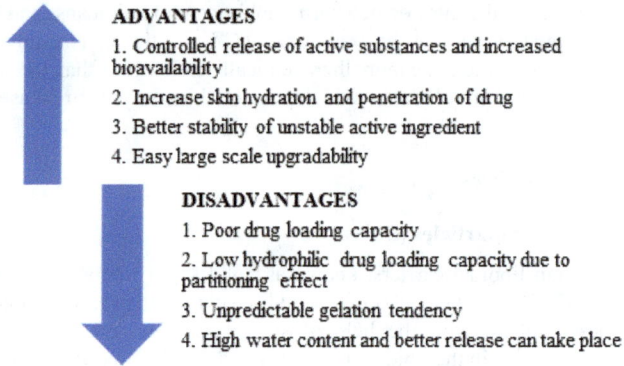

ADVANTAGES

1. Controlled release of active substances and increased bioavailability
2. Increase skin hydration and penetration of drug
3. Better stability of unstable active ingredient
4. Easy large scale upgradability

DISADVANTAGES

1. Poor drug loading capacity
2. Low hydrophilic drug loading capacity due to partitioning effect
3. Unpredictable gelation tendency
4. High water content and better release can take place

Figure 5 Advantages and Disadvantages of SLNs [70-72]

4.3.2 Nanostructured lipid carriers (NLCs)

Nanostructured lipid carriers are the second generation of lipid nanoparticles. To overcome the problems with SLNs, NLCs have been developed. The blend of liquid and solid lipids that make up NLCs gives them a less organized structure that allows them to hold more active ingredients in their pockets. NLCs are divided into amorphous, numerous, and imperfect subtypes based on structural variations, formulation techniques, and ingredients employed [73]. There is more scholarly and business interest now for NLCs during the last few years due to the reduced systemic side effects risk. NLCs in contrast to SLNs demonstrate increased medication loading for entrapped bioactive due to the compound's twisted structure, which aids in expanding the area. Additional drawbacks of SLNs include lowering drug expulsion and particle concentration during storage. The creation of NLCs have solved these problems. They are created by physiological lipids that are biodegradable and exhibit very low toxicity [74]. They have many beneficial qualities, including enhanced skin moisture brought on by their occlusive abilities and their small size, which guarantees close contact with the stratum corneum. There are improved UV protection systems with fewer adverse effects and steady drug integration during storage [75].

The first lipid nanoparticle-containing cosmetic product, Dr. Rimpler GmbH's NanoRepair Q10 cream, and serum, which offer greater skin penetration, was released onto the market in October 2005. More than 30 cosmetic items are on the market right now that contain NLCs [68,76]. Figure 6 [77-78] illustrates a few of NLC's advantages. Table 4 [79-80] contains a list of manufactured goods, their manufacturers, and their applications.

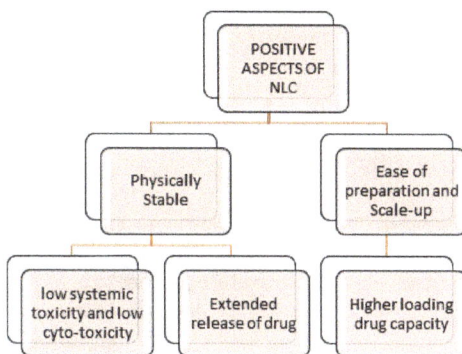

Figure 6 Advantages of NLC's [77-78]

Table 4 List of marketed products, manufacturers, and uses of NLCs

S.No.	Product Name	Uses	Promoted By
1.	Olivenol Augenpfegebalsam	eliminates wrinkles, eye puffiness, and eye rings	Dr. Teiss/Medipharma cosmetics
2.	Olivenol Anti Falten Pfegekontrat	skin tightening and wrinkle prevention	Dr. Teiss/Medipharma cosmetics
3.	Regenerations Cream Intensive Ampoules	smoothest wrinkles and encourages cell regeneration	Scholl
4.	Swiss Cellular White Illuminating Eye Essence	eliminates discoloration and darkness under the eyes	La Prairie

5.	Surmer Creme Leg ` ere Nano-Protection	assists with hydrating	Isabelle Lancray
6.	Cutanova-Cream Nanorepair Q1	Getting rid of small wrinkles and aiding in restructuring	Dr. Rimpler
7.	Intensive Serum Nanorepair Q10	Helps in fighting the sign of aging and anti-wrinkle serum	Dr. Rimpler
8.	Cutanova Cream Nanovital Q10	UV protection for anti-aging treatment	Dr. Rimpler
9.	Iope Supervital Extra Moist Sofner	helps to moisturize dry, rough skin	Amore Pacifc
10.	Iope Supervital Extra Moist Eye Cream	lack of suppleness around the eyes.	Amore Pacifc
11.	Surmer Masque Creme Nano-Hydratant	Reduction of crease, and helps in restricting dry skin	Isabelle Lancray

4.4 Nanosphere

Nanospheres are spherical, core-and-shell-arranged particles. The medicine is entrapped, solution connected or encapsulated in nanospheres, where it is shielded from enzymatic and chemical deterioration. The medication is physically and evenly disseminated throughout the polymer matrix structure. The nanospheres makeup might be either crystalline or amorphous [81]. It holds great potential that this technology can transform chemicals with poor absorptivity, solubility, and physiological activity into the desired delivered medication. Different enzymes, DNA, and medications can be encapsulated inside nanospheres' centers [82].

The cosmetics industry uses nanospheres in skin protection yields to dispatch active chemicals to the skin's deep layers and dispatch their beneficial benefits to the skin's afflicted area more precisely and successfully. These tiny components aid in the defense against actinic aging. Nanospheres are increasingly being used in the cosmetics sector, particularly in skin care items including anti-aging, hydrating, and anti-acne creams [83]. Biodegradable nanospheres and nonbiodegradable nanospheres are the two categories into which nanospheres can be separated. Examples of biodegradable nanospheres include gelatin, starch, and albumin nanospheres, while polylactic acid is an example of a nonbiodegradable nanosphere. Figure 7 illustrates a pictorial illustration of nanospheres' positive attributes and Table 5 lists the names of marketed product makers and their intended usage.

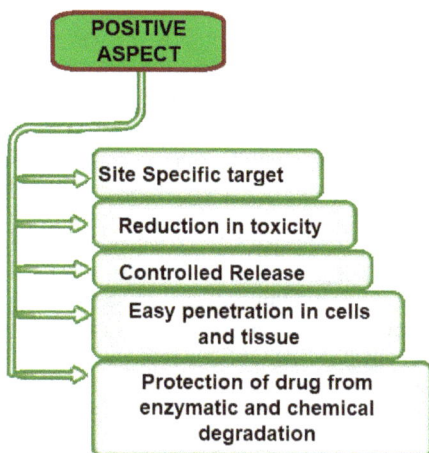

Figure 7 Positive Aspects of Nanosphere [83]

Table 5 Formulation of Nanosphere

S.No.	Product Name	Uses	Marketed By
1.	Cell act DNA Filler Intense Cream	It helps into Reduced firms Skin and wrinkles	Cell Act Switzerland
2.	Clear It! Complex Mist	Helps in Antiacne	Kara Vita
3.	Hydralane Ultra Moisturizing Day Cream	Helps in hydration and deep moisturizing	Hydralane Paris
4.	Fresh as a Daisy Body Lotion	Lotion for the Body	Kara vita
5.	Lip Tender	Moisturizing lips	Kara vita
6.	Nanosphere Plus	Antiwrinkle, antiaging	Dermaswiss
7.	Coryse Salome competence hydration ultra-Moisturizing Cream	hydrating cream	Coryse Salome Paris
8.	Eye tender	Antiwrinkle	Kara vita
9.	Nano salt moisture key	hydrating cream	Salvon

4.5   Nanoemulsions

Nanoemulsions are transparent or translucent liquids that have exceptional kinetic and thermodynamic stability. There are several different forms of nanoemulsion systems, including the oil in water (O/W), water in oil (W/O), and systems with many repeated layers i.e. oil/water/oil or water/oil/water. They differ in preparation technique used, which results in formulations with various consistencies, depending on the need, such as watery, creamy, or gel-like [83–88].

They are typically made by combining phase inversion, high-pressure homogenization, sonication, microfluidization, and other methods with co-surfactants and surfactants to give the formulation stability. Consequently, nanoemulsions have been widely used in the pharmaceutical business to create creams, lotions, and other cosmetic products. shampoos, gel-based products, sunscreens, sprays, lotions, deodorants, and lotions [89]. They are ideal cosmetic vehicles because of their attractive appearance and rich blending texture [90]. Figure 8 illustrates the merits of nanoemulsions. The nano gel Kemira is one of many systems based on nanoemulsions that have received patent protection. Similarly, to this, L'Oreal has obtained patents for nanoemulsions based on fatty acid esters and phosphoric acid [91-92].

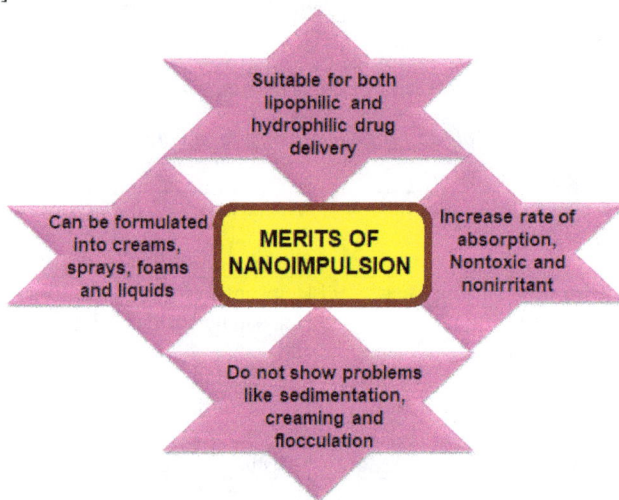

Figure 8 Merits of Nanoemulsions [89]

4.6   Dendrimers

A category of macromolecular organic chemical nanocarriers known as dendrimers comprises a central core and a web of long, symmetric branches that are joined to functional groups at their ends [93]. They are very small particles (2-20 nm), with a high permeability rate, and can easily distribute the chemical present in the core. are employed in the formulation of nail polishes, hair care products, and cosmetics for the skin and hair that call for extremely thin films. The usage of dendrimers as vitamin conjugates has also been suggested in several kinds of research.

Additionally, they are employed as a carrier to improve the penetration of vitamins into the deep layers of skin. A large number of products based on dendrimers were patented by cosmetic corporations, like L'Oreal, Unilever, and Wella. Carbo siloxanexane dendrimer is patented which provides the skin and hair a glossy appearance and can withstand both water and oil [94]. Resveratrol-containing dendrimer formulations may also improve the substance's solubility, stability, and transdermal permeability; as a result, they may be used in anti-aging goose. The advantages of dendrimers are illustrated in Figure 9. In addition, they are allegedly utilized in the creation of shampoos, hair gels, lotions, spray gels, sunscreens, and anti-acne medications.

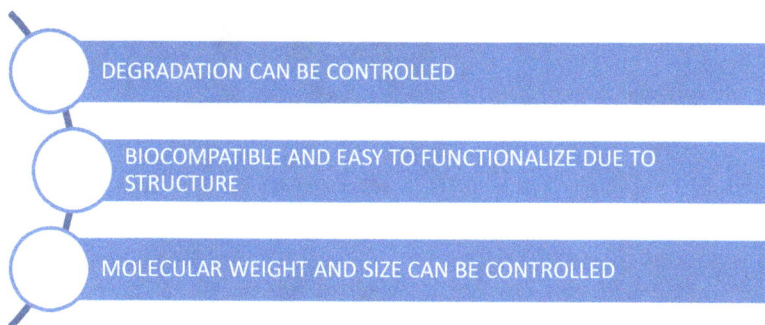

Figure 9 Advantages of Dendrimers [95]

5. Green nanomaterials for cosmetics

Green synthesized nanomaterials (NMs) have a lot of applications in various fields as they have amazing physicochemical properties. Moreover, green-synthesized NMs have outstanding advantages over chemically and physically synthesized NMs [96-97]. Green synthesized NMs are extensively used for antioxidant, medical, and clinical purposes like anticancer, derma-pharmaceutical, wound healing, anti-inflammatory, antiaging antimicrobial, and so forth [98-100]. Likewise, green NMs have an interesting application in the cosmetics industry. Nowadays, all over the globe, the different green NMs have been widely used for the cosmetic, treatment of cancerous and human skin diseases (Figure 10) [101]. Basically, for the last thirty years, green NMs have been utilized in the cosmetics business [102–104]. The cosmetics industry is a godsend for the country as it contributes huge economy.

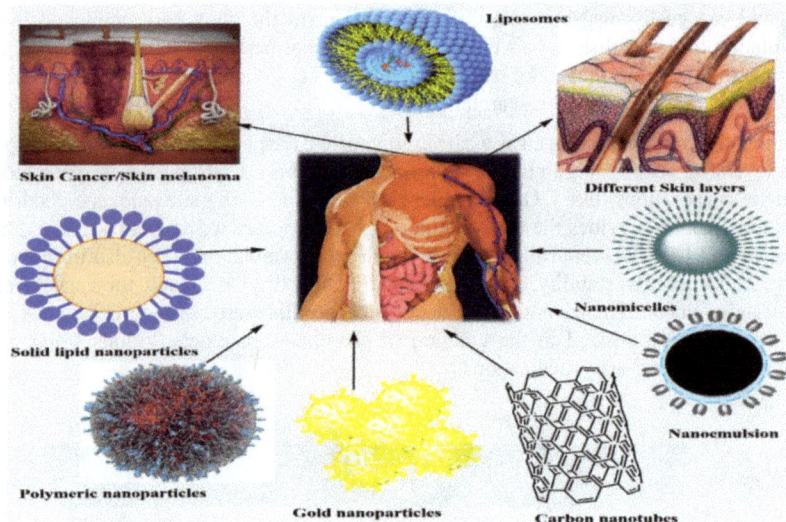

Figure 10 Applications of green NMs in the treatment of skin diseases and tumors [101]

The green NMs are always showing long-lasting effects, and increasing stability towards the uses of cosmetics. It is because of its high surface area and microporosity behaviors [105]. For instance, green NMs are currently used in eye-brow, anti-acne, lipsticks, sunscreens, nail polishes, makeup, hair protection, face-mask, anti-aging cream, etc (Figure 11). Moreover, it can use for body care, sun care, oral care, etc. The widespread uses of nanoparticles (NPs) in cosmetics products especially in skin care and sunscreens. L'Oréal et al. obtained nanotechnology-related patents which were based on green NMs for cosmetic applications. The most useful nano-ingredients like titanium oxide, zinc oxide, silica, and carbon black have been used for cosmetic preparation. Moreover, it is popularly known that the nanostructured TiO_2 NPs and ZnO NPs have great advantages over many product formations [107]. Even, microscale sizes of TiO and ZnO NPs are used as ingredients in various sunscreens ointment due to their wonderful absorption potential. Similarly, lipid nanoparticles are also one of the best remedies for skin treatment.

Other microscale lipid NPs can be employed for transdermal delivery. These types of drug delivery are used due to enhanced skin perforation with lower side effects [108]. Likewise, chitin-derived nanoparticles also have been used in the cosmetic and biomedical fields [109-110].

Cosmetics and nanotechnology

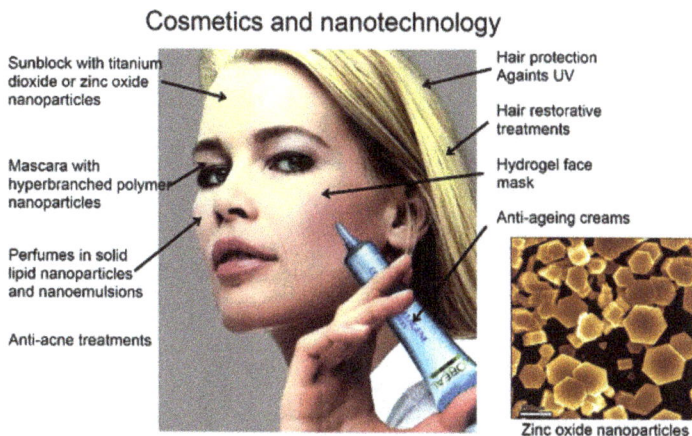

Sunblock with titanium dioxide or zinc oxide nanoparticles

Mascara with hyperbranched polymer nanoparticles

Perfumes in solid lipid nanoparticles and nanoemulsions

Anti-acne treatments

Hair protection Againts UV

Hair restorative treatments

Hydrogel face mask

Anti-ageing creams

Zinc oxide nanoparticles

Figure 11 The green NMs for beautification [106]

6.   Roles and mechanism of cosmetics

There are two types of nanomaterials used in cosmetics: soluble nanoparticles also called biodegradable nanoparticles and insoluble or non-biodegradable nanoparticles. They serve as nanocarriers and increase the stability of products and permeability of active ingredients. They reduce the significant toxicity of the product and rashes on the skin. Applications in cosmetics and cosmeceuticals must include the physicochemical properties of nanoparticles as well as the fundamental interactions between human tissues (skin conditions). Passive diffusion is a prominent transport method for delivering active compounds via the skin. The only areas where nanoparticles can enter the body normally are through the skin pores and hair follicle openings [112].

Fig. 12 shows a schematic representation of the structure of the skin, which is a multilayered organ made up of the epidermis, dermis, and hypodermis, and table 6 represents the mechanism study of nanomaterials in cosmetics. The principal physical-chemical barrier to the penetration of bioactive chemicals is represented by the epidermis, which is the most superficial layer and is made up of a multilayered, tight epithelium. Keratinocytes make up the epidermis and are constantly moving from deeper to shallower layers. Desmosomes, adherent junctions, and tight junctions connect them and prevent chemicals from diffusing into the dermis.

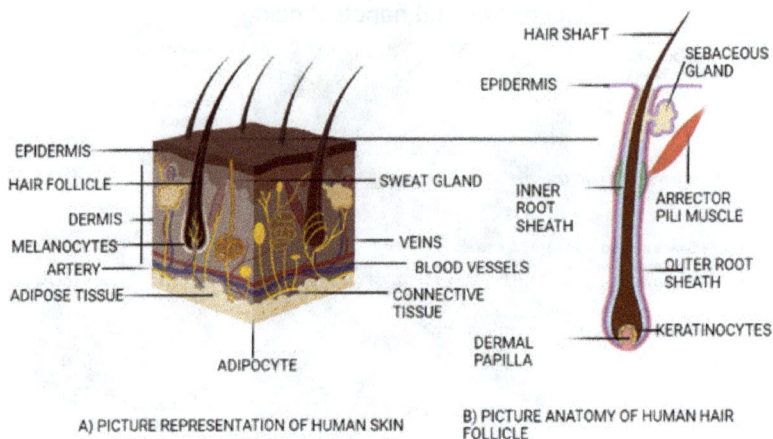

Figure 12 Picture representation of the human skin and human hair follicle

These deeper layers contain blood vessels and innervations. According to their physical-chemical characteristics, molecules can pass through the epidermal layer in different ways. The molecular weight (MW) limit for molecules that can pass through the skin has been proposed as 500 Da [64]. Skin metabolism, its location, and tissue condition at the application site are additional factors that affect skin penetration. Other factors that affect skin penetration are how substances bind with the tissue structure and the degree of vehicle incorporation, and how substances interact with the tissue [113]. The latter is especially significant when used in cosmetics; in fact, the formulation can function via altering the polarity and lipophilic/hydrophilic ratio of substances as well as through mechanisms such as hydration and/or epidermal barrier adjustment. Phase behavior and viscosity also appear to have a significant impact on the partitioning of bioactive molecules [114-115].

Finally, it is important to note that the ultimate target determines whether a bioactive substance needs to partially pass the epidermis, penetrate the dermis, or enter the systemic circulation. Since the action is frequently limited to the top layers of the skin when used for cosmetic purposes.

Applications of Emerging Nanomaterials and Nanotechnology
Materials Research Foundations 148 (2023) 127-169

Materials Research Forum LLC
https://doi.org/10.21741/9781644902554-5

Table 6 Roles and mechanism study of nanomaterials in cosmetics

S.No.	ROLES IN COSMETICS	TYPES OF NMs	MECHANISMS	REFERENCE
1.	Antiaging	Fullerenes	Squalane's fullerene contains anti-aging properties. Extraordinary antioxidant activities of 2,4-nonadienal	[116-117]
		Retinol	Amplify the manufacture of collagen, cell renewal, epidermal hyperplasia, and water content in the skin.	[118]
		Gold	The rigidity of skin is enhanced and rejuvenated by protein nanofiber gold.	[119-120]
2.	Whitening agent	Arbutin	A naturally occurring hydroquinone derivative that reduces age spots, dullness, and undesirable pigmentation by interfering with tyrosinase activity.	[121]
3.	Moisturizers	Hyaluronic acid	To prevent dehydration, a significant amount of highly viscous water is covered by hyaluronic acid.	[122]
4.	Concealer	Iron oxide	To conceal wrinkles and fine lines, it maintains transparency while diffusing light.	[123]
5.	Lip care	Gold-silica	The lips' thin line is kept from migrating or bleeding, and greasiness brought on by sebum secretion is inhibited by the pigments' homogeneous distribution.	[119]
		Gold and Silver	To prevent pigment deterioration and maintain color for a longer amount of time	[119]
6.	Sunscreen	ZnO and TiO	Create a skin barrier that is less irritative to deflect UVA and UVB rays. Free radicals may develop as a result of photocatalytic activity.	[124-126]

7.	Delivery active ingredients	Liposomes	Able to interact with cell membrane bilayers to promote delivery and discharge. capable of distributing both hydrophobic and hydrophilic chemicals and protecting the encapsulated medications from the environment.	[29,78,127]
		Ceramides	The maintenance and construction of the skin's water permeability barrier Reduce aging signs like dryness, lines, and wrinkles.	[128]
8.	Antibacterial agent/ skin cleanser	Ag and metal oxides	By cleaning/infecting wounds, beauty soap can treat decubitus, gangrene pimples, and acne. shown bacterial activity against gram-positive and negative bacteria	[129]
9.	Dental care	Calcium fluoride	preventing the brain from receiving pain signals by long-lasting remineralization	[130]
10.	Hair care	Liposome	change the color of the hair follicle's melanin by using the bulge area Hair color with a long-lasting impact and little toxicity	[29,131]
		Dendrimers	Improved tactile sensitivity, glossiness, features of water resistance, sebum resistance, and hair adhesion	[132-133]
		Silicone	Capable of diffusing into hair fibers to increase lubrication, gloss, and moisture	[21,134]
		Nanoemulsions	Rejuvenate your dry hair to make it look shinier, softer, and less oily.	[121,135]
		Metal oxides	Control hair grease	[136-137]

| | | Fullerene | Stop the free radicals and oxidative stress that lead to the death of hair follicles to encourage the creation of new hair follicles within the dermis of human skin, and aging | [138-140] |
| | | Carbon nanotubes | Nanoparticles' high surface area to volume ratio increases their affinity for and contact with hair fibres, leading to a long-lasting effect. | [141-142] |

7.   Toxicity of nanoparticles

The theme of the toxicological effects of "cosmeceuticals" made with a nanoformulation on the skin barrier. This implies that each nanoformulation must be carefully considered. NPs' permeability through the live tissues, is known to have a substantial impact on dermatological safety. Insoluble nanoparticles are supposed to be less harmful than soluble nanomaterials since they have a lower diffusion coefficient, Because the latter are typically made up of substances that are widely used in cosmetics (such as lipids and surfactants), their use has always raised more questions. Nanoparticle toxicity is highly dependent on a range of parameters, including surface characteristics. Smaller the size of particle more will be the surface area and thus shows more toxicity [143]. Toxicity is also influenced by the chemical composition of nanoparticles that are absorbed through the skin [144]. The level of exposure and the pathway through which nanoparticles enter the body determine the health risk they pose to people. Ingesting, inhaling, and dermal these are some of the potential ways that humans could be exposed to nanoparticles (Figure 13) [145].

7.1   Ingestion

The purposeful or unintentional transfer of nanomaterials from the hands to the mouth can result in the ingestion of such materials. After ingestion, the majority of nanoparticles quickly leave the body; however, some of the research we looked at suggested that a tiny proportion of nanoparticles might be absorbed and then reach into organs. Nanoparticles may be present in lipsticks, lip balms, lip glosses, and other cosmeceuticals and enter the body when used on the lips or mouth [146]. Certain nanomaterials can permeate the skin's layers within 24 hours of exposure. Mice subjected to 20 nm and 120 nm zinc oxide nanoparticles at different doses all exhibited problems in their spleen, hearts, livers, bones, and pancreas [147]. A range of widely marketed cosmeceuticals also contain copper nanoparticles. Mice treated with copper nanoparticles showed toxicological effects and severe internal organ damage [114]. Silver nanoparticles are utilized in several cosmeceuticals, including soaps, face creams, and toothpaste, as well as in wound dressing and antibacterial compositions. Silver nanoparticles are used in cosmetic industry due to its antimicrobial effects. But the same silver concentration which is fatal for bacteria can also be harmful for keratinocytes and fibroblasts [148].

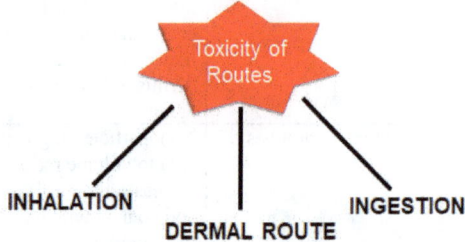

Figure 13 Picture representation of toxicity route [145]

7.2 Inhalation

The National Institute of Occupational Health and Safety claims that inhalation is the most frequent way by which people are exposed to airborne nanoparticles. Workers might breathe in nanoparticles, for instance. Customers who use products containing nanoparticles, such as spray versions of sunscreens containing nanoscale titanium dioxide, may inhale them if the proper safety measures are not implemented during utilization. According to the National Institutes of Health authorities, the majority of inhaled particles migrate to the lungs. However, research on laboratory animals refer that some of the nanoparticals may go to the brain through the nasal passage and can also enter the blood, neurological system, and other organs.

According to carbon nanotube tests, long-term exposure can cause interstitial inflammation and also results in epithelioid granulomatous lesions. Some carbon-based fullerenes may cause cell oxidation or pose a risk to health if inhaled [149]. The pulmonary administration of TiO_2 ultrafine particles caused higher lung damage than TiO_2 fine particles. Gold nanoparticles with sizes of 2, 40, and 100 nm were found in the liver and macrophages after intratracheal exposure. There is evidence that even low doses of exposure to TiO_2 can destroy DNA, but TiO_2 with a particle size of 500 nm only slightly damages DNA strands. [150]

7.3 Dermal route

Three different pathways, including intracellular, transcellular, and trans follicular, are used to penetrate cosmeceuticals in the skin. The consequences of cutaneous exposure to particles smaller than 10 nm can be devastating. For particles larger than 30 nm, there is a chance that skin barrier modifications including cuts, wounds, and dermatitis diseases will affect nanoparticle penetration [151]. Currently, fullerenes are utilized in cosmeceuticals like moisturizers and face creams, although it is unclear how harmful they are. According to a report fullerene-containing face creams have been shown to harm fish brain tissue and have harmful effects on human liver cells [152]. According to several research, peptides based on fullerene can penetrate into the skin and can easily cross the dermal layer when subjected to mechanical stress. Some studies have demonstrated that nanoparticles such as titania, quantum dots with surface coatings, and single- or multi-wall carbon nanotubes are capable of killing epidermal keratinocytes and fibroblasts by altering gene or protein expression [153]. Nanoparticles such as TiO_2 and ZnO present in sunscreens and their effects on human health safety, and the environment are currently not widely debated. Reactive oxygen species (ROS), including free radicals, are used more frequently due to their smaller size, higher surface area, and more reactive chemical makeup. The main mechanism for nanoparticle toxicity is the generation of free radicals and reactive oxygen species. TiO_2 and ZnO both produce reactive oxygen species as well as free radicals when exposed to ultraviolet (UV) light, which can cause oxidative stress and seriously disrupt cell membranes, proteins, RNA, and lipids [154]. A research revealed that when TiO_2 nanoparticle were delivered subcutaneously to pregnant mice, the nanoparticles were transferred to the offspring and caused lower sperm production in male offspring as well as brain damage. Cobalt-chromium nanoparticles can penetrate the skin's protective layer and harm human fibroblasts [155].

8. Safety assessment of nanomaterials in the cosmetic industry

Nanotechnology-based cosmetic firms face an uncertain future in terms of customer response and the regulatory landscape. Cosmetic nanomaterials may serve a variety of purposes (Nano-preservatives, UVA and UVB filters in sunscreen, etc.). Consumers may also be at risk from the distinctive properties of any specific nanomaterial that could give the cosmetic product the desired function or functionality. In light of this, a general assessment of the safety of all nanomaterials is required, including tests addressing the nano-characteristics. And Nanomaterial synthesis may result in exposure by eating, skin contact, and inhalation. The manufacturing of nanoparticles includes skin contact, which results in unintentional exposure. Inhalation exposure may be caused by the direct release of nanoscale airborne reactants or products at work. Nevertheless, airborne nanoparticles may inevitably be in all manufacturing lines as a result of product recovery, extra processing, and cleaning.

Intentional cutaneous exposure to nanoparticles can occur through skin contact with cosmetics and direct application. In the form of sunscreen, TiO_2 and ZnO are in touch with the skin. Dermal exposure very certainly leads to swallowing exposure due to hand-to-mouth contact, and exposure to ingesting brought on by hand-to-mouth contact [156]. Because these particles are tiny and have strong hypersensitivity, they could become potentially fatal elements in causing unfavorable cellular toxicity and damaging effects. Nanomaterials with known harmful effects can enter living organisms through the mouth or nose and then travel to various organs and tissues throughout the body [157]. Based on several types of nanoparticles used in cosmetics, table 7 illustrates the effects of nanomaterials on people and figure 14 represents the safety assessments in cosmetics.

Cosmetic ingredients

Physicochemical characterization consideration nano -related aspects

Is the ingredients a NM according to EC No 1223/2009

Exposure assessment considering possible routes

Local effect ← → Systematic exposure

Physicochemical characterization

Hazard identification and dose -response effect characterization

Nanomaterial Risk Assessment

General principles for risk assessments

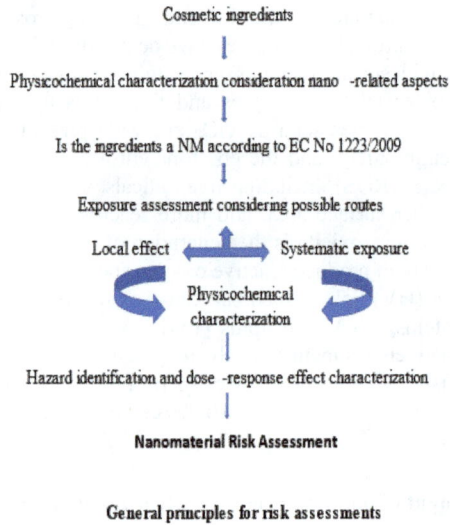

Figure 14 Outline for the NMs safety Assessment in cosmetics[158]

Table 7 Implication of nanomaterials in cosmetics to human

S.No.	Nanomaterials in cosmetics	The implication of NMs on Humans	References
1	Metallic Nanomaterials	Tumors and inflammatory reactions membrane deterioration and DNA damage	[159-161]
2	Carbon Nanomaterials	Potential for cancer if breathed fibrotic lesions that are inflammatory and present in the lungs and may cause granulomas inflammation and skin sensitivity	[12,162]
3	Polymer-based Nanomaterials	Inflammation of the airways momentarily, Zeolites suspended in an alkaline solution may cause skin or mucosal irritation, impede the immune system's reaction to common allergens	[163]
4	Lipid-based Nanomaterials	Nontoxic, biodegradable, and biocompatible; less harmful	[164]

Sprays and aerosols that could contain NMs should be subject to a more thorough safety assessment because inhalation exposure is a possibility.

An incomplete list of factors that are contained in the Scientific Committee on Consumer Safety, (SCCS/1602/18) is required in a case involving exposure to these nanoparticles. The concentration of nanomaterials is expressed in terms of particle number concentration and the surface area should be used to express the concentration in addition to the weight-based concentration of the NM. The European Parliament gave its approval to the revised European Union Cosmetics Directive, which included a reference to "nanomaterials" in EU law. According to European Parliament, the new law establishes a process for evaluating the safety of all nanomaterial-containing products, which could result in the banning of a chemical if they are caising a risk to human health. [165] The following are some of the most significant quotes from the act [166]

According to the ruling, a nanomaterial is "a purposefully created insoluble or bio-persistent material with one or more exterior dimensions, or an interior structure, on the scale from 1 to 100 nm."

• The accountable person makes sure for safety compliances, Good Manufacturing Practices, safety assessment, file containing product information, annexure list for restricted substances, regular sampling, analysis, labeling, animal testing, notification, restrictions for substances listed in annexes, carcinogenic, mutagenic and toxic substances for reproduction and public information.

• Communication of SUE, and information on substances are all followed.

• The responsible party must provide the Commission with the following information before releasing the cosmetic product for sale.

A) Substances in nanomaterial form present in product.
Their identifying information, which includes the chemical name and further details.
B) The exposure circumstances that were conceivably foreseen.

• The SCCS opinion on the safety of the nanomaterial must be promptly requested by the Commission if it has any reservations about its suitability.

• All components included in the form of nanomaterials shall be explicitly mentioned in the ingredients list for the relevant categories of cosmetic goods and the reasonably foreseeable exposure conditions. The word "nano" in brackets must come after the names of these substances.

• Any potential effects on the toxicological profile caused by

A) Particle sizes, including nanomaterials;

B) Impurities of the chemicals and raw materials employed; and

C) Substance Interactions shall be given particular care.

9. Advantages and disadvantages of nanocosmetics

There is a debate that NPs have a significant impact on human health. Both positive and negative impacts are there (Fig.15) [167].

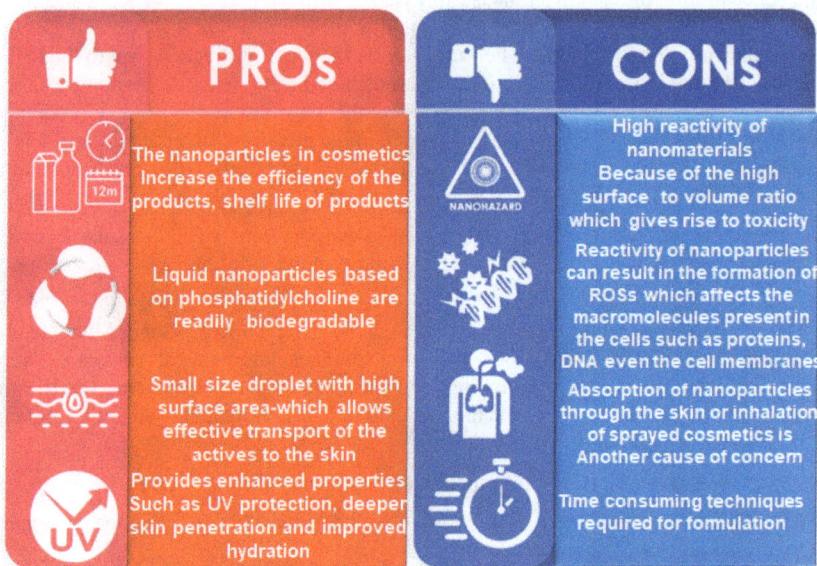

Fig.15 Advantages and disadvantages of nano systems in cosmetic formulations

Conclusions

The cosmetics industry has seen an upsurge in interest in nanotechnology. Cosmetic goods now contain a variety of nanoparticle and nanomaterial types. Thanks to the use of nanotechnology, active compounds can reach deeper layers of the skin and can be successfully supplied to the desired place. However, rather than becoming complacent just in the financially lucrative aesthetic arena, there is a need for more research into the health ramifications of these skin applications. Although the amount of information on the toxicity of nanoparticles is still growing, it is crucial. Advancements in the nano-system preparation techniques, improvement of evaluations/tests for quality check, and efficacy are some areas of advancement as businesses strive to introduce products that are not tested on animals and to create "cruelty-free" products that are completely recyclable and environmentally friendly. Scientists have developed new methods of distribution and technological advancements that are being used in the production of cosmeceuticals. Due to the rise in nano cosmeceuticals traditional delivery methods are being replaced by unique ones.

Liposomes, niosomes, nanostructured lipid carriers, soild lipid nanoparticles, gold nanoparticles, nanoemulsions, and nanosomes are novel nanocarriers which are now used in a variety of cosmeceuticals. These innovative drug delivery methods offer great potential to achieve several goals, including site attentiveness, improved stability, biocompatibility, extended action, and increased drug-loading capacity. Because there aren't enough strong arguments to support effectiveness claims, businesses must offer them. There are significant disagreements over the toxicity and safety of nanomaterials, and numerous studies are being conducted to identify the potential effect, toxins, and health risks. Careful research is needed on the safety potential of nanomaterials. Nanoproducts ought to be developed in such a way as to improve both their worth and the health of their consumers. Cosmetics are exempt from the need for clinical trials; thus, producers take advantage of this exemption and forego expensive and time-consuming trials.

References

[1] A. Mihranyan, N. Ferraz, and M. Strømme, Current status and prospects of nanotechnology in cosmetics. Progress in Materials Science, 57(5) (2012) pp.875-910. https://doi.org/10.1016/j.pmatsci.2011.10.001

[2] N.A. Ibrahim, and M.A.A. Zaini, Nanomaterials in detergents and cosmetics products: the mechanisms and implications. Handbook of nanomaterials for manufacturing applications, Elsevier. (2012) pp.23-49. https://doi.org/10.1016/B978-0-12-821381-0.00002-8

[3] R. Yadwade, S. Gharpure, and B. Ankamwar, Nanotechnology in cosmetics pros and cons. Nano Express, 2(2) (2021), pp.022003. https://doi.org/10.1088/2632-959X/abf46b

[4] A. Gautam, D. Singh, and R. Vijayaraghavan, Dermal exposure of nanoparticles: an understanding. Journal of Cell and Tissue Research, 11(1) (2011), pp.2703-2708.

[5] R. Saha, Cosmeceuticals and herbal drugs: practical uses. International Journal of Pharmaceutical Sciences and Research, 3(1) (2012), pp.59-65.

[6] K. Srinivas, The current role of nanomaterials in cosmetics. Journal of Chemical and Pharmaceutical Research, 8(5) (2016), pp.906-914.

[7] S. Kaul, N. Gulati, D. Verma, S. Mukherjee, and U. Nagaich, Role of nanotechnology in cosmeceuticals: a review of recent advances. Journal of pharmaceutics, 2 (2018), pp.1-19. https://doi.org/10.1155/2018/3420204

[8] B.H. Abbasi, H. Fazal, N. Ahmad, M. Ali, N. Giglioli-Guivarch, and C. Hano, Nanomaterials for cosmeceuticals: nanomaterials-induced advancement in cosmetics, challenges, and opportunities. Nanocosmetics, (2020) pp.79-108. https://doi.org/10.1016/B978-0-12-822286-7.00005-X

[9] S. Mukta, and F. Adam, Cosmeceuticals in day-to-day clinical practice. Journal of Drugs in Dermatology: JDD, 9(5 Suppl ODAC Conf Pt 1), (2010), pp.s62-6.

[10] A. Lohani, A. Verma, H. Joshi, N. Yadav, and N. Karki, Nanotechnology-based cosmeceuticals. International Scholarly Research Notices, (2014), http://dx.doi.org/10.1155/2014/843687 https://doi.org/10.1155/2014/843687

[11] L.M. Katz, K. Dewan, and R.L. Bronaugh, Nanotechnology in cosmetics. Food and Chemical Toxicology, 85, (2015), pp.127-137. https://doi.org/10.1016/j.fct.2015.06.020

[12] S. Raj, S. Jose, U.S. Sumod, and M. Sabitha, Nanotechnology in cosmetics: Opportunities and challenges. Journal of pharmacy & bioallied sciences, 4(3), (2012), p.186-193. https://doi.org/10.4103/0975-7406.99016

[13] L. Rigano, and N. Lionetti, Nanobiomaterials in galenic formulations and cosmetics. Applied Nanobiomaterials, 10, (2016), pp.121-148. https://doi.org/10.1016/B978-0-323-42868-2.00006-1

[14] C.C. Chang, C.P. Chen, T.H. Wu, C.H. Yang, C.W. Lin, and C.Y. Chen, Gold nanoparticle-based colorimetric strategies for chemical and biological sensing applications. Nanomaterials, 9(6), (2019), https://doi.org/10.3390/nano9060861 https://doi.org/10.3390/nano9060861

[15] G. Ghodake, M. Kim, J.S. Sung, S. Shinde, J. Yang, K. Hwang, and D.Y. Kim, Extracellular synthesis and characterization of silver nanoparticles-Antibacterial activity against multidrug-resistant bacterial strains. Nanomaterials, 10(2), (2020), doi: 10.3390/nano10020360 https://doi.org/10.3390/nano10020360

[16] N. Sharma, S. Singh, N. Kanojia, A.S. Grewal, and S. Arora, Nanotechnology: a modern contraption in cosmetics and dermatology. Applied Clinical Research, Clinical Trials and Regulatory Affairs, 5(3), (2018), pp.147-158. https://doi.org/10.2174/2213476X05666180528093905

[17] G. Miller, L. Archer, E. Pica, D. Bell, R. Senyen, and G. Kimbrell, Nanomaterials, sunscreens and cosmetics: small ingredients, big risks: Friends of the Earth Australia & Friends of the Earth United States. (2006)

[18] T.G. Smijs, and S. Pavel, Titanium dioxide and zinc oxide nanoparticles in sunscreens: focus on their safety and effectiveness. Nanotechnology, science and applications, 4, (2011), pp.95-112. https://doi.org/10.2147/NSA.S19419

[19] D.A. Glaser, Anti-aging products and cosmeceuticals. Facial Plastic Surgery Clinics, 12(3), (2004), pp.363-372. https://doi.org/10.1016/j.fsc.2004.03.004

[20] J. Rosen, A. Landriscina, and A.J. Friedman, Nanotechnology-based cosmetics for hair care. Cosmetics, 2(3), (2015), pp.211-224. https://doi.org/10.3390/cosmetics2030211

[21] Z. Hu, M. Liao, Y. Chen, Y. Cai, L. Meng, Y. Liu, N. Lv, Z. Liu, and W. Yuan, A novel preparation method for silicone oil nanoemulsions and its application for coating

hair with silicone. International journal of nanomedicine, 7, (2012), pp.5719-5724. https://doi.org/10.2147/IJN.S37277

[22] C. Fox, Cosmetic and pharmaceutical vehicles: skin care, hair care, makeup and sunscreens. Cosmetics and Toiletries, 113(1), (1998), pp.45-56.

[23] H. Bethany, Zapping nanoparticles into nail polish. Laser Ablation Method Makes Cosmetic and Biomedical Coatings in a Flash, 95(12), (2017), pp.9.

[24] L. Pereira, N. Dias, J. Carvalho, S. Fernandes, C. Santos, and N. Lima, Synthesis, characterization and antifungal activity of chemically and fungal-produced silver nanoparticles against T richophyton rubrum. Journal of applied microbiology, 117(6), (2014), pp.1601-1613. https://doi.org/10.1111/jam.12652

[25] P.T. Sundari, and H. Anushree, Novel delivery systems: current trend in cosmetic industry. European Journal of Pharmaceutical and Medical Research, 4(8), (2017), pp.617-627.

[26] N. Arora, S. Agarwal, and R.S.R. Murthy, Latest technology advances in cosmaceuticals. Int. J. Pharm. Sci. Drug Res, 4(3), (2012), pp.168-182.

[27] M.J. Hope, and C.N. Kitson, Liposomes: a perspective for dermatologists. Dermatologic clinics, 11(1), (1993), pp.143-154. https://doi.org/10.1016/S0733-8635(18)30291-2

[28] B.G. Prajapati, N.K. Patel, N.M. Panchal, and R.P. Patel, Topical liposomes in drug delivery: a review. International Journal of Pavement Research and Technology, 4(1), (2012), pp.39-44.

[29] I.P. Kaur, and R. Agrawal, Nanotechnology: a new paradigm in cosmeceuticals. Recent patents on drug delivery & formulation, 1(2), (2007), pp.171-182. https://doi.org/10.2174/187221107780831888

[30] S.K. Sriraman, and V.P. Torchilin, Recent advances with liposomes as drug carriers. Advanced Biomaterials and Biodevices, 1, (2014), pp.79-120. https://doi.org/10.1002/9781118774052.ch3

[31] D.D. Lasic, Novel applications of liposomes. Trends in biotechnology, 16(7), (1998) pp.307-321. https://doi.org/10.1016/S0167-7799(98)01220-7

[32] C.C. Müller-Goymann, Physicochemical characterization of colloidal drug delivery systems such as reverse micelles, vesicles, liquid crystals and nanoparticles for topical administration. European Journal of Pharmaceutics and Biopharmaceutics, 58(2), (2004), pp.343-356. https://doi.org/10.1016/j.ejpb.2004.03.028

[33] K. Egbaria, and N. Weiner, Liposomes as a topical drug delivery system. Advanced drug delivery reviews, 5(3), (1990), pp.287-300. https://doi.org/10.1016/0169-409X(90)90021-J

[34] M.M. Rieger, Skin lipids and their importance to cosmetic science. Cosmetics and Toiletries, 102(7), (1987), pp.45-49.

[35] G. Blume, E. Teichmüller, and E. Teichmüller, New evidence of the penetration of actives by liposomal carrier system. Cosmetics & Toiletries Manufacture Worldwide, (1997), pp.135-139.

[36] M. Ghyczy, H.P. Nissen, and H. Biltz, The treatment of acne vulgaris by phosphatidylcholine from soybeans, with a high content of linoleic acid. Journal of applied cosmetology, 14, (1996), pp.137-146.

[37] G. Blume, Flexible liposomes for topical applications in cosmetics. Science and applications of skin delivery systems,(2008), pp.269-282.

[38] A. Akbarzadeh, R. Rezaei-Sadabady, S. Davaran, S.W. Joo, N. Zarghami, Y. Hanifehpour, M. Samiei, M. Kouhi, and K. Nejati-Koshki, Liposome: classification, preparation, and applications. Nanoscale research letters, 8(1), (2013), pp.1-9. https://doi.org/10.1186/1556-276X-8-102

[39] Dermosome, https://www.ulprospector.com/en/la/personalcare/Detail/1832/44044/ Dermosomes.

[40] Decorte, https://www.decortecosmetics.com/skincare/liposome

[41] K.M. Kazi, A.S. Mandal, N. Biswas, A. Guha, S. Chatterjee, M. Behera, and K. Kuotsu, Niosome: a future of targeted drug delivery systems. Journal of Advanced Pharmaceutical Technology & Research, 1(4), (2010), pp.374-380. https://doi.org/10.4103/0110-5558.76435

[42] S. Duarah, K. Pujari, R.D. Durai, and V.H.B. Narayanan, Nanotechnology-based cosmeceuticals: a review. International Journal of Applied Pharmaceuticals, 8(1), (2016), pp.8-12.

[43] A. Sankhyan, and P. Pawar, Current trends in niosome as vesicular drug delivery system. Asian Journal of Pharmacy and Life Science, 02(06), (2012) pp.20-32.

[44] P. Sudheer, and K. Kaushik, Review on niosomes-a novel approach for drug targeting. Journal of Pharmaceutical Research, 14(1), (2015), pp.20-25. https://doi.org/10.18579/jpcrkc/2015/14/1/78376

[45] V. Pola Chandu, A. Arunachalam, S. Jeganath, K. Yamini, and K. Tarangini, Niosomes: a novel drug delivery system. International Journal of Novel Trends in Pharmaceutical Sciences, 2(1), (2012), pp. 2277-2782.

[46] G.P. Kumar, and P. Rajeshwarrao, Nonionic surfactant vesicular systems for effective drug delivery-an overview. Acta Pharmaceutica Sinica B, 1(4), (2011), pp.208-219. https://doi.org/10.1016/j.apsb.2011.09.002

[47] S. Biswal, P.N. Murthy, J. Sahu, P. Sahoo, and F. Amir, Vesicles of non-ionic surfactants (niosomes) and drug delivery potential. International journal of Pharmaceutical Sciences and Nanotechnology, 1(1), (2008), pp.1-8. https://doi.org/10.37285/ijpsn.2008.1.1.1

[48] A. Nasir, S.L. Harikumar, and K. Amanpreet, Niosomes: An excellent tool for drug delivery. International Journal of Research in Pharmacy and Chemistry, 2(2), (2012), pp.479-487.

[49] M. Gandhi, P. Sanket, and S. Mahendra, Niosomes: novel drug delivery system. International Journal of Pure & Applied Bioscience, 2(2), (2014), pp.267-274.

[50] S. Verma, S.K. Singh, N. Syan, P. Mathur, and V. Valecha, Nanoparticle vesicular systems: a versatile tool for drug delivery. Journal of Chemical and Pharmaceutical Research, 2(2), (2010), pp.496-509.

[51] T. Varun, A. Sonia, P. Bharat, V. Patil, P.O. Kumharhatti, and D. Solan, Niosomes and liposomes-vesicular approach towards transdermal drug delivery. International Journal of Pharmaceutical and Chemical Sciences, 1(3), (2012), pp.632-644.

[52] A. Gupta, S.K. Prajapati, M. Balamurugan, M. Singh, and D. Bhatia, Design and development of a proniosomal transdermal drug delivery system for captopril. Tropical Journal of Pharmaceutical Research, 6(2), (2007), pp.687-693. https://doi.org/10.4314/tjpr.v6i2.14647

[53] S. Waghmare, A. Patil, and P. Patil, Novasome: Advance in Liposome and Niosome. The Pharma Innovation, 5(5), (2016), pp.34.

[54] S. Mosallam, M.H. Ragaie, N.H. Moftah, A.H. Elshafeey, and A.A. Abdelbary, Use of novasomes as a vesicular carrier for improving the topical delivery of terconazole: In vitro characterization, in vivo assessment and exploratory clinical experimentation. International Journal of Nanomedicine, 16, (2021) pp.119-132. https://doi.org/10.2147/IJN.S287383

[55] A. Singh, R. Malviya, and P.K. Sharma, Novasome-a breakthrough in pharmaceutical technology a review article. Advances in Biological Research, 5(4), (2011), pp.184-189.

[56] B. Gorain, B.E. Al-Dhubiab, A. Nair, P. Kesharwani, M. Pandey, and H. Choudhury, Multivesicular liposome: A lipid-based drug delivery system for efficient drug delivery. Current Pharmaceutical Design, 27(43), (2021), pp.4404-4415. https://doi.org/10.2174/1381612827666210830095941

[57] D. Puri, A. Bhandari, P. Sharma, and D. Choudhary, Lipid nanoparticles (SLN, NLC): A novel approach for cosmetic and dermal pharmaceutical. Journal of Global Pharma Technology, 2(9), (2010), pp.1-15.

[58] N. Hasan, M. Imran, P. Kesharwani, K. Khanna, R. Karwasra, N. Sharma, S. Rawat, D. Sharma, F.J. Ahmad, G.K. Jain, and A. Bhatnagar, Intranasal delivery of Naloxone-loaded solid lipid nanoparticles as a promising simple and non-invasive approach for the management of opioid overdose. International Journal of Pharmaceutics, 599, (2021), doi: 10.1016/j.ijpharm.2021.120428. https://doi.org/10.1016/j.ijpharm.2021.120428

[59] G.S. Bhagwat, R.B. Athawale, R.P. Gude, S. Md, N.A. Alhakamy, U.A. Fahmy, and P. Kesharwani, Formulation and development of transferrin targeted solid lipid nanoparticles for breast cancer therapy. Frontiers in Pharmacology, 11, (2020), doi: 10.3389/fphar.2020.614290. https://doi.org/10.3389/fphar.2020.614290

[60] R.K. Khurana, R. Kumar, B.L. Gaspar, G. Welsby, P. Welsby, P. Kesharwani, O.P. Katare, K.K. Singh, and B. Singh, Clathrin-mediated endocytic uptake of PUFA enriched self-nanoemulsifying lipidic systems (SNELS) of an anticancer drug against triple negative cancer and DMBA induced preclinical tumor model. Materials Science and Engineering: C, 91, (2018), pp.645-658. https://doi.org/10.1016/j.msec.2018.05.010

[61] R.K. Khurana, S. Beg, A.J. Burrow, R.K. Vashishta, O.P. Katare, S. Kaur, P. Kesharwani, K.K. Singh, and B. Singh, Enhancing biopharmaceutical performance of an anticancer drug by long chain PUFA based self-nanoemulsifying lipidic nanomicellar systems. European Journal of Pharmaceutics and Biopharmaceutics, 121, (2017), pp.42-60. https://doi.org/10.1016/j.ejpb.2017.09.001

[62] A. Jain, G. Sharma, V. Kushwah, N.K. Garg, P. Kesharwani, G. Ghoshal, B. Singh, U.S. Shivhare, S. Jain, and O.P. Katare, Methotrexate and beta-carotene loaded-lipid polymer hybrid nanoparticles: A preclinical study for breast cancer. Nanomedicine, 12(15), (2017), pp.1851-1872. https://doi.org/10.2217/nnm-2017-0011

[63] A. Zur Mühlen, C. Schwarz, and W. Mehnert, Solid lipid nanoparticles (SLN) for controlled drug delivery-drug release and release mechanism. European Journal of Pharmaceutics and Biopharmaceutics, 45(2), (1998), pp.149-155. https://doi.org/10.1016/S0939-6411(97)00150-1

[64] J. Pardeike, A. Hommoss, and R.H. Müller, Lipid nanoparticles (SLN, NLC) in cosmetic and pharmaceutical dermal products. International Journal of Pharmaceutics, 366(1-2), (2009), pp.170-184. https://doi.org/10.1016/j.ijpharm.2008.10.003

[65] A. Patidar, D.S. Thakur, P. Kumar, and J. Verma, A review on novel lipid based nanocarriers. International Journal of Pharmacy and Pharmaceutical Sciences, 2(4), (2010), pp.30-35.

[66] C. Song, and S. Liu, A new healthy sunscreen system for human: Solid lipid nannoparticles as carrier for 3, 4, 5-trimethoxybenzoylchitin and the improvement by adding Vitamin E. International Journal of Biological Macromolecules, 36(1-2), (2005), pp.116-119. https://doi.org/10.1016/j.ijbiomac.2005.05.003

[67] R.H. Müller, M. Radtke, and S.A. Wissing, Solid lipid nanoparticles (SLN) and nanostructured lipid carriers (NLC) in cosmetic and dermatological preparations. Advanced Drug Delivery Reviews, 54, (2002), pp.131-S155. https://doi.org/10.1016/S0169-409X(02)00118-7

[68] M. Hassan Hany, O.N. El Gazayerly, Rice bran solid lipid nanoparticles: Preparation and characterization. International Journal of Research in Drug Delivery, 1(2), (2011), pp.6-9.

[69] R. López-García, and A. Ganem-Rondero, Solid lipid nanoparticles (SLN) and nanostructured lipid carriers (NLC): occlusive effect and penetration enhancement ability. Journal of Cosmetics, Dermatological Sciences and Applications, 5(2), (2015), pp.62-72. https://doi.org/10.4236/jcdsa.2015.52008

[70] E.B. Souto, and R.H. Müller, Cosmetic features and applications of lipid nanoparticles (SLN®, NLC®). International Journal of Cosmetic Science, 30(3), (2008), pp.157-165. https://doi.org/10.1111/j.1468-2494.2008.00433.x

[71] P. Ekambaram, A.H. Sathali and K. Priyanka, Solid lipid nanoparticles: a review. Scientific Reviews and Chemical Communications, 2(1), (2012), pp. 80-102.

[72] K.H. Ramteke, S.A. Joshi, and S.N. Dhole, Solid lipid nanoparticle: a review. IOSR Journal of Pharmacy, 2(6), (2012), pp.34-44. https://doi.org/10.9790/3013-26103444

[73] R. Costa, and L. Santos, Delivery systems for cosmetics-From manufacturing to the skin of natural antioxidants. Powder Technology, 322, (2017), pp.402-416. https://doi.org/10.1016/j.powtec.2017.07.086

[74] D.K. Purohit, Nano-lipid carriers for topical application: Current scenario. Asian Journal of Pharmaceutics, 10(1), (2016), https://doi.org/10.22377/ajp.v10i1.544

[75] K. Westesen, H. Bunjes, and M.J.H. Koch, Physicochemical characterization of lipid nanoparticles and evaluation of their drug loading capacity and sustained release potential. Journal of Controlled Release, 48(2-3), (1997), pp.223-236. https://doi.org/10.1016/S0168-3659(97)00046-1

[76] S. Khan, S. Baboota, J. Ali, S. Khan, R.S. Narang, and J.K. Narang, Nanostructured lipid carriers: an emerging platform for improving oral bioavailability of lipophilic drugs. International Journal of Pharmaceutical Investigation, 5(4), (2015), pp.182-191. https://doi.org/10.4103/2230-973X.167661

[77] B. Sharma, and A. Sharma, Future prospect of nanotechnology in development of anti-ageing formulations. International Journal of Pharmacy and Pharmaceutical Sciences, 4(3), (2012), pp.57-66.

[78] N. Naseri, H. Valizadeh, and P. Zakeri-Milani, Solid lipid nanoparticles and nanostructured lipid carriers: structure, preparation and application. Advanced Pharmaceutical Bulletin, 5(3), (2015), pp.305-313. https://doi.org/10.15171/apb.2015.043

[79] S. Kaur, U. Nautyal, R. Singh, S. Singh, and A. Devi, Nanostructure lipid carrier (NLC): the new generation of lipid nanoparticles. Asian Pacific Journal of Health Sciences, 2(2), (2015), pp.76-93. https://doi.org/10.21276/apjhs.2015.2.2.14

[80] C.L. Fang, S.A. Al-Suwayeh, and J.Y. Fang, Nanostructured lipid carriers (NLCs) for drug delivery and targeting. Recent patents on nanotechnology, 7(1), (2013), pp.41-55. https://doi.org/10.2174/187221013804484827

[81] A. Singh, G. Garg, and P.K. Sharma, Nanospheres: a novel approach for targeted drug delivery system. International Journal of Pharmaceutical Sciences Review and Research, 5(3), (2010), pp.84-88.

[82] B. Mamo, W. Abebe, and T.D. Gabriel, Literature review on biodegradable nanospheres for oral and targeted drug delivery. Asian Journal of Biomedical and Pharmaceutical Sciences, 5(51), (2015), pp.01-12. https://doi.org/10.15272/ajbps.v5i51.761

[83] S.S. Guterres, M.P. Alves, and A.R. Pohlmann, Polymeric nanoparticles, nanospheres and nanocapsules, for cutaneous applications. Drug Target Insights, 2, (2007), pp.147-157. https://doi.org/10.1177/117739280700200002

[84] B. Gorain, H. Choudhury, A.B. Nair, S.K. Dubey, and P. Kesharwani, Theranostic application of nanoemulsions in chemotherapy. Drug Discovery Today, 25(7), (2020), pp.1174-1188. https://doi.org/10.1016/j.drudis.2020.04.013

[85] H. Choudhury, N.F.B. Zakaria, P.A.B. Tilang, A.S. Tzeyung, M. Pandey, B. Chatterjee, N.A. Alhakamy, S.K. Bhattamishra, P. Kesharwani, B. Gorain, and S. Md, Formulation development and evaluation of rotigotine mucoadhesive nanoemulsion for intranasal delivery. Journal of Drug Delivery Science and Technology, 54, (2019), 101301. doi: 10.1016/j.jddst.2019.101301. https://doi.org/10.1016/j.jddst.2019.101301

[86] H. Choudhury, M. Pandey, B. Gorain, B. Chatterjee, T. Madheswaran, S. Md, K.K. Mak, M. Tambuwala, M.K. Chourasia, and P. Kesharwani, Nanoemulsions as effective carriers for the treatment of lung cancer. In Nanotechnology-based targeted drug delivery systems for lung cancer, (2019), pp. 217-247. https://doi.org/10.1016/B978-0-12-815720-6.00009-5

[87] M. Pandey, H. Choudhury, O.C. Yeun, H.M. Yin, T.W. Lynn, C.L. Tine, N.S. Wi, K.C. Yen, C.S. Phing, P. Kesharwani, and S.K. Bhattamisra, Perspectives of nanoemulsion strategies in the improvement of oral, parenteral and transdermal chemotherapy. Current pharmaceutical biotechnology, 19(4), (2018), pp.276-292. https://doi.org/10.2174/1389201019666180605125234

[88] H. Choudhury, B. Gorain, M. Pandey, L.A. Chatterjee, P. Sengupta, A. Das, N. Molugulu, and P. Kesharwani, Recent update on nanoemulgel as topical drug delivery system. Journal of Pharmaceutical Sciences, 106(7), (2017), pp.1736-1751. https://doi.org/10.1016/j.xphs.2017.03.042

[89] M.F.R.G. Dias, Educational Website for Cosmetologists.

[90] O. Sonneville-Aubrun, M. Yukuyama, and A. Pizzino, Application of nanoemulsions in cosmetics. Nanoemulsions: Formulation, Applications and

Characterization. Elsevier Inc; (2018). https://doi.org/10.1016/B978-0-12-811838-2.00014-X

[91] T.J. Ashaolu, Nanoemulsions for health, food, and cosmetics: A review. Environmental Chemistry Letters, 19(4), (2021), pp.3381-3395. https://doi.org/10.1007/s10311-021-01216-9

[92] Z.A.A. Aziz, H. Mohd-Nasir, A. Ahmad, S.H. Mohd. Setapar, W.L., Peng, S.C. Chuo, A. Khatoon, K. Umar, A.A. Yaqoob, and M.N. Mohamad Ibrahim, Role of nanotechnology for design and development of cosmeceutical: application in makeup and skin care. Frontiers in Chemistry, 7, (2019), https://doi.org/10.3389/fchem.2019.00739 https://doi.org/10.3389/fchem.2019.00739

[93] A. Sheikh, S. Md, and P. Kesharwani, RGD engineered dendrimer nanotherapeutic as an emerging targeted approach in cancer therapy. Journal of Controlled Release, 340, (2021), pp.221-242. https://doi.org/10.1016/j.jconrel.2021.10.028

[94] S.K. Dubey, A. Dey, G. Singhvi, M.M. Pandey, V. Singh, and P. Kesharwani, Emerging trends of nanotechnology in advanced cosmetics. Colloids and Surfaces B: Biointerfaces, (2022), doi: 10.1016/j.colsurfb.2022.112440. https://doi.org/10.1016/j.colsurfb.2022.112440

[95] E.B. Souto, A.R. Fernandes, C. Martins-Gomes, T.E. Coutinho, A. Durazzo, M. Lucarini, S.B. Souto, A.M. Silva, and A. Santini, Nanomaterials for skin delivery of cosmeceuticals and pharmaceuticals. Applied Sciences, 10(5), (2020), https://doi.org/10.3390/app10051594. https://doi.org/10.3390/app10051594

[96] C. Yang, M. Wang, J. Zhou, and Q. Chi, Bio-synthesis of peppermint leaf extract polyphenols capped nano-platinum and their in-vitro cytotoxicity towards colon cancer cell lines (HCT 116). Materials Science and Engineering: C, 77, (2017), pp.1012-1016. https://doi.org/10.1016/j.msec.2017.04.020

[97] S. Joglekar, K. Kodam, M. Dhaygude, and M. Hudlikar, Novel route for rapid biosynthesis of lead nanoparticles using aqueous extract of Jatropha curcas L. latex. Materials Letters, 65(19-20), (2011), pp.3170-3172. https://doi.org/10.1016/j.matlet.2011.06.075

[98] A.K. Potbhare, R.G. Chaudhary, P.B. Chouke, S. Yerpude, A. Mondal, V.N. Sonkusare, A.R. Rai, and H.D. Juneja, Phytosynthesis of nearly monodisperse CuO nanospheres using Phyllanthus reticulatus/Conyza bonariensis and its antioxidant/antibacterial assays. Materials Science and Engineering: C, 99, (2019), pp.783-793. https://doi.org/10.1016/j.msec.2019.02.010

[99] A.C. Paiva-Santos, A.M. Herdade, C. Guerra, D. Peixoto, M. Pereira-Silva, M. Zeinali, F. Mascarenhas-Melo, A. Paranhos, and F. Veiga, Plant-mediated green synthesis of metal-based nanoparticles for dermopharmaceutical and cosmetic applications. International Journal of Pharmaceutics, 597, (2021), DOI:10.1016/j.ijpharm.2021.120311. https://doi.org/10.1016/j.ijpharm.2021.120311

[100] R. Alsubki, H. Tabassum, M. Abudawood, A.A. Rabaan, S.F. Alsobaie, and S. Ansar, Green synthesis, characterization, enhanced functionality and biological evaluation of silver nanoparticles based on Coriander sativum. Saudi Journal of Biological Sciences, 28(4), (2021), pp.2102-2108. https://doi.org/10.1016/j.sjbs.2020.12.055

[101] F. Sabir, M. Barani, A. Rahdar, M. Bilal, and M. Nadeem, How to face skin cancer with nanomaterials: A review. Biointerface Research in Applied Chemistry, 11, (2021), pp.11931-11955. https://doi.org/10.33263/BRIAC114.1193111955

[102] H. Pastrana, A. Avila, and C.S. Tsai, Nanomaterials in cosmetic products: The challenges with regard to current legal frameworks and consumer exposure. Nanoethics, 12(2), (2018), pp.123-137. https://doi.org/10.1007/s11569-018-0317-x

[103] F. Carrouel, S. Viennot, L. Ottolenghi, C. Gaillard, and D. Bourgeois, Nanoparticles as anti-microbial, anti-inflammatory, and remineralizing agents in oral care cosmetics: a review of the current situation. Nanomaterials, 10(1), (2020), doi: 10.3390/nano10010140. https://doi.org/10.3390/nano10010140

[104] R.A. Revia, B.A., Wagner, and M. Zhang, A portable electrospinner for nanofiber synthesis and its application for cosmetic treatment of alopecia. Nanomaterials, 9(9), (2019), doi: 10.3390/nano9091317. https://doi.org/10.3390/nano9091317

[105] U. Ahmad, Z. Ahmad, A.A. Khan, J. Akhtar, S.P. Singh, and F.J. Ahmad, Strategies in development and delivery of nanotechnology based cosmetic products. Drug Research, 68(10), (2018), pp.545-552. https://doi.org/10.1055/a-0582-9372

[106] C. Ngô, and M.H. Van de Voorde, Nanomaterials and cosmetics. Nanotechnology in a Nutshell (2014), pp. 311-319. https://doi.org/10.2991/978-94-6239-012-6_18

[107] J. Cornier, C.M. Keck, and M. Van de Voorde, Nanocosmetics: From Ideas to Products. Springer (2019). https://doi.org/10.1007/978-3-030-16573-4

[108] L.A. Huber, T.A. Pereira, D.N. Ramos, L.C. Rezende, F.s. Emery, L.M. Sobral, A.M. Leopoldino, and R.F. Lopez, Topical skin cancer therapy using doxorubicin-loaded cationic lipid nanoparticles and iontophoresis. Journal of Biomedical Nanotechnology, 11(11), (2015), pp.1975-1988. https://doi.org/10.1166/jbn.2015.2139

[109] M.R. Santos, A.C. Fonseca, P.V. Mendonça, R. Branco, A.C. Serra, P.V. Morais, and J.F. Coelho, Recent developments in antimicrobial polymers: A review. Materials, 9(7), (2016), doi:10.3390/ma9070599. https://doi.org/10.3390/ma9070599

[110] P. Morganti, New horizon in cosmetic dermatology. J. Appl. Cosmetol, 34, (2016), pp.15-24.

[111] N. Dragicevic, and H.I. Maibach, Percutaneous penetration enhancers physical methods in penetration enhancement. New York, NY, USA: Springer (2017). https://doi.org/10.1007/978-3-662-53273-7

[112] A. Singh, G. Garg, and P.K. Sharma, Nanospheres: a novel approach for targeted drug delivery system. International Journal of Pharmaceutical Sciences Review and Research, 5(3), (2010), pp.84-88.

[113] H. Idrees, S.J.Z. Zaidi, A. Sabir, R.U. Khan, X. Zhang, and S. Hassan, A review of biodegradable natural polymer based nanoparticles for drug delivery applications. Nanomaterials, 10(10), (2020), https://doi.org/10.3390/nano10101970 https://doi.org/10.3390/nano10101970

[114] S. Deng, M.R. Gigliobianco, R. Censi, and P.D. Martino, Polymeric Nanocapsules as Nanotechnological Alternative for Drug Delivery System: Current Status, Challenges and Opportunities. Nanomaterials, 10(5), (2020), https://doi.org/10.3390/nano10050847 https://doi.org/10.3390/nano10050847

[115] D.A. Fernandes, Review on the applications of nanoemulsions in cancer theranostics. Journal of Materials Research, 37, (2022), pp.1953-1977. https://doi.org/10.1557/s43578-022-00583-5

[116] S. Kato, H. Aoshima, Y. Saitoh, and N. Miwa, Defensive effects of fullerene-C60 dissolved in squalane against the 2, 4-nonadienal-induced cell injury in human skin keratinocytes HaCaT and wrinkle formation in 3D-human skin tissue model. Journal of Biomedical Nanotechnology, 6(1), (2010), pp.52-58. https://doi.org/10.1166/jbn.2010.1091

[117] T.M. Benn, P. Westerhoff, and P. Herckes, Detection of fullerenes (C60 and C70) in commercial cosmetics. Environmental Pollution, 159(5), (2011), pp.1334-1342. https://doi.org/10.1016/j.envpol.2011.01.018

[118] Z.D. Draelos, Retinoids in cosmetics. Cosmetic Dermatology, 18(1), (2005), pp.3-5.

[119] B.H. Jung, Y.T. Lim, J.K. Kim, J.Y. Jeong, and T.H. Ha, Cosmetic pigment composition containing gold or silver nano-particles. (2008), European Patent 1909745A1.

[120] S.M. Hirst, A.S. Karakoti, R.D. Tyler, N. Sriranganathan, S. Seal, and C.M. Reilly, Anti-inflammatory properties of cerium oxide nanoparticles. Small, 5(24), (2009), pp.2848-2856. https://doi.org/10.1002/smll.200901048

[121] C. Couteau, and L. Coiffard, Overview of skin whitening agents: Drugs and cosmetic products. Cosmetics, 3(3), (2016). https://doi.org/10.3390/cosmetics3030027

[122] G.M. Turino, and J.O. Cantor, Hyaluronan in respiratory injury and repair. American Journal of Respiratory and Critical Care Medicine, 167(9), (2003), pp.1169-1175. https://doi.org/10.1164/rccm.200205-449PP

[123] P. Maitra, S. Carlo, and R.A. Ranade, Nanoparticle Compositions Providing Enhanced Color for Cosmetic Formulations. (2014), U.S. Patent Application 14/040,800.

[124] J.D. Dreher, G.J. Stepniewski, Optical makeup composition. Patent no. CA 2366619 C; (2008).

[125] A.S. Barnard, One-to-one comparison of sunscreen efficacy, aesthetics and potential nanotoxicity. Nature nanotechnology, 5(4), (2010), pp.271-274. https://doi.org/10.1038/nnano.2010.25

[126] D. Schlossman, E,.Bartholomey, and C. Orr, Kobo Products Inc, Uv protective cosmetic product incorporating titanium dioxide and transparent iron oxide. (2010) U.S. Patent Application 12/515,395.

[127] J. Vollhardt, N. Malkan, and R.P. Manzo, Dragoco Gerberding and Co GmbH. Process for producing cosmetic and pharmaceutical formulations, and products comprising same. (2002), U.S. Patent 6,387,398.

[128] M. Herstein,, Cosmetic skin-renewal-stimulating composition with long-term irritation control. (1997), U.S. Patent 5,616,332.

[129] Z.V.Q. Jurado, M.A.W. Mendoza, L.M.C. Covarrubias, H.M.A. Bandin, and J.E.P. Lopez, Antibacterial composition of silver nanoparticles bonded to a dispersing agent. Patent no. WO 2016 122995 A1; 2016 https://doi.org/10.1155/2016/1641352

[130] H.W. Lee, OROSCIENCE Inc, Process to form nano-sized materials, the compositions and uses thereof. (2009), U.S. Patent Application 11/933,397.

[131] R.M. Hoffman, Topical liposome targeting of dyes, melanins, genes, and proteins selectively to hair follicles. Journal of Drug Targeting, 5(2), (1998), pp.67-74. https://doi.org/10.3109/10611869808995860

[132] M. Franzke, K. Steinbrecht, T. Clausen, S. Baecker, J. Titze, and G.M.B.H. Wella, Cosmetic compositions for hair treatment containing dendrimers or dendrimer conjugates. (2000), U.S. Patent 6,068,835.

[133] H. Furukawa, and T. Limura, Copolymer having carbosiloxane dendrimer structure, and composition and cosmetic containing the same. (2012) Patent no. US 2012 0263662 A1;

[134] F. Giroud, H. Samain, and I. Rollat, Loreal, S.A., Cosmetic composition based on nanoparticles and on water-soluble organic silicon compounds. (2013), U.S. Patent 8,377,427.

[135] J.L. Gesztesi, L.M. Santos, P.D.T. Hennies, and K.A. Macian, Natura Cosmeticos SA, Oil-in-water nanoemulsion, a cosmetic composition and a cosmetic product comprising it, a process for preparing said nanoemulsion. (2015), U.S. Patent 8,956,597.

[136] S.R. Fogg, T.R. Kapsner, and P. Matravers, Hair treatment composition and method. (1999), Patent no. WO 1999 055293 A1.

[137] S. Dickhof, J. Franklin, P. Busch, C. Kropf, and D. Fischer, Cosmetic composition, for preventing greasy appearance on hair, contains nanoparticles of oxide, oxide-

hydrate, hydroxide, carbonate, silicate or phosphate of calcium, magnesium, aluminum, titanium, zirconium or zinc. (2001), Patent DE19946784 A, 12001, p.19.

[138] M. Mellul, S.A. LOreal, Cosmetic make-up composition containing a fullerene or mixture of fullerenes as a pigmenting agent. (1997), U.S. Patent 5,612,021.

[139] P.J. Borm, D. Robbins, S. Haubold, T. Kuhlbusch, H. Fissan, K. Donaldson, R. Schins, V. Stone, W. Kreyling, J. Lademann, and J. Krutmann, The potential risks of nanomaterials: a review carried out for ECETOC. Particle and Fibre toxicology, 3(1), (2006), pp.1-35. https://doi.org/10.1186/1743-8977-3-11

[140] Z. Zhou, R. Lenk, A. Dellinger, D. MacFarland, K. Kumar, S.R. Wilson, and C.L. Kepley, Fullerene nanomaterials potentiate hair growth. Nanomedicine: Nanotechnology, Biology and Medicine, 5(2), (2009), pp.202-207. https://doi.org/10.1016/j.nano.2008.09.005

[141] F. Giroud, and V. Favreau, LOreal SA, Cosmetic composition for volumizing keratin fibers and cosmetic use of nanotubes for volumizing keratin fibers. (2004), U.S. Patent Application 10/455,499.

[142] X. Huang, R.K. Kobos, and G. Xu, EI Du Pont de Nemours and Co, Hair coloring and cosmetic compositions comprising carbon nanotubes. (2007), U.S. Patent 7,276,088.

[143] V. Jeanne-Rose, Cosmetic composition comprising a dyestuff said dyestuff and cosmetic treatment process. Patent no. WO 2012 035029 A2; 2012.

[144] C. Buzea, I.I. Pacheco, and K. Robbie, Nanomaterials and nanoparticles: sources and toxicity. Biointerphases, 2(4), (2007), pp.17-71. https://doi.org/10.1116/1.2815690

[145] C.S. Yah, S.E. Iyuke, and G.S. Simate, A review of nanoparticles toxicity and their routes of exposures. Iranian Journal of Pharmaceutical Sciences, 8(1), (2012), pp.299-314.

[146] M.T. Zhu, W.Y. Feng, Y. Wang, B. Wang, M. Wang, H. Ouyang, Y.L. Zhao, and Z.F. Chai, Particokinetics and extrapulmonary translocation of intratracheally instilled ferric oxide nanoparticles in rats and the potential health risk assessment. Toxicological Sciences, 107(2), (2009), pp.342-351. https://doi.org/10.1093/toxsci/kfn245

[147] P.H. Hoet, I. Brüske-Hohlfeld, and O.V. Salata, Nanoparticles-known and unknown health risks. Journal of Nanobiotechnology, 2(1), (2004), pp.1-15. https://doi.org/10.1186/1477-3155-2-12

[148] A. Nel, T. Xia, L. Madler, and N. Li, Toxic potential of materials at the nanolevel. Science, 311(5761), (2006), pp.622-627. https://doi.org/10.1126/science.1114397

[149] V.K. Poon, and A. Burd, In vitro cytotoxity of silver: implication for clinical wound care. Burns, 30(2), (2004), pp.140-147. https://doi.org/10.1016/j.burns.2003.09.030

[150] T.G. Singh, and N. Sharma, Nanobiomaterials in cosmetics: current status and future prospects. Nanobiomaterials in Galenic Formulations and Cosmetics, (2016), pp.149-174. https://doi.org/10.1016/B978-0-323-42868-2.00007-3

[151] G. Wakefield, S. Lipscomb, E. Holland, and J. Knowland, The effects of manganese doping on UVA absorption and free radical generation of micronised titanium dioxide and its consequences for the photostability of UVA absorbing organic sunscreen components. Photochemical & Photobiological Sciences, 3(7), (2004), pp.648-652. https://doi.org/10.1039/b403697b

[152] A. Mavon, C. Miquel, O. Lejeune, B. Payre, and P. Moretto, In vitro percutaneous absorption and in vivo stratum corneum distribution of an organic and a mineral sunscreen. Skin Pharmacology and Physiology, 20(1), (2007), pp.10-20. https://doi.org/10.1159/000096167

[153] C.M. Sayes, J.D. Fortner, W. Guo, D. Lyon, A.M. Boyd, K.D. Ausman, Y.J. Tao, B. Sitharaman, L.J. Wilson, J.B. Hughes, and J.L. West, The differential cytotoxicity of water-soluble fullerenes. Nano letters, 4(10), (2004), pp.1881-1887. https://doi.org/10.1021/nl0489586

[154] B. Arvidson, A review of axonal transport of metals. Toxicology, 88(1-3), (1994), pp.1-14. https://doi.org/10.1016/0300-483X(94)90107-4

[155] H. Shi, R. Magaye, V. Castranova, and J. Zhao, Titanium dioxide nanoparticles: a review of current toxicological data. Particle and Fibre Toxicology, 10(1), (2013), pp.1-33. https://doi.org/10.1186/1743-8977-10-15

[156] O.M. Posada, R.J. Tate, and M.H. Grant, Toxicity of cobalt-chromium nanoparticles released from a resurfacing hip implant and cobalt ions on primary human lymphocytes in vitro. Journal of Applied Toxicology, 35(6), (2015), pp.614-622. https://doi.org/10.1002/jat.3100

[157] R.J. Aitken, M.Q. Chaudhry, A.B.A. Boxall, and M. Hull, Manufacture and use of nanomaterials: current status in the UK and global trends. Occupational Medicine, 56(5), (2006), pp.300-306. https://doi.org/10.1093/occmed/kql051

[158] I. Khan, K. Saeed, and I. Khan, Nanoparticles: properties, applications and toxicities. Arabian Journal of Chemistry, 12(7), (2017), pp. 908-931. https://doi.org/10.1016/j.arabjc.2017.05.011

[159] J. Chen, X. Dong, J. Zhao, and G. Tang, In vivo acute toxicity of titanium dioxide nanoparticles to mice after intraperitioneal injection. Journal of applied toxicology, 29(4), (2009), pp.330-337. https://doi.org/10.1002/jat.1414

[160] F. Gottschalk, T. Sonderer, R.W. Scholz, and B. Nowack, Modeled environmental concentrations of engineered nanomaterials (TiO2, ZnO, Ag, CNT, fullerenes) for different regions. Environmental Science & Technology, 43(24), (2009), pp.9216-9222. https://doi.org/10.1021/es9015553

[161] J. Sun, S. Wang, D. Zhao, F.H. Hun, L. Weng, and H. Liu, Cytotoxicity, permeability, and inflammation of metal oxide nanoparticles in human cardiac microvascular endothelial cells. Cell Biology and Toxicology, 27(5), (2011), pp.333-342. https://doi.org/10.1007/s10565-011-9191-9

[162] E. Hood, Fullerenes and fish brains: nanomaterials cause oxidative stress (2004). https://doi.org/10.1289/ehp.112-a568a

[163] C. Fruijtier-Pölloth, The safety of synthetic zeolites used in detergents. Archives of toxicology, 83(1), (2009), pp.23-35. https://doi.org/10.1007/s00204-008-0327-5

[164] Z. Fakhravar, P. Ebrahimnejad, H. Daraee, and A. Akbarzadeh, Nanoliposomes: Synthesis methods and applications in cosmetics. Journal of Cosmetic and Laser Therapy, 18(3), (2016), pp.174-181. https://doi.org/10.3109/14764172.2015.1039040

[165] S. Tekmen, and S. Öksüz, Nanomaterials and Human Health. In Nanotoxicology and Nanoecotoxicology 1 (2021), (pp. 21-55). Springer, Cham. https://doi.org/10.1007/978-3-030-63241-0_2

[166] A. Gergely, and L. Coroyannakis, Nanotechnology in the EU cosmetics regulation. Household and Personal Care Today, 3, (2009), pp.28-30.

[167] C. Cardoza, V. Nagtode, A. Pratapa and S. N. Mali, Emerging applications of nanotechnology in cosmeceutical health science: Latest updates, Health Sciences Review 4(2022)100051 https://doi.org/10.1016/j.hsr.2022.100051

Applications of Emerging Nanomaterials and Nanotechnology   Materials Research Forum LLC
Materials Research Foundations 148 (2023) 170-199   https://doi.org/10.21741/9781644902554-6

Chapter 6

Nanomaterials for Leather Production

Uswatun Hasanah, Muhammed Shah Miran, and Md. Mominul Islam*

Department of Chemistry, Faculty of Science, University of Dhaka, Dhaka 1000, Bangladesh

* mominul@du.ac.bd

Abstract

Leather is used to make various daily life goods including footwear, clothing, automobile seats, furniture, etc. from prehistoric time. Various defects can occur either due to failure of controlling the factors of processing or due to low grade of raw material or both. The tanned leathers are, before use, necessarily being passed through different processes including re-tanning, filling, coating the surface with dyes, and other functional materials for protecting the surface of leather. Nanoparticles of natural and synthetic polymers, metal ions, metal oxides, metal etc. have been used for finishing tanned leather in providing them adequate quality as per the demand of buyers. In this chapter, the potential of using nanomaterials in different steps of leather processing and finishing are focused.

Keywords

Leather Processing, Nanoparticles, Nano-preservation, Nano-tannage, Nano-dying, Nanofinishing

Contents

1.   Introduction

Leather has been used as a clothing substance from ancient times. Being one of the oldest trades of mankind, nowadays leather is produced from hides and skins on a large scale, in a multibillion-dollar industry to make various goods including footwear, clothing, automobile seats, furniture, and different technical applications and so on. It is produced in a wide variety of types and styles and decorated by a wide range of techniques. Leather production is largely based on the utilization of the raw hides and skins which occur as a 'waste' product in the slaughtering of domesticated animals that are kept for meat. Leather production contributes to the socio-economic development of a country by earning currency for her. We live in the age of synthetic materials, plastics and metals, but leather has been sustained for its unique aesthetics, superior quality and exceptional properties.

Leather being a natural polymer is a durable and flexible material created from animal raw hides, mostly bovine hide. Raw hides and skins undergo several stages of processing, e.g., tanning, finishing, coating, dyeing etc. to become useable. In fact, the processing of hides and skins for making leather accompanies various defects either due to failure of controlling the factors of processing or due to low grade of raw material. One of the most common defects is the looseness in leather that has been causing around 5-10% of finished bovine leather produced currently to be deemed second-grade lower priced materials. Generally, the tanned leathers are not directly used and passed through different processes including re-tanning, filling, and coating the surface with dyes and other functional materials for protecting surface of leather from mechanical damage, moisture, microbial attack and so on.

The leather sector is one of the most environmental pollution creating sectors, the researchers have been focusing on sustainability centered and environment-friendly research goals and objectives for the leather industry. Therefore, over a decade and more now several sustainable technologies addressing each unit operation during leather processing have been developed and showcased. The main focus of this chapter is to showcase various sustainable technological options for leather processing operations in terms of utilization of nanomaterials.

2. Leather: History, importance and production

One of man's first and most useful inventions is leather. As part of a 'waste to worth' strategy to prevent animal parts from going to waste, our predecessors utilized leather to defend themselves. To make footwear, clothes, and crude tents out of the skins of the wild animals that they hunted for survival, primitive people smeared fatty substances to make the hides last longer. Leather was used for footwear, clothing, gloves, buckets, bottles, burial shrouds, and military gear as early as 5000 B.C., according to wall murals and objects found in Egyptian tombs. The development of tanning recipes utilizing particular tree bark and leaves soaked in water to preserve the leather is attributed to the ancient Greeks. Around 500 B.C., Greece developed a thriving market for vegetable-tanned leather, and this was the earliest mention of its existence [1]. Even in the present, leathers developed using vegetable tannins are still in demand. The Romans used leather extensively for their boots, clothing, and military equipment including shields, saddles and tack, and harnesses. Leather has been used for upholstery throughout the history of transportation and furniture due to its durability and comfort. Due to its ease of maintenance and resistance to food odor absorption, leather became the material of choice for dining chair covers during the middle ages. In the 18th and 19th centuries, as industrialization expanded, there was a growing need for new types of leather, such as belting leathers for propelling machines. The invention of the automobile, the need for softer, lighter footwear with a trendy appearance, and an overall improvement in living standards all contributed to the desire for soft, supple, vibrant leather. Because the original vegetable-tanned leather was too thick and hard for these uses, chromium salt was used instead, and chrome tanning evolved to be the standard in contemporary clothing, footwear, and upholstery leather. Modern technology initiated the key step for the evolution of science and technology providing technological support for the development of leather industry. Therefore, the appearance of leather, feel, and range of potential uses have all been considerably enhanced by the invention of chemicals and advanced processing techniques. Not only for commercial and household furnishings but also for automobile, aviation, and marine uses, leather is still the material of choice.

Leather is a non-putrescible material that is manufactured from spoilable hides and skins through tanning process. According to International Union of Leather Technologists and Chemist's Society, general terminology for hide or skin with its unique fibrous structure more or less fixed, tanned to be non-putrescible with the hair or wool may or may not be removed. Fur skin refers to a variety of skins that have been similarly dressed or treated but have not had their hair removed [1]. Leather is made from hide or skin that is segmented or divided into layers either before or after tanning. A tanned hide or skin that is mechanically, chemically, or both into fiber particles, tiny pieces, or powder and then

Applications of Emerging Nanomaterials and Nanotechnology   Materials Research Forum LLC
Materials Research Foundations 148 (2023) 170-199   https://doi.org/10.21741/9781644902554-6

formed into sheets or various forms with or without the addition of a binding agent is not leather. If the leather has a surface coating, the thickness of the applied layer cannot exceed 0.15 mm.

3.   Leather making process

3.1   Raw materials: Hides and skins

Hide refers to the skin of bigger animals (e.g., cowhide or horsehide), whereas skin refers to the skin of smaller animals (e.g., calfskin or kidskin). The method of preservation used is a chemical process known as tanning, which turns animal hides and skins into putrescible, stable materials known as leather [2]. Fish or animal oils, mineral salts like chromium sulfate, and vegetable tannins (from sources like tree bark) are all tanning agents. There are seven main categories from which the more common leathers come: cattle, including calves and oxen; sheep and lambs; goats and kids; equine animals, including horses, mules, and zebras; buffalo; pigs and hogs; and such aquatic animals as seals, walrus, whales, and alligators. The hide of mammals is made up of three layers: the epidermis, a corium or dermis-thin outer layer, a thick core layer, and a subcutaneous fatty layer. After removing the two sandwiching layers, the corium is used to produce leather. Fresh hides have a weight percentage of between 60 and 70 percent water and between 30 and 35 percent protein. Collagen, a fibrous protein bound together by chemical bonds, makes up around 85% of the protein [2]. Fundamentally, the science of creating leather involves dissolving lipids and non-fibrous proteins with the use of acids, bases, salts, enzymes, and tannins while also strengthening the links between the collagen fibers.

3.2   Collagen and skin structures

The term "collagen" refers to a family of at least 28 different collagen types, each of which has a distinctive purpose in animals, most notably as a component of connective tissues. In collagen-based biological tissues such as skin, leather and pericardium, multiple collagen fibrils bundle together in various bundle-sizes to form collagen fibers. The major component of skin is type I collagen: so, unless otherwise specified, the term collagen will always refer to type I collagen [3]. Collagen fibers can range in diameter from 30-300 nm and lengths up to the millimetre range depending on the number of fibrils present [4]. Collagens are fibrous proteins that are the polymer of amino acids: α- amino acids and β-amino acids. Protein molecules are formed through peptide link between terminal amino group and a terminal carboxyl group. When the amino acids are linked together to form proteins in fibrils (Figure 1), they create an axis or backbone to the polymer, from which the side chains extend. Most of the characteristics of any protein are determined by the content and distribution of its side chains. For collagen, it is the side chains that largely define its reactivity and its ability to be modified by the stabilizing reactions of tanning when leather is made [5]. Then fibrils bundle up to form fibers that are held in place through the existence of crosslinks. These can be either physical lysine-based crosslinks or proteoglycan crosslinks composed of glycosaminoglycans (GAGs) [6]. In nature, the cross-links between fibrils are provided by proteoglycan bridges. These proteoglycan bridge are mostly made of decoran, a proteoglycan that contains either the glycosaminoglycans, dermatan sulphate or chrondroitin sulphate [6,7].

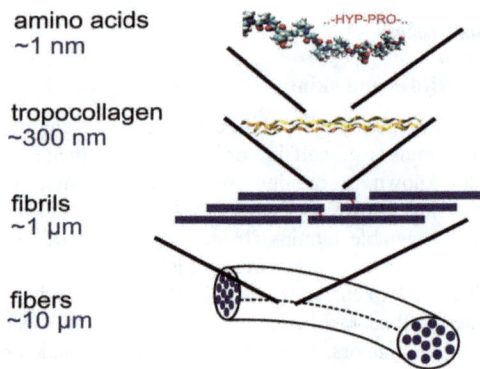

Figure 1: The hierarchical features of collagen, ranging from the amino acid sequence level at the nanoscale up to the scale of collagen fiber with length on the order of 10 [8].

3.3   Collagen–nanoparticle interaction

As collagen structure contains both carboxyl and amino group, the impregnated nanoparticles may interact with reactive functional groups of protein in collagen under favorable condition which is gravely related to the pH of leather and the isoelectric point (IEP) of tanned leather. Castaneda et al. showed that the carboxyl group present on the surface of Tiopronin (*N*-(2-mercaptopropionyl) glycine)-protected gold nanoparticles facilitated the formation of multiple cross-links with lysine moieties of collagen fibers, thus reducing the pore size from 140 to 1 μm which in turn improved stability and biocompatibility [9].

The naturally occurring holes in leather with the sizes ranging from micro (2 nm), meso (2-50 nm), to macro (>50 nm) tend to remain open in the acidic situation. Therefore, there is also a chance that solely physical deposition of nanoparticles will occur through the pores of collagen matrix [10]. Due to the collagen alienated fibers and layers of tape or cord-like shape and the fact that each of these fibers is made up of a bundle of fibrils with diameters ranging from 30 to 300 nm depending on where in the body they are located, such nano-sized particles can exist within the alienated fibers and layers without stiffening the collagen matrix [11]. Sangeetha et al. have studied the stability of nano-particulate interaction with collagen by observing metal oxide nanoparticles, preferably chromium (III) oxide nanoparticles, encapsulated in a polymeric matrix sized 50–70 nm which have been treated against collagen solution [12]. The triple helix structure of collagen that is unmodified and stable in its three-dimensional conformation has been demonstrated by a positive circular dichroism spectrum at a peak of about 220 nm and ratio of positive to negative peak intensity close to 0.1. The involvements of amino acids present in the collagen enhanced the cross-linking of the core–shell nanostructures through hydrophilic poly(acrylic acid) surface groups conferring thermomechanical stability.

The interaction of Ag nanoparticles placed on TiO_2 with collagen under simulated circumstances has been studied by Gaidau *et al.* utilizing Fourier transform infrared (FT-IR) spectroscopy, atomic absorption, and fluorescence spectroscopy. Although FT-IR analysis showed an alteration in the amide I band correlating to altered secondary collagen structure through hydrogen bonds in the carbonyl group, no collagen structural deformation has been seen during treatment with nanoparticle and/or chromium salt [13]. **Figure 2** illustrates the use of metal nanocolloids (Ag/Au/TiO_2, ZnO, etc.) and the effectiveness of their preservation through an antibacterial mechanism.

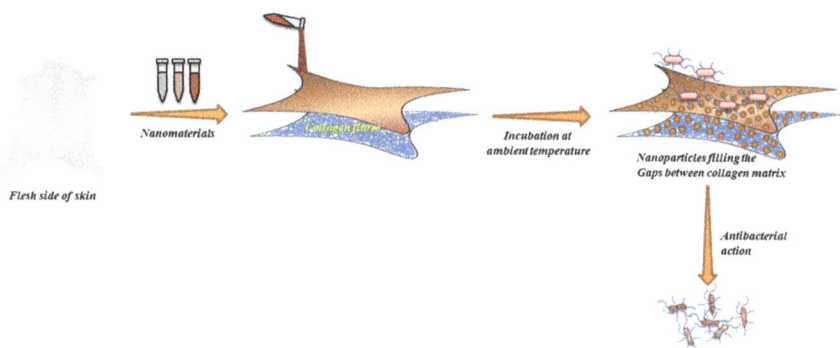

Figure 2: Application of metal nanocolloids (Ag/Au/TiO_2, ZnO, etc.) and their preservation efficacy demonstrated through antibacterial mechanism.

Studies have been carried out to investigate the conformational changes inside microbial cell during the utilization of nanomaterials in leather preservation [14,15]. Membrane permeability enables the uptake and accumulation of nanomaterials in bacterial cell membranes following electrostatic interaction with the microbial cell wall, leading to cell surface depression. Most of the ions produced from the nanoparticles have a strong affinity for thiol (−SH−) groups, which inactivates the enzyme activities and causes a significant proton leak that disrupts the cellular membrane. As the nanomaterial undergoes oxidation on its surface, ROS (H_2O_2, $\cdot$OH, $O_2{}^{\cdot-}$ etc.) are generated from the oxidative stress-induced nanomaterials, which eventually damage proteins and DNA, is responsible for antibacterial activity.

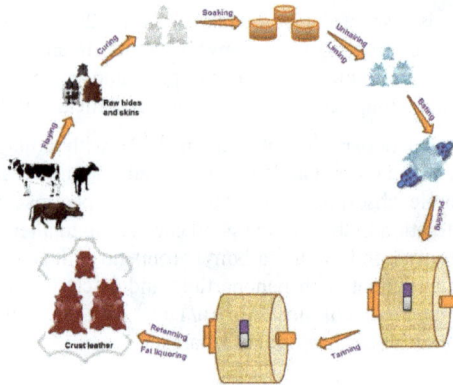

Figure 3: Steps involved in leather processing starting beam house operations till finished crust [16].

3.4. Leather preparation process– a bird's eye view

Steps involved in leather processing starting beam house operations till finished crust are shown in Figure 3. The operation flow is generally to follow the preparatory → tanning → crusting → surface coating sub-process order without deviation (see in Table 1). The preparatory stages are when the hides/skins are prepared for tanning [16]. In the preparatory stages, many of the unwanted components of raw hides/skins are removed. These operations also include chemical and mechanical treatment and processing respectively.

Table 1. Steps involved in processing of skins and hides to leather

Stages of Processing	Agents	Objectives
Beam house operations [2,17]		
Soaking	Water, Detergents, Soaking Enzymes, Biocides, Soda Ash, Polyphosphate	Removal of the curing salt in the case of salted skin Rehydrate the skin proteins.
Liming (unhairing and scudding)	Lime $(Ca(OH)_2)$, sodium sulfide, sodium hydroxide, sodium hydrosulfide, dimethyl amine	Removal of the hairs, nails, hooves and other keratinous matters. Removal of some interfibrillar soluble proteins and extraneous flesh. To swell and split up the fibers to the desired extent.

De-liming	Ammonium sulfates and ammonium chlorides	Removal of alkali from the pelt Removal of all swelling and plumping
Bating	Bating enzyme	Removal of all swelling and plumping Removal of scud or dirt, short hairs, grease and lime soaps, dark colored pigments to obtain clean, flat, slippery, non-elastic or flaccid pelt.
Pickling	Sulfuric acid, formic acid, hydrochloric acid and salt	The pelt is brought to the desired pH before tanning to facilitate even penetration of tanning agents
De-greasing	Lipases, detergents or solvents	Removal of grease and fat from the interfibrillary space
Tanning [3]		
Vegetable tanning	Chestnut, myrobalan, oak, tanoak, wattle, hemlock, mimosa, quebracho, and mangrove	Increasing the resistance to bacterial attack Increasing shrinkage temperature, hydrothermal stability and durability is increased
Mineral tanning	Chromium sulfate	
Post-tanning operations [18]	Water, tanning agents, fats/oils, waxes, dyes, filling agents etc.	Decreasing thickness, retanning, fatliquoring and dying the skins and hides to convert them into crust leather
Finishing	Pigment, lacquer, wax, emulsifier etc.	Levelness of color on the leather. Improving the light fastness, scuff resistance, heat fastness, fastness to acids and alkali etc. of the leather. Filling the gaps of loose leather to improve the break of the grain.

4. The advent of nanotechnology

A nanoparticle is a physicochemical structure that is larger than an atomic or molecular size but smaller than 100 nanometers [19]. Nanoparticles can have different shapes, such as being spherical or tube-shaped, at the nanoscale in one, two, or three dimensions, in a single, combined, or accumulating form [20]. Because of their charge, nanoparticles have greater reactivity toward living cells and proteins [21]. Nearly all inorganic and organic nanoparticles are solubilized by the pH of living systems, whereas metals, once within the cell, accumulate and increase oxidative stress [22]. Based on these factors, researches have been carried out on nanotechnological interventions to address some of the prevalent problems with leather processing and environmental effects for a cleaner leather manufacturing in the future. Some of the nanotechnology-based patented products for leather application are shown in Table 2.

Table 2. Nanotechnology-based patents filed in leather applications

Product	Product Specifications	Country	Patent ID/year
Leather nanosilver	The natural leather is evenly coated with a high concentration of colloidal silver (Ag). The antibacterial characteristics of the fabric are retained for a long time on both the outer and inner sides. The permanent protection including scratch resistance, superhydrophobic effect, weather-, UV-, and temperature resistance (withstanding 750 °C) is provided by SiO_2 nanocolloids of diameter of 2 nm.	South Korea	KR10075 0196B1/2 006
Leather nanopigment paste	Nanoscale versions of TiO_2, ZnO, and SiO_2 are present in nanopigment paste. This nanometer-based paste exhibits antibacterial, UV radiation resistant, mildewproof, and flame retardant characteristics, and is used to coat leather for its durability.	China	CN10141 2869B/20 08
Polyacrylate/nan ometer ZnO composite finishing agent	This finishing product is appropriate for base coating on instep, corrected, and garment leathers. The benefits include increased elongation at break, water vapor permeability, air permeability, and water tolerance.	China	CN10230 4316B/20 11
Nanoparticle-coated genuine leather	This genuine leather has leather pigment and a nanopowder comprised of a mixture of nanosilver and sub-nanometer SiO_2. This has strong resistance to wear and tear, wetness, and antibacterial activity.	United States	US20130 078451A 1/2013
Polyacrylate/nan o-ZnO composite leather finishing Agent	The polyacrylate/nano-ZnO composite emulsion made by the in situ approach has the drawback of having low polymerization stability, which is fixed by this method of preparation. Therefore, water resistance and tensile strength are enhanced.	China	CN10311 3804B/20 13

Antibacterial casein-based nano-ZnO composite leather finishing agent	For the purpose of creating a finishing agent for leather, nano-ZnO is added to the casein system. Excellent antibacterial qualities are displayed by this inorganic/casein leather finishing product.	China	CN10483 0231A/20 15
Graphene/alumin um oxide nanocomposite tanning agent	The shrinkage temperature is raised to 112 °C with 95% absorption while using a tanning agent based on a graphene/aluminum oxide nanocomposite at 5%. Like chrome-tanned leather, it has similar levels of softness, elasticity, and fullness.	China	CN10474 5736A/20 15
Leather product cleaning agent	A self-cleaning leather product made of anatase-type nanotitanium dioxide, acrylic resin, lemon juice, sodium silicate, etc. It simultaneously works as an excellent antibacterial agent and a crack filler.	China	CN10636 7233A/20 17
Semiconductor-oxides nanotubes-based composite for dye-removal	This novel technique uses an ion-exchange mechanism that operates in aqueous solution and in the dark. On the surface of fly ash and metal oxide nanoparticles, semiconductor oxide nanotubes (hydrogen titanate, anatase TiO_2, and Ag- doped anatase TiO_2) are deposited. Through surface-adsorption, this technology eliminates organic and synthetic dye from wastewater.	United States	US20180 290135 A1/2018
Poly-octavinyl polyhedral oligomeric silsesquioxane–acrylic acid nanocomposite auxiliary agent	Octavinyl silsesquioxane, an organic solvent, ammonium persulfate, distilled water, and a surfactant are combined in a cocktail to fill in the gaps between collagen fibers and facilitates easy penetration. Ideal for better filling and minimizing the leakage of chromium content in water bodies.	United States	US 20190284 650 A1/2019

4.1   Nano-preservation

Through a difference in osmotic pressure, salt (NaCl) curing eliminates water from hides and skins, making the environment unfavorable for bacterial development. Although the salt curing process is efficient and affordable, it suffers greatly from an environmental standpoint due to high TDS (> 2100 mg/L), which accounts for 40% of the effluents, and chloride content, which accounts for 55% [23]. Nanomaterials, such as polymers, metals, bimetals, and graphene oxide, have been the subject of many studies into more environmentally friendly curing techniques [24]. These materials not only have antimicrobial capabilities but may also act as transporters for antimicrobial additives and chemicals. The most commonly used nanomaterials with their antimicrobial mechanism are shown in Table 3.

Table 3. Nanomaterials with their antimicrobial mechanism.

Nanomaterials	Antimicrobial mechanism	Refs
Silica (SiO_2)	Creating an environment that is adverse for bacterial growth Minimizing bacterial adherence and growth	[25]
Calcium oxide (CaO)	Strong alkalinity Reactive oxygen species formation (ROS) Cell membrane degeneration and cellular content leaking	[27]
Silver (Ag)	Effect of oligodynamic damage to the cell membrane Interfering with protein disulfide linkages Generating free radicals Genotoxic effects cause DNA damage	[28]
Magnesium oxide (MgO)	Availability of oxygen gaps Damage to cell membrane Peroxidation of lipids	[28]
Aluminum oxide (Al_2O_3)	Disruption of the bacterial cell wall ROS formation Cell wall intrusion and accumulation	[29]
Copper oxide (CuO)	Easy migration across the cell membrane Degradation of the bacterial enzyme system	[30]
Gold (Au)	Photothermal impact and ROS production DNA damage and bacterial cell wall disruption Osmotic imbalance and cell wall integrity disturbance	[31]
Iron oxide (Fe_3O_4)	Cell membrane permeabilization Electron transfer interference	[32]

Titanium dioxide (TiO$_2$)	Peroxidation of lipids Damage to the cell membrane DNA alterations	[33]
Zinc oxide (ZnO)	Decreased microbial viability H$_2$O$_2$ formation Accumulation on bacterial membrane through electrostatic attraction ROS generation	[34]

4.2 Nano-unhairing

One of the most crucial phases in the production of leather is liming, which involves the removal of non-collagenous components such as hairs, flesh, and interfibrillar proteins etc. Chemical usage in this stage as a whole accounts for 50–60% of the TDS, COD, and BOD in the effluent [35]. This advocated for a paradigm change in which the dehairing process would employ enzymes rather than lime and sodium sulfide [36].

In a research, enzymatic fiber opening and unhairing have been accomplished using an enzyme cocktail immobilized on metal oxide nanoparticles (nanozyme). The results have been contrasted with those obtained using a control cocktail of amylase and protease (fibrozyme). Based on the raw skin weight, the enzyme level has been kept at 3% where drumming has been chosen as the application methodology. To evaluate the effectiveness of the nanozyme for unhairing and fiber opening, the treated leather samples were assessed, showing satisfactory results within 7 hours and fiber opening in skins without any adverse effects attributing high membrane permeability in comparison to conventional leathers as shown by the organoleptic and histological studies [37].

In a different study, the protease-amylase enzyme combination has been immobilized using CuO nanoparticles to facilitate the dehairing and fiber opening processes. The buffering effect of CuO stopped the activity of protease antagonistic toward amylase [38]. Recently, the potentiality of enzyme immobilized on ZnO has been investigated. A 1.0% concentration of enzyme-immobilized nanoparticles has been optimized for hair removal and contrasted with the traditional sodium sulfide approach. The use of a nano-biocatalyst has been seen to reduce the time needed for matching efficiencies and also did not cause any deformation of the fibers, as shown by histopathology data. The study therefore suggests that the prospective application of enzyme-immobilized nanoparticles in the production of leather may, in theory, be a sustainable approach on account of reduction in emission in terms of COD, BOD and chromium content in spent liquor [39].

4.3 Nano-tannage

During the process of tanning, cross-linking causes a permanent change in the collagen matrix making the leather more durable. In comparison to other tanning techniques, chrome tannage still produces leather from 90% of skins and hides worldwide [40]. Aside from maintaining the versatility of chrome tanning, the leather industry has recently been challenged by the environmental effects and is in need of an alternative sound eco-benevolent method. Secondly, the substantial chemical absorption (including 60% chromium) and hexavalent chromium release in the effluent are creating a notwithstanding environmental danger. The paradigm shifting toward the implications of nanotechnology is necessary for cleaner leather manufacturing.

To increase chromium absorption by lowering its consumption, nanomaterials are making their way into the leather industry. In a similar manner, the microemulsion polymerization method has been used to effectively create an acrylic retanning emulsion from methyl methacrylate, n-butyl acrylate, and methacrylic acid. They have the ability to penetrate the skins and fill the interfibrillar gaps with these 100 nm-sized nanoscale polymers. Because of the higher surface area to volume ratio, proton transfer might create stable complexes with either collagen or chromium. The shrinkage temperature in wet blue goat skins might rise to 118°C using this nanotechnology-based method [41]. There has been a higher risk of denaturing the retanned leather due to the excessive aggregation of the emulsifiers. In order to avoid this, semicontinuous microemulsion copolymerization has been employed to reduce the size in the16–64 nm which led to the improvement of tensile and tear strength by 68% and 44%, respectively, in retanned pig bottom splits after treatment with 2% dose [42]. Additionally, a montmorillonite nanocomposite produced by in situ polymerization of an aldehyde-acid copolymer has been used in the tanning of leather. They have shown exceptional dispersion along collagen fibers. The amino groups of the collagen side chains have been assumed to be connected to the carboxyl and aldehyde groups in the nanocomposite via electrovalent and covalent interactions, respectively. In the inside of the chrome complex, this combinatorial technique and tanning agents (2% chrome) build a tightly knit network that eventually cross-links with the collagen fibers to impart stability. The temperature at which the tanned sheep skin shrank significantly increased to 90°C [44, 45].

On the other hand, coordinate covalent bonds are used in chrome tanning to create a cross-link between bi-/polynuclear chromium ions and the carboxylate side chains of collagen fibers [46-48]. Similar to this, SiO_2 nanoparticles have been employed to create a vinyl polymer-based nanocomposite with a mean size of 10 nm, which has been then used as a tanning agent. They use an appropriate coupling agent on SiO_2 to operate at the interface, separating the organic from the inorganic phase. These nanostructures played a key role in filling the interfibrillar gap, which is made possible by hydrogen bonds generated between the hydroxyl groups on the surface and the carboxylate/amino groups of collagen. Through extensive cross-linking with the collagen strands, these interactions have provided hydrothermal stability [49]. In addition, in situ nanohybridization using tetrabutyl titanate or tetraethoxysilane as a precursor has been suggested for efficient tanning. In order to easily penetrate the skins or hides, these precursor molecules are first mixed with a polymer or oil-based supporter. Inorganic particles like SiO_2 and TiO_2 are produced in situ as a result of the subsequent reduction of precursors under specified triggering circumstances

Applications of Emerging Nanomaterials and Nanotechnology   Materials Research Forum LLC
Materials Research Foundations 148 (2023) 170-199   https://doi.org/10.21741/9781644902554-6

(radiation, pH, heat, etc.). Critical triggering determines homogeneity and size reduction (90 nm), allowing for simpler penetration through skins and hides. These highly reactive nanoparticles show a high level of selectivity for collagen fiber cross-linking [50]. It is believed that hydrolyzed and condensed tetraethoxysilane interacts with pendant groups of arginine, histidine, tryptophane, and hydroxyl groups of collagen to generate -Si-C- bonds and hydrogen bonds (with C=O and N-H) that provide stability and raise the shrinkage temperature to 95 °C [50].

Gao et al. [51] have studied the diallyl dimethyl ammonium chloride copolymer-mediated production of SiO_2 nanoparticles for tanning operations which cut down chromium utilization to 2% when used in conjunction with polymer/nanocomposite. The intramolecular bonding that confers stability between the free -NH_3 group of nano-SiO_2 and the -COOH of the polymer was discovered by spectroscopic research. Diallyldimethyl ammonium chloride-acrylamide-acrylic acid with nanosilicon dioxide nanocomposite has been used to tan the leather. This process leads to high levels of hydrothermal stability, tensile strength, and flexibility of leather. According to Li et al. [52], ZnO nanocomposite may be produced using vinyl polymers and used as a tanning agent in the leather sector. Free radical polymerization has been used to create the polymer matrix, and ultrasonic processing has been employed to create the polymer-ZnO nanocomposite. They have used a polymer/ZnO nanocomposite with a commercial-grade tanning chemical, chromate B (33% basicity), to study the effectiveness of chromium absorption in the tanning process of suede leather. Studies have shown that the tanning and retanning processes significantly absorbed chromium, lowering the chromium concentration in the effluent with a BOD: COD ratio of 0.37. The leather treated has shown excellent hydrothermal stability, biodegradability, softness, and cleaner approach-related improvements [53].

A simple approach has been followed to create surface modified spherical SiO_2 nanoparticles from tetraethyl silicate [54]. Modifiers with hydrophobic, amphiphilic, and hydrophilic qualities were utilized, including poly (methacrylic acid), dimethyl diallyl ammonium chloride, and methacryloxy(propyl) trimethoxysilane. In comparison to pure polymer-mediated processes, the tanning appropriateness of surface-modified SiO_2 demonstrated improved tanning efficacy. Agitation in the spinning drum made it easier for SiO_2 to enter the pores of the sheep skin due to their tiny size. After being absorbed into the skin, the nanoparticles-polymer conjugate combines with the tanning agent to produce a complex and creates cross-links with the collagen fiber network. The shrinkage temperature of the wet white sheep skin is raised to 76 °C by this thick cross-link, which also boosts thermal stability. By using this method of surface modification, SiO_2 nanoparticles with great dispersibility may penetrate and fill in between collagen fibers, enhancing the physical and mechanical qualities of leather.

To further boost chromium absorption and decrease chromium release in cowhide shoe top leather, nanocomposites have been utilized as pre-tanning agents. An effective method for chromium reduction has been put forth by Lyu et al. [55] via the production of polymer-ZnO nanocomposites. The amount of chromium as a pretanning agent has been decreased from 6 to 4%. Moreover, the chromium content in the leather dramatically rose from 13,489 to 15,030 mg kg^{-1}; this led to a decrease in the concentration in the effluent, which is what these nanostructures' effective absorption is attributed to. When 2% of polymer/nanocomposites and 4% basic chromium sulfate have been mixed, the shrinkage temperature (Ts) is significantly increased to 99.8 °C and is only marginally higher than the Ts (99.9 °C) of the sample treated using the standard approach by employing 6% basic chromium sulfate.

For even more effective chromium distribution in situ, self-assembling mineralized nanoparticles loaded with chromium have been created which remain stable at high salt concentrations and pH levels greater than 3 and can be efficiently administered into pickled pelt without interference with collagen bundles during penetration. When the pH is lowered to 2.5 inside the pelt, the chromium core ($Cr(OH)_3$) from the nanocomposite complex disintegrates, releasing Cr^{3+} ions that are then evenly distributed throughout the pelt. Higher thermal stability and durability are given by the cross-linking of the liberated Cr^{3+} ions during the basification process with the -COOH groups on collagen fibers and copolymers [56].

Beyond its potential use in tissue engineering and medication administration, laponite, a synthetic form of smectite clay, is presently used as a tanning ingredient in the leather industry. Using nanoparticles of laponite clay and tetrakis(hydroxymethyl)phosphonium sulfate (THPS), a unique wet-white tanning product has been created. By taking this strategy, the risk of formaldehyde has been marginally minimized, which has improved leather's functionality. Notably, the Ts is over 85 °C in this system of THPS (2.5%-laponite (3%) at pH 4.5. What is more intriguing is that these laponite nanoparticles have been discovered evenly distributed and bonded between the collagen fibers without changing the collagen's natural structure. According to reports, this unusual mixture has better strength, light fastness, and resistance to yellowing [57].

There have lately been suggestions for high-exhaustion chrome tanning and chromeless tanning. In brief, chromium (III)-loaded nanoparticles have been employed to self-assemble poly (ethylene glycol) methyl ether acrylateco-acrylic acid-co-glycidyl methacrylate polymers to be used as chrome-free tanning agents. This core-shell approach stops Cr^{3+} from interacting with the pelt during penetration while yet successfully delivering Cr^{3+} inside the pelt. Following basification, the liberated Cr^{3+} forms cross-links with the polymer and collagen fibrils via the -COOH group, producing high exhaustion chrome tanning effect. In addition, the synergistic impact of chrome and organic tanning significantly lowered the utilization of chromium [58].

4.4   Retanning / Looseness removal

Studies to improve chromium collagen cross-linking have been carried out by pre-treating complexing agents (sodium formate and disodium phthalate) with nanoclay in order to promote uniformity across the collagen matrix of bovine hide (sodium montmorillonite). The molecular level insights can improve the measurements to indicate a route to more sustainable technologies [59].

A simple soaking approach in an aqueous solution with a pH of 5.7 may efficiently penetrate citrate-stabilized MnO_2 nanoparticles into the loose collagen matrix (**Figure 4**), therefore reducing the looseness of the leather matrix. Strong electrostatic interactions between MnO_2 nanoparticles and citrate ions result in a coordination complex that forms cross-links with functional protein groups of collagen, changing the collagen's loose architecture as seen by FT-IR spectra [60].

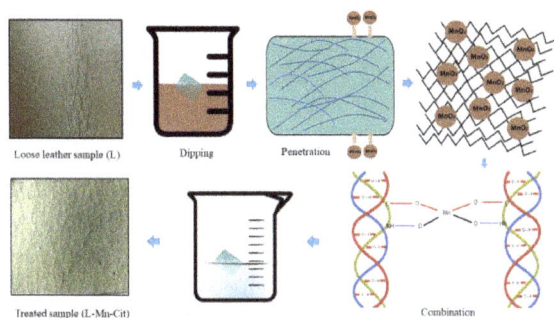

Figure 4: Incorporation and possible binding of MnO_2 nanoparticles with functional group of protein of collagen fiber of leather [60].

4.5   Nano-dying

Protein-based nanoparticle polymers in post-tanning trialsare found to show a considerable dye absorption, which has been supported by modeling using estimations of binding energy and percent dye depletion [61]. For a sustainable leather dyeing procedure, encapsulated nanoparticles composed of polymeric material with homogenized particle size and simple functionalization have been utilized. The copolymer has 50 nm-sized particles and a pH range of 5-7. This is actually an effective dye absorption and fixing during the leather-dying process. This process is both economically and environmentally feasible because the amount of dye is reduced, and auxiliaries are avoided [62].

In order to dye leather, the usage of an organic cationic nano-based colorant that has been emulsified in silica matrices has been investigated. The resistance of dye to extreme optical and thermal conditions has been noted. The dyeing technique is eco-friendly since the colorant enters and fixes into the leather matrix on its own without the need of any auxiliary [63].

4.6   Functional leather finishes

Finishing that is one of the most important steps in the leather manufacturing is the process of crafting coatings on leather surface in the leather business. Pigments, binders, auxiliaries, and other materials are applied to the surface of the leather during this procedure. Finishing compounds may also easily provide leather with superior mechanical characteristics in addition to improving its look [64-68]. Currently, the use of nanotechnology in finished products aids in the development and improvement of intelligent leathers with distinctive characteristics and properties, such as antibacterial properties, self-cleaning abilities, water resistance, flame retardant abilities, and more [69-76].

4.6.1 Antibacterial leather finishes

The wear-feeling of clothing may be impacted by the rapid growth of germs and fungus that occurs when clothing, for example, garment leather is in touch with the body for an extended period of time. $Ag-TiO_2$ and $Ag-N-TiO_2$ nanoparticles have been encapsulated in binders and used in leather finishing [77] where under visible light irradiation, the final leather demonstrated outstanding antibacterial and self-cleaning characteristics. An anionic polymer called PA30 has been chosen to alter ZnO in a polyacrylate-based ZnO nanocomposite [78]. Improved antibacterial properties have been added to the final leather by capping PA30 on ZnO surface.

ZnO nanoparticles of various shapes have been added to the polyacrylate binder [79]. When a composite emulsion comprising hollow, columnar-shaped ZnO has been applied to leather, the end product displayed increased hygienic and antibacterial characteristics. Different structural ZnO nanoparticles have been added into polyacrylate emulsion in order to study the impact of ZnO nanoparticle structure on the performance of composite [80]. The results showed that adding sphere- and flower-shaped ZnO nanoparticles might provide the composite films with improved antibacterial characteristics. ZnO nanoparticles that resemble flowers have been added, increasing the water vapor permeability of the resulting film by 122.17%. In contrast, adding sphere-shaped ZnO nanoparticles to polyacrylate film is advantageous for improving its mechanical characteristics. The twofold in situ approach has also been used to create a TiO_2 composite emulsion based on polyacrylate. Nano TiO_2 has been added to the polyacrylate coating to enhance its antibacterial characteristics and thermal stability [81]. The natural protein casein is prone to bacterial development, therefore, in a recent research, an in situ approach has been employed to manufacture a casein-based ZnO nanocomposite to address this issue [82]. SEM images of a composite film based on casein revealed presence of ZnO nanoparticles on the surface and in the cross-section demonstrating exceptional antibacterial power against *S. aureus* and *Escherichia coli*.

4.6.2 Water resistant leather finishes

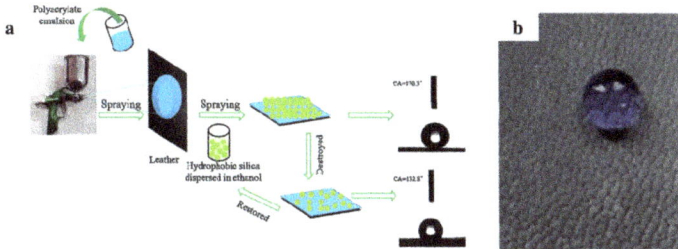

Figure 5: Fabrication of superhydrophobic leather coating via spraying method (a); Optical image of water droplet on the superhydrophobic leather coating surface (b) [86].

Since collagen-based materials include several hydrophilic groups such as -NH₂, -COOH, and -OH, leather can be referred to as a naturally hydrophilic substance. Leather goods are readily damaged by bacteria or water. The aforementioned problems can be prevented if the leather product is water resistant. In order to polish leather, Cu nanoparticles have been added to a regularly used leather finishing chemical. On the base and top layer, a leather finishing compound containing Cu nanoparticles was applied. Because Cu nanoparticles are rather hydrophobic in comparison to other finishing chemicals, adding them to leather can give it good wet and rub fastness as well as color fastness to water [83]. In order to polish leather, an acrylic resin-based SiO_2 emulsion polymerization emulsifier-free emulsion polymerization process has been adopted [84] which showed nanocomposite-finished leather absorbed water at a rate that has been comparatively 17.89% higher. Acrylic resin-based SiO_2 composite for polishing leather that has been made by mixing nano-SiO_2 with acrylic resin has been used [85]. Meanwhile, in 2015, super hydrophobic coatings on leather surfaces have been created using a simple layer-by-layer (LBL) spraying technique [86]. As shown in Figure 5, the coating is created by simply spraying hydrophobic silica nanoparticles and polyacrylate emulsion over a leather surface. By adjusting the amount of SiO_2 nanoparticles sprayed on coated leather surfaces, the hydrophobicity of the surface may be adjusted. Notably, re-spraying the coated leather surface with a hydrophobic SiO_2 dispersion helps repair cracks or scratches. This technique had great potential for practical applications since it is easy to use and affordable.

4.6.3 Self-cleaning leather finishes

In everyday living, dirt may readily damage a clothing's surface. Leather garments cannot be cleaned directly like clothing made of other materials. Intelligent surfaces have been created to change leather surfaces in order to solve this problem. Self-cleaning coatings for leather surfaces have recently been researched. In this industry, creating superhydrophilic coatings that repel stains is a popular technique. As is well known, TiO_2 is a strong choice for creating superhydrophilic self-cleaning coatings due to its inherent hydrophilicity and unique photocatalytic capabilities. TiO_2 nanoparticles doped with N and Fe have been created in research of Petica. Under visible light, the produced nanoparticles display enhanced photocatalytic activity [87].

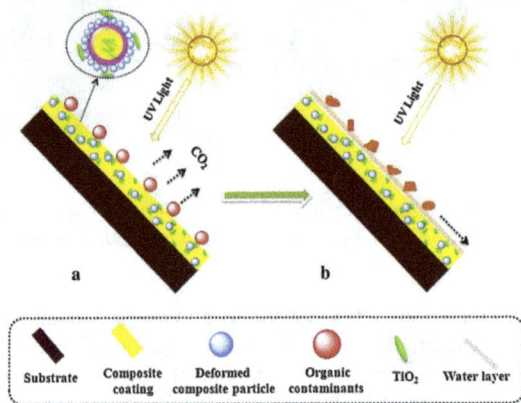

Figure 6: self-cleaning behavior of casein-based TiO_2 composite coatings in the initial (a) and intermediate (b) stage [89].

Doped TiO_2 nanoparticles have been encapsulated in an acrylic binder and used to polish leather. Methylene blue (MB) clearly degraded when leather treated with Fe-N co-doped TiO_2 nanoparticles have been exposed to visible light. MB spots and ballpoint pen ink lines on the leather surface have been practically undetectable after 30 hours of exposure to visible light. Tests on contact angle of water is revealed that the deterioration of surface of leather is mostly caused by the photo-induced hydrophilicity process. Additionally, silica-doped TiO_2 nanoparticles have been employed for polishing leather [88]. Due to the hydrophilicity of the leather surface, leather treated with the composites exhibits evident photocatalytic capabilities against MB during UV and visible light exposure.

Casein integrated TiO_2 nanocomposite film prepared from polyacrylate and commercially available TiO_2 nanoparticles exhibit an efficient self-cleaning ability to stains including, dye, coffee, red wine and oil as potential functional coatings on various substrates [89]. The self-cleaning behavior is schematically shown in Figure 6. TiO_2 may produce electron-hole pairs under UV illumination, which can then transfer to the surface of the fabric. The formed hole then interacts with the adsorbed water or hydroxyl ion to produce hydroxyl radicals. These hydroxyl radicals are known as the most potent oxidant and are what causes the photocatalytic breakdown of organic stains.

4.6.4 Other functional leather finishes

Studies have been initiated to develop other functional leather to provide a continuous performance such as resistance to yellowing, high water vapor permeability, fragrance/mildew preventive sustained releasing properties, etc. To improve the water vapor permeability of polished leather, hollow silica spheres have been mixed with a polymer binder and applied to the surface of leather [90, 91]. The hollow size and shell thickness of hollow silica spheres significantly contributes to the improvement of the permeability of water vapor inside the finished leather. The insertion of hollow silica spheres has improved water vapor permeability and may provide a more free area for the composite film. Furthermore, because ZnO nanoparticles may absorb UV radiation and decrease its effects on polymeric coatings and leather itself, polyacrylate/ZnO nanocomposite has been employed as a leather finish. In particular, it helps keep leather goods with bright colors from yellowing [82].

Chitosan-coated silica nanocapsules with a double-shelled structure were added to casein binder as leather finishes in order to provide prolonged aromatic coatings [92, 93]. The polished leather has persistent scent release patterns. In the meanwhile, to give leather long-term mildew resistance, casein-based silica hollow spheres have been employed as a carrier for a mildew preventative and then sprayed onto the leather surface [94]. Encapsulation techniques, however, can lead to expensive and labor-intensive processes. As a result, a simple LBL spraying technique has been used to create distinct sustained aromatic coatings [95] where the fragrance is coated by a chitosan and SiO_2 layer that prevented its quick dispersion. Due to its ease of use, this LBL spraying technique shows great potential in the creation of practical leather finishes.

Methyl methacrylate and 2-ethyl hexyl acrylate have been used to make polyacrylic resin using microemulsion polymerization, which has then been followed by the sol-gel method to add SiO_2. These nano-formulations and acrylic resins have been considered in various combinations to coat the surface of leather. Utilizing SiO_2 improves the tensile, tear strength, and elongation percentage of leather by rising to 3%. As a result, durable, green leather finishing technique was developed [96].

As water-based polyurethane formulations proved to be both practical and secure for the leather industry and its workers it was used for leather finishing operations instead of organic solvent-based ones. Then, 1–5% SiO_2 nanoparticles were added to the polyurethane binder [97]. Similar to the earlier study, TiO_2- SiO_2 nanocomposite has been used as the finishing coat on leathers. The physical-mechanical and fastness features, such as dry/wet rubbing fastness, color fastness under UV light, and finish adhesion of the leathers finished with the TiO_2- SiO_2 nanocomposite as top coat, are enhanced compared to conventionally treated leathers [98]. ZnO has been applied to the top layer of leathers with cationic finishes in order to improve their fastness properties. When the ZnO nanoparticle dose was up to 24 g/L during the season, it has been demonstrated that wet/dry rub fastness and flexing resistance significantly improved [99].When polishing leather, copper nanoparticles have been employed in the base and top coats. Wet and rub resistance, color stability in water, and adhesive strength of leather have been significantly increased [100].

5.   Advantages and disadvantages of nanomaterials

The use of nanomaterials in leather processing provides a number of benefits, including an improvement in the desired leather product's stability, quality, and durability. Without the application of severe osmotic pressures, metal oxide nanoparticles with magnetic characteristics (e.g., Fe_3O_4) have been combined with enzymes to be employed for skin unhairing and fiber opening [37].

Secondly, to eradicate the bacteria that cause skin putrefaction, Ag, ZnO, CuO, etc. with broad-spectrum antimicrobial activity may be employed [75, 101]. Nanomaterials play a significant role in attributing self-cleaning capabilities to the finished products.

Nanocomposites have been initially used in leather processing as tanning agents, and reports concerning its use in leather finishing only began in the early twenty-first century [102]. They increased the mechanical and hygienic properties of the finished leather by incorporating nano TiO_2 into polyacrylate emulsion and applying the composite emulsion to the leather surface. Since then, several polymer matrices have been functionalized with nanoparticles, especially Ag, TiO_2, ZnO, SiO_2, graphene oxide, etc., to impart heat and chemical resistance, block UV radiation, and extend the lifespan of the product. The antibacterial property that most nanoparticles possess stops the growth of germs and fungus when they come into contact with leather. Additionally, antimicrobial activity of nano Ag slows the spread of infections and shields the body against odor, itching, and sores. TiO_2 and SiO_2 displaying hydrophobicity have been created employing novel nanotechnological implications in the production of self-cleaning leathers [103, 104].

Despite the fact that nanotechnology, its foray into the leather sector, and its use on an industrial scale are yet in their infancy, there are several restrictions that prevent its widespread implementation. For instance, the large-scale synthesis and post-processing of nanoparticles utilized in various leather production processes are required. Second, regular inspection is required for cleanup and remediation procedures owing to the toxicity risk caused by the reduced size and altered physicochemical characteristics. While using the ideal solvent system and natural systems, the 'green synthesis' approach, on the contrary, prevents the generation of undesirable or dangerous by-products, making it a dependable, sustainable, and environmentally friendly method [105].

Organic-inorganic nanocomposite has received a lot of attention up to this point and exhibits exceptional performance for practical leather treatments. Functional organic-inorganic nanocomposite-based leather finishes are, however, the subject of very few studies, particularly in the leather sector. The following list of problems includes a few that still require further attention. First and foremost, the stability of the organic-inorganic nanocomposite is crucial for delivering the desired performance. Before being used in industry, inorganic nanoparticles can still aggregate when being stored or transported in a water-based polymer binder. In addition to hydrophilic surface treatments for nanoparticles, this issue can be solved by looking for innovative synthesis techniques to create nanocomposites with improved compatibility between inorganic and organic phases. In-depth and systematic research on the interactions at the interfaces between various phases should also receive significant focus.

Secondly, according to recent research advances, there are just a few options for polymers or nanoparticles, which makes it difficult to create leather finishes with a variety of uses. As different sizes of nanoparticles exhibit different characteristics, researches into additional inorganic particles or polymer binder is necessary. For instance, while developing green leather finishes, natural polymers like chitosan, zein, or others should be attempted. Quantum dot materials can be used to create innovative finishing agents for inorganic particles, providing leather goods with a unique feature.

Finally, there has to be more research done on the impact of nanocomposite microstructure on the performance of final leather. The consequences on the functionality of polished leather may be complex because leather is a natural substance. In order to explore the microstructure changes of leather surfaces before and after finishing, a number of advanced characterization procedures should be employed. In the meanwhile, the interaction between nanocomposite and leather surface may be studied using computer modeling techniques, which may make it easier to identify a link between the microstructure of nanocomposites and the functionality of final leather. The hefty cost of manufacturing nanomaterials is another obstacle that prevents wide-scale use of these materials.

Conclusions

Despite being attractive and interesting, the manufacturing procedures for leather raise major environmental concerns. It is high time to create and suggest alternative nanotechnological techniques for processing leather. Without a doubt, nanotechnology is the most sophisticated and futuristic cutting-edge technology in the near future, perhaps influencing nearly all operations. The industries would eventually move toward a high-efficient, cost-effective, and ecologically sustainable future, which would open up prospects for creative and competitive goods. As huge investment is needed in order to make this nanotechnology compatible with leather production, researchers from various branches of science and engineering must concentrate their efforts in order to address the inevitable related problems described above to pave way for nanomaterials toward their practical applications.

References

[1] T. C. Thorstensen, Practical Leather Technology, Third edition (1985).

[2] S. S. Dutta, An introduction to the principles of leather manufacture, Fourth edition (1999).

[3] A. Covington, Modern tanning chemistry, Chem Soc Rev 26(1997)111–126. https ://doi.org/10.1039/cs997 26001 11

[4] B. R. Williams, R. A. Gelman, D. C. Poppke, K. A. Piez, Collagen fibril formation, J Bio Chem 253(1978)6578–6585.

[5] M. M. Taylor, M. B. Medina, J. Lee, L. P. Bumanlag, N. P. Latona, E. M. Brown, C. K. Liu, Treatment of hides with tara-modified protein products, J Am Leather Chem Assoc 108(2013)438–444.

[6] A. Gautieri, S. Vesentini, A. Redaelli, M. J. Buehler, Hierarchical structure and nanomechanics of collagen microfibrils from the atomistic scale up. Nano Lett 11(2011)757–766. https://doi.org/10.1021/nl103943u

[7] Leather Facts, 3rd edition, New England Tanners Club, Peabody, MA, (1994)21.

[8] J. M. Buehler, Nature designs tough collagen, explaining the nanostructure of collagen fibrils, PNAS 103(33)(2006)12285–12290. https://doi.org/10.1073/pnas.0603216103

[9] L. Castaneda, J. Valle, N. Yang, S. Pluskat, K. Slowinska, Collagen cross-linking with Au nanoparticles, Biomacromol 9(2008)3383–3388. https://doi.org/10.1021/bm800793z

[10] Y. Li, B. Wang, Z. Li, L. Li, Variation of pore structure of organosilicone-modified skin collagen matrix, J Appl Polym Sci 134(2017)44831 https://doi.org/10.1002/app.44831

[11] T. Ushiki, The three-dimensional ultrastructure of the collagen fibers, reticular fibers and elastic fibers: a review, Kaibogaku Zasshi 67(1992)186–199.

[12] S. Sangeetha, U. Ramamoorthy, K. J. Sreeram, B. U. Nair, Enhancing collagen stability through nanostructures containing chromium (III) oxide, Colloids Surf B 100(2012)36–41. https ://doi.org/10.1016/j. colsu rfb.2012.05.015

[13] C. Gaidau, M. Guirginca, T. Dragomir, A. Petica, W. Chen, Study of collagen and leather functionalization by using metallic nanoparticles. J Optoelectron Adv Mater 12(10)(2010)2157–2163.

[14] R. Wahab, I. Hwang, H. S. Shin, Y. S. Kim, J. Musarrat, M. Siddiqui, In: Tiwari A, Mishra AK, Kobayashi H, Turner APF (eds) Intelligent nanomaterials: processes, properties, and applications, Scrivener Publishing LLC, Salem, (2012)183–212.

[15] C. You, C. Han, X. Wang, et al. The progress of silver nanoparticles in the antibacterial mechanism, clinical application and cytotoxicity, Mol Biol Rep 39(2012)9193–9201. https ://doi.org/10.1007/ s1103 3-012-1792-8

[16] C. F. Carter, H. Lange, S. V. Ley, I. R. Baxendale, B. Wittkamp, J. G. Goode, N. L. Gaunt, React IR Flow Cell: A new analytical tool for continuous flow chemical processing, Organic Process Research 14(2010)393–404. https://doi.org/10.1021/op900305v

[17] J. H. Sharphouse, Leather technician's handbook. Leather Producer's Association, (1983)104. ISBN 0-9502285-1-6.

[18] E. Heidemann, Fundamentals of leather manufacture. In: Eduard Roether KG (ed), (1993)296. ISBN 3-7929-0206-0.

[19] G. Oberdörster, A. Maynard, K. Donaldson, V. Castranova, J. Fitzpatrick, K. Ausman, D. Lai, Principles for characterizing the potential human health effects from exposure to nanomaterials: elements of a screening strategy. Part. Fibre Toxicol 2(2005)8.

[20] C. Buzea, I. I. Pacheco, K. Robbie, Nanomaterials and nanoparticles: sources and toxicity, Biointerphases 2(2007) MR17–MR71. https://doi.org/10.1116/1.2815690

[21] Z. J. Deng, M. Liang, I. Toth et al., Plasma protein binding of positively and negatively charged polymer-coated gold nanoparticles elicits different biological responses, Nanotoxicology 7(2013)314–322. http://dx.doi.org/10.3109/17435390.2012.655342

[22] M. Sajid, M. Ilyas, C. Basheer, M. Tariq, M. Daud, N. Baig, F. Shehzad, Impact of nanoparticles on human and environment: review of toxicity factors, exposures, control strategies, and future prospects, Environ Sci Pollut Res 22(2015)4122–4143. https://doi.org/10.1007/s11356-014-3994-1

[23] J. Kanagaraj, P. Sastry, C. Rose, Effective preservation of raw goat skin for the reduction of total dissolved solids, J Clean Prod 13(2005)959–996. https://doi.org/10.1016/j.jclepro.2004.05.001

[24] B. S. Gholizadeh, F. Buazar, S. M. Buazar, S. M. Mousavi, Enhanced antibacterial activity, mechanical and physical properties of alginate/ hydroxyapatite bionanocomposite film. Int J Biol Macromol 116(2018)786–792. https ://doi.org/10.1016/j.ijbio mac.2018.05.104

[25] B. G. Cousins, H. E. Allison, P. J. Doherty, C. Edwards, M. J. Garvey, D. S. Martin et al., Effects of a nanoparticulate silica substrate on cell attachment of *Candida albicans*, J Appl Microbiol 102(2007)757–765. https://doi.org/10.1111/j.1365-2672.2006.03124.x

[26] O. Yamamoto, T. Ohira, K. Alvarez, M. Fukuda, Antibacterial characteristics of $CaCO_3$-MgO composites, Mater Sci Eng B 173(2010)208–212. https ://doi.org/10.1016/j.mseb.2009.12.007

[27] N. Duran, P. D. Marcato, R. D. Conti, O. L. Alves, F. T. M. Costa, M. Brocchi, Potential use of silver nanoparticles on pathogenic bacteria, their toxicity, and possible mechanisms of action, Braz Chem Soc 21(2010)949–959. https://doi.org/10.1590/S0103-50532010000600002

[28] T. Jin, Y. He, Antibacterial activities of magnesium oxide (MgO) nanoparticles against foodborne pathogens, J Nanopart Res 13(2011)6877–6885. https://doi.org/10.1007/s11051-011-0595-5

[29] M. A. Ansari, H. M. Khan, A. A. Khan, R. Pal, S. S. Cameotra, Antibacterial potential of Al_2O_3 nanoparticles against multidrug resistance strains of *Staphylococcus aureus* isolated from skin exudates, J. Nanoparticle Res 15 (2013)1970. https ://doi.org/10.1007/ s1105 1-013-1970-1

[30] A. P. Ingle, N. Duran, M. Rai, Bioactivity, mechanism of action and cytotoxicity of copper-based nanoparticles: a review, Appl Microbiol Biotechnol 98(2014) 1001–1009. https ://doi.org/10.1007/ s0025 3-013-5422-8

[31] M. Shah, V. Badwaik, Y. Kherde, H. K. Waghwani, T. Modi, Z. P. Aguilar, *et al.* Gold nanoparticles: various methods of synthesis and antibacterial applications, Front Biosci 19(2014)1320–1344. https://doi.org/10.2741/4284

[32] M. E. El-Zowalaty, S. H. H. Al-Ali, M. I. Husseiny, B. M. Geilich, T. J. Webster, M. Z. Hussein, The ability of streptomycin-loaded chitosan coated magnetic nanocomposites to possess antimicrobial and antituberculosis activities, Int J Nanomed 10 (2015)3269–3274. https ://doi.org/10.2147/ijn.s7446 9.

[33] H. M. Yadav, J. Kim, S. H. Pawar, Developments in photocatalytic antibacterial activity of nano TiO_2: A review, Korean J Chem Eng 33(2016)1989–1998. https ://doi.org/10.1007/s1181

[34] A. B. Moghaddam, M. Moniri, S. Azizi, R. A. Rahim, A. B. Ariff, W. Z. Saad, F. Namvar, M. Navaderi, Biosynthesis of ZnO nanoparticles by a new *Pichia kudriavzevii* yeast strain and evaluation of their antimicrobial and antioxidant activities, Molecules 22(2017)872. https://doi.org/10.3390/molecules22060872

[35] P. Thanikaivelan, J. R. Rao, B. U. Nair, T. Ramasami, Progress and recent trends in biotechnological methods for leather processing, Trend Biotechnol 22 (2004)181–188. https ://doi.org/10.1016/j.tibtech.2004.02.008

[36] N. George, P. S. Chauhan, V. Kumar, N. Puri, N. Gupta, Approach to eco-friendly leather: characterization and application of an alkaline protease for chemical free dehairing of skins and hides at pilot scale, J Clean Prod 79 (2014)249–257. https ://doi.org/10.1016/j.jclep ro.2014.05.046

[37] G. Murugappan, M. J. Zakir, G. C. Jayakumar, Y. Khambhaty, K. J. Sreeram, J. R. Rao, A novel approach to enzymatic unhairing and fiber opening of skin using enzymes immobilized on magnetite nanoparticles, ACS Sustain Chem Eng 4 (2016)828–834. https://doi.org/10.1021/acssuschemeng.5b00869

[38] G. Murugappan, K. J. Sreeram, Effective use of enzymatic processes in beamhouse through nanoparticle immobilization, XXXV. Congress of IULTCS (2019).

[39] G. Murugappan, Y. Khambhaty, K. J. Sreeram, Protease immobilized nanoparticles: a cleaner and sustainable approach to dehairing of skin, Appl Nanosci 10 (2020)213–221. https ://doi.org/10.1007/s1320 4-019-01113 -2

[40] M. A. Eid, E. A. Al-Ashkara, Speciation of chromium ions in tannery effluents and subsequent determination of Cr(VI) by ICPAES, J Am Leather Chem Assoc 97(2007)451. https://doi.org/10.3126/jncs.v23i0.2102

[41] G. Mallikarjun, P. Saravanan, G. V. R. Reddy, Microemulsion solutions of acrylic copolymers for retanning applications on chrome tanned goat skins, J Am Leather Chem Assoc 97(2002)215. http://hdl.handle.net/123456789/9525

[42] X. C. Wang, H. R. An, M. Sun, Y. H. Luo, J. Y. Feng, An acrylic resin retanning agent with a reinforcing effect: Synthesized by high solids content microemulsion copolymerization, J Soc Leath Tech Ch 89(2005)164–168.

[43] J. Z. Ma, X. J. Chen, Y. Chu, Z. S. Yang, The preparation and application of a montmorillonite-based nanocomposite in leather making, J Soc Leath Tech Ch. 87(2003)131.

[44] Y. Bao, J. Z. Ma, The interaction between collagen and aldehyde acid copolymer/MMT nano-composite, J Soc Leath Tech Ch 94(2010)53.

[45] R. Usha, T. Ramasami, Effect of crosslinking agents (basic chromium sulphate and formaldehyde) on the thermal and thermomechanical stability of rat tail tendon collagen fiber, Thermochim Acta 356(1–2)(2000)59–66. https ://doi.org/10.1016/S0040-6031(00)00518 -9

[46] Š. Rýglová, M. Braun, T. Suchý, Collagen and its modifications—crucial aspects with concern to its processing and analysis, Macromol Mater Eng 302(6)(2017) 1600460. https ://doi.org/10.1002/mame.201600460

[47] Š. Rýglová, M. Braun, T. Suchý, Collagen and its modifications— crucial aspects with concern to its processing and analysis, Macromol Mater Eng 302 (2017)1600460.

[48] H. Pan, Z. J. Zhang, J. X. Zhang, H. X. Dang, The preparation and application of a nanocomposite tanning agent-MPNS/SMA, J Soc Leath Tech Ch 89(2005)153.

[49] Y. S. Liu, Y. Chen, J. Yao, H. J. Fan, B. Shi, B. Y Peng, An environmentally-friendly leather-making process based on silica chemistry, J Am Leather Chem Assoc 105(2010)84.

[50] H. J. Fan, L. Li, B. Shi, Q. He, B. Y. Peng, Characteristics of leather tanned with nano-SiO$_2$, J Am Leather Chem Assoc 100(2005)22.

[51] D. G. Gao, J. Z. Ma, D. Gao, B. Lv, Study on diallyldimethyl ammonium chloride copolymer/nano SiO$_2$ composite tannage, Leather Sci Eng 20(2010)45–48.

[52] Y. Li, D. Gao, J. Ma, B. Lu, Synthesis of vinyl polymer/ZnO nano-composite and its application in leather tanning agent, Mater Sci Forum 694(2011)103–107. https ://doi.org/10.4028/www.scien tific .net/MSF.694.103

[53] J. Ma, X. Lu, D. Gao, Y. Li, B. Lv, J. Zhang, Nanocomposite-based green tanning process of suede leather to enhance chromium uptake, J Clean Prod 72(2014)120–126. https ://doi.org/10.1016/j.jclepro.2014.03.016

[54] H. Pan, L. Guang-L., L. Rui-Qi, W. Su-Xia, W. Xiao-D., Preparation, characterization and application of dispersible and spherical Nano-SiO$_2$@copolymer

nanocomposite in leather tanning. Appl Surf Sci., 426(2017) 376–385. https ://doi.org/10.1016/j.apsus c.2017.07.106

[55] B. Lyu, R. Chang, D. Gao, J. Ma, Chromium footprint reduction: nanocomposites as efficient pretanning agents for cowhide shoe upper leather, ACS Sustain Chem Eng 6(4) (2018)5413–5423. https ://doi.org/10.1021/acssu schem eng.8b002 33

[56] L. Kaijun, R. Yu, Z. Ruixin, L. Ruifeng, L. Gongyan, P. Biyu, pH-sensitive and chromium-loaded mineralized nanoparticles as a tanning agent for cleaner leather production, ACS Sustain Chem Eng, 7(9) (2019)8660–8669. https ://doi.org/10.1021/acssu schem eng.9b004 82

[57] S. Jiabo, W. Chunhua, L. Hu, X. Yuanhang, L. Wei, Novel wet-white tanning approach based on Laponite clay nanoparticles for reduced formaldehyde release and improved physical performances, ACS Sustain Chem Eng 7(1) (2019)1195–1201. https ://doi.org/10.1021/acssu schem eng.8b048 45

[58] R. Zhu, C. Yang, K. Li, R. Yu, G. Liu, B. Peng, A smart high chrome exhaustion and chrome-less tanning system based on chromium (III)-loaded nanoparticles for cleaner leather processing, J Clean Prod 277(2020)123278. https ://doi.org/10.1016/j.jclep ro.2020.12327 8

[59] Y. Zhang, T. Snow, A. J. Smith, G. Holmes, S. Prabakar, A guide to high-efficiency chromium (III)-collagen cross-linking: synchrotron SAXS and DSC study, Int J Biol Macromol 126(2019)123–129.

[60] U. Hasanah, M. S. Miran, M. M. Rahman, M. M. Islam, Simultaneous reductions of production loss and environmental burden through the treatment of loose leather with non-toxic manganese dioxide nanoparticles, J. Cleaner Prod 318(2021) 128541. https://doi.org/10.1016/j.jclepro.2021.128541

[61] J. Kanagaraj, R. C. Panda, Modeling of dye uptake rates, related interactions, and binding energy estimation in leather matrix using protein, Ind Eng Chem Res 50(22)(2011)12400–12408.

[62] S. Ramalingam, J. R. Rao, Tailoring nanostructured dyes for auxiliary free sustainable leather dyeing application, ACS Sustain Chem Eng 5(6)(2017)5537–5549.

[63] S. Ramalingam, K. J. Sreeram, J. R. Rao, B. U. Nair, Organic nanocolorants: self-fixed, optothermal resistive, silica-supported dyes for sustainable dyeing of leather, ACS Sustain Chem Eng 4(5) (2016)2706–2714.

[64] O. A. Mohamed, A. B. Moustafa, M. A. Mehawed, N. H. El-Sayed, Styrene and butyl methacrylate copolymers and their application in leather finishing, J Appl Polym Sci 111(2009)1488–95.

[65] Q. Fan, J. Ma, Q. Xu, J. Zhang, D. Simion, G. Carmen, C. Guo. Animal-derived natural products review: focus on novel modifications and applications, Colloids Surf B 128(2015)181–90.

[66] R. Kothandam, M. Pandurangan, R. Jayavel, S. Gupta, A novel nano-finish formulation for enhancing performance properties in leather finishing applications, J Clust Sci 27(2016)1263–72.

[67] J. Ma, Q. Xu, J. Zhou, D. Gao, J. Zhang, L. Chen, Nano-scale core–shell structural casein based coating latex: synthesis, characterization and its biodegradability, Prog Org Coat 76(2013)1346–55.

[68] S. Sundara, N. Vijayalakshmia, S. Guptab, R. Rajaramc, G. Radhakrishnan, Aqueous dispersions of polyurethane-polyvinyl pyridine cationomers and their application as binder in base coat for leather finishing, Prog Org Coat 56(2006) 178–84.

[69] I. P. Fernandes, J. S. Amaral, V. Pinto, Development of chitosan-based antimicrobial leather coatings, Carbohydr Polym 98(2013)1229–35.

[70] L. Fang, Y. Honglei, M. Shulu, Antibacterial activity of chitosan-metal ion complexes in leather top-finishing, China Leather 38(2009)9–12.

[71] J. Xiang, L. Ma, H. Su, J. Xiong, K. Li, Q. Xia, G. Liu, Layer-by-layer assembly of antibacterial composite coating for leather with cross-link enhanced durability against laundry and abrasion, Appl Surf Sci 458(2018)978–87.

[72] H. Shi, Y. Chen, H. Fan, J. Xiang, B. Shi, Thermosensitive polyurethane film and finished leather with controllable water vapor permeability, J Appl Polym Sci 117(2010)1820–7.

[73] O. A. Mohamed, F. A. Abdel-Mohdy, Preparation of flame-retardant leather pretreated with pyrovatex CP, J Appl Polym Sci 99(2006)2039–43.

[74] H. Fan, L. Gao, Q. Wang, Research on the application of nanometer TiO_2 in the leather industry, Leather Chem 28(2011)22–5.

[75] T. Wang, Y. Bao, Advances on functional polyacrylate/inorganic nanocomposite latex for leather finishing, Mater Rev 31(2017)64–71.

[76] P. Velmurugan, M. Cho, S. M. Lee, J. H. Park, S. Bae, B. T. Oh, Antimicrobial fabrication of cotton fabric and leather using green-synthesized nanosilver, Carbohydr Polym 106(2014)319–25.

[77] C. Gaidau, A. Petica, M. Ignat, O. Iordache, L. Ditu, M. Ionescu, Enhanced photocatalysts based on ag-TiO_2 and ag-N-TiO_2 nanoparticles for multifunctional leather surface coating, Open Chem 14(2016)383–92.

[78] J. Liu, J. Ma, Y. Bao, J. Wang, H. Tang, L. Zhang, Polyacrylate/surface-modified ZnO nanocomposite as film-forming agent for leather finishing, Int J Polym Mater Polym Biomater 63(2014)809–14.

[79] Y. Bao, C. Feng, C. Wang, J. Ma, C. Tian, Hygienic, antibacterial, UV-shielding performance of polyacrylate/ZnO composite coatings on a leather matrix, Colloids Surf A Physicochem Eng Asp 518(2017)232–40.

[80] J. Liu, J. Ma, Y. Bao, J. Wang, Z. Zhu, H. Tang, L. Zhan, Nanoparticle morphology and film-forming behavior of polyacrylate/ZnO nanocomposite, Compos Sci Technol 98(2014)64–71.

[81] W. Chen, L. Feng, B. Qu, In situ synthesis of poly (methyl methacrylate)/MgAl layered double hydroxide nanocomposite with high transparency and enhanced thermal properties, Solid State Commun 130(2004)259–63.

[82] Y. Wang, J. Ma, Q. Xu, J. Zhang, Fabrication of antibacterial casein-based ZnO nanocomposite for flexible coatings, Mater Des 113(2017)240–5.

[83] R. Kothandam, M. Pandurangan, R. Jayavel, S. Gupta, A novel nano-finish formulation for enhancing performance properties in leather finishing applications, J Clust Sci 27(2016)1263–72.

[84] H. Jing, J. Ma, W. Deng, Properties of acrylic resin/nano-SiO$_2$ leather finishing agent prepared via emulsifier-free emulsion polymerization, Mater Lett 62(2008)2931–4.

[85] W. Zhang, J. Ma, D. Gao, Y. Zhou, C. Li, J. Zha, J. Zhang, Preparation of aminofunctionalized graphene oxide by Hoffman rearrangement and its performances on polyacrylate coating latex, Prog Org Coat 94(2016)9–17.

[86] J. Ma, X. Zhang, Y. Bao, J. Liu, A facile spraying method for fabricating superhydrophobic leather coating, Colloids Surf A Physicochem Eng Asp 472(2015)21–5.

[87] A. Petica, C. Gaidau, M. Ignat, C. Sendrea, L. Anicai, Doped TiO$_2$ nanophotocatalysts for leather surface finishing with self-cleaning properties, J Coat Technol Res 12(2015)1153–63.

[88] C. Gaidau, A. Petica, M. Ignat, L. M. Popescu, R. M. Piticescu, I. A. Tudor, Preparation of silica doped titania nanoparticles with thermal stability and photocatalytic properties and their application for leather surface functionalization, Arab J Chem 10(2017)985–1000.

[89] Q. Xu, Q. Fan, J. Ma, Z. Yan, Facile synthesis of casein-based TiO$_2$ nanocomposite for self-cleaning and high covering coatings: insights from TiO$_2$ dosage, Prog Org Coat 99(2016)223–9.

[90] Y. Bao, Y. Yang, J. Ma, Fabrication of monodisperse hollow silica spheres and effect on water vapor permeability of polyacrylate membrane. J Colloid Interface Sci 407(2013)155–63.

[91] Y. Bao, Y. Yang, C. Shi, J. Ma, Fabrication of hollow silica spheres and their application in polyacrylate film forming agent, J Mater Sci 49(2014)8215–25.

[92] Q. Fan, J. Ma, Q. Xu, J. Wang, Y. Ma, Facile synthesis of chitosan-coated silica nanocapsules via interfacial condensation approach for sustained release of vanillin, Ind Eng Chem Res 57(2018)6171–9.

[93] Q. Fan, J. Ma, Q. Xu, Facile synthesis of chitosan-based silica nanocapsules for fragrance-controlled release leather finishes, XXXIII IULTCS congress proceedings, (2017) India.

[94] F. Zhang, J. Ma, Q. Xu, J. Zhou, D. Simion, G. Carmen, A facile method for fabricating room-temperature-film-formable casein-based hollow nanospheres, Colloids Surf A Physicochem Eng Asp 484(2015)329–35.

[95] Q. Fan, J. Ma, Q. Xu, W. An, R. Qiu, Multifunctional coatings crafted via layer-by-layer spraying method, Prog Org Coat 125(2018)215–21.

[96] O. Mohamed, H. Elsayed, R. Attia, A. Haroun, N. El-Sayed, Preparation of acrylic silicon dioxide nanoparticles as a binder for leather finishing, Adv Poly Technol 37(8)(2018)3276–3286.

[97] H. Elsayed, R. Attia, O. Mohamed, A. Haroun, N. El-Sayed, Preparation of polyurethane silicon oxide nanomaterials as a binder in leather finishing, Fibers Polym 19(2018)832–842.

[98] M. K. Kaygusuz, M. Meyer, A. Aslan, The effect of TiO_2-SiO_2 nanocomposite on the performance characteristics of leather, Mat Res 20(4) (2017)1103–1110.

[99] K. R. Kumar, R. Jayavel, G. Sanjeev, Zinc oxide (ZnO) nanoparticles for enhancement of fastness properties in cationic finishing, J Am Leather Chem Assoc 112 (5) (2017)162–167.

[100] K. Ramkumar, P. Muthuraman, R. Jayavel, G. Sanjeev, Some novel nano-finish formulations for enhancing performance properties in leather finishing applications, J Cluster Sci 27(4)(2016)1263–1272.

[101] L. Muthukrishnan, M. Chellappa, A. Nanda, Bio-engineering and cellular imaging of silver nanoparticles as weaponry against multidrug resistant human pathogens, J Photochem Photobiol B Biol 194(2019)119–127. https ://doi.org/10.1016/j.jphot obiol .2019.03.021

[102] J. Kanagaraj, P. Sastry, C. Rose, Effective preservation of raw goat skin for the reduction of total dissolved solids, J Clean Prod 13(2005)959–964.

[103] L. Kaijun, R. Yu, Z. Ruixin, L. Ruifeng, L. Gongyan, P. Biyu, pH-sensitive and chromium-loaded mineralized nanoparticles as a tanning agent for cleaner leather production, ACS Sustain Chem Eng 7(9)(2019)8660–8669. https ://doi.org/10.1021/acssu schem eng.9b004 82

[104] D. Gurera, B. Bhushan, Fabrication of bioinspired superliquiphobic synthetic leather with self-cleaning and low adhesion, Colloids Surf A Physicochem Eng Asp 545(2018)130–137. https ://doi.org/10.1016/j.colsu rfa.2018.02.052.

[105] J. Singh, T. Dutta, K. H. Kim et al., Green synthesis of metals and their oxide nanoparticles: applications for environmental remediation, J Nanobiotechnol 16 (2018)84. https ://doi.org/10.1186/s12951-018-0408-4

Applications of Emerging Nanomaterials and Nanotechnology
Materials Research Foundations 148 (2023) 200-228

Materials Research Forum LLC
https://doi.org/10.21741/9781644902554-7

Chapter 7

Production of Nanomaterials from Forest Resources

M. Mahbubur Rahman[1,2], M. Mostafizur Rahman[1], Md. Abu Bin Hasan Susan[2],
M. Sarwar Jahan[1*]

[1]Bangladesh Council of Scientific and Industrial Research, Dr. Qudrat-i-Khuda Road, Dhaka 1205, Bangladesh

[2]Department of Chemistry, University of Dhaka, Dhaka-1000, Bangladesh

* sarwar2065@hotmail.com

Abstract

Renewable resources such as lignocellulose are prospective alternatives for the development of different products in connection with climate change. Cellulose, lignin, and hemicellulose can be extracted from wood and non-wood through a chemical process, subsequently, nanocellulose, nanolignin, and nanohemicellulose can be obtained through mechanical, chemical, and a combination of these two processes. Nanocellulose is suggested for application in improving barrier properties, drug delivery, energy storage, composite film, scaffolds for tissue regeneration, and other smart materials due to its nanoscale dimension, hydrogen bond formation capability, and high surface area. This chapter presents various methods for the extraction of lignin nanoparticles and their applications. Due to the high reactivity, large surface area, and homogeneity, nanolignin is applied in the preparation of nanocomposites, and so far, various thermally stable composites have been suggested. Though very little information is available on nanohemicellulose, it is a very promising nanomaterial from forest resources to show a definite improvement in the tensile strength of biofilm.

Keywords

Lignocellulose, Nanocellulose, Nanolignin, Nanohemicellulose, Barrier Properties, Biofilm

Contents

1.   Introduction

The key forest product is wood, which comprises cellulose, hemicellulose, and lignin. These three are the major components of the plant, which represent cellulose 35–50%, hemicellulose 10–25%, and lignin 20–35% of dry mass respectively. The lignin contributes to the strength of the plant through the binding with cellulose and hemicellulose in the cell wall of fiber. Cellulose is the most abundant renewable natural biopolymer in the world that exists generally in plant cell walls as well as a small amount in several marine animals (e.g., tunicates), bacteria, fungi, invertebrates, and algae [1]. Cellulose is a polymer of glucose monomer linked by 1-4 β glucoside linkage (Fig. 1a) and the hydroxyl groups in glucose monomer in the same and adjacent cellulose chains form intra- and inter-molecular hydrogen bonds respectively [2]. The elementary fibrils of cellulose can be 3-4 nm in width and a few micrometers long. The fibrils are interlinked in a regular and irregularly ordered manner that creates a crystalline and amorphous region in the fiber. Cellulose is typically fibrous, structurally rigid, and water-insoluble, which is very important to maintain the structural integrity of plant cells [3]. The hydrogen bonding in the cellulose network (Fig. 1b) gives strength and common solvent insolubility. Xylans and glucomannans are the main components of hemicellulose that makes a branched network with a short chain length. The main constituents of different lignocellulosic biomasses are shown in Table 1.

a)

b)

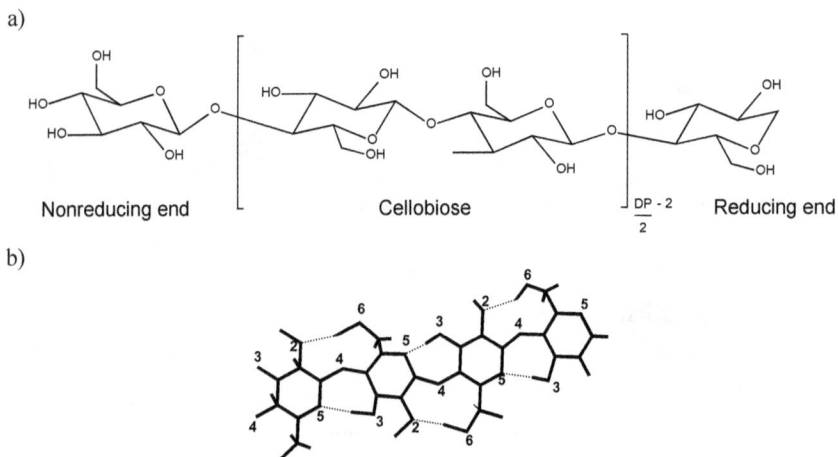

Fig.1. a) Chemical structure of cellulose, b) Intramolecular hydrogen bonding network in cellulose.[2]

Table 1. Chemical composition of different lignocellulosic biomasses

Raw materials	Composition (%)			References
	Cellulose	Pentosan	Lignin	
Hardwood	40–50	15–25	21–34	[4]
Softwood	40–44	25–29	25–31	[5]
Jute	63	14	13	[6]
Bagasse	39	17	20	[7]
Corn stalk	35	18	20	[8]
Wheat straw	37	18	16	[9]

The wood and other lignocellulosic biomasses are converted to pulp and paper in forest product industry. This industry faces challenges due to document digitalization and is in search of alternative products. The successful production of high-value biochemicals and biomaterials from lignocellulosic biomasses would shift the present global economy to a bio-based one, and raise the revenue of the forest product industry. Van Heiningen et al proposed an integrated forest biorefinery concept, where lignocellulosic biomass was pre-extracted prior to pulping, and the dissolved biomass was modified to biofuel and biochemical along with pulp [10-13]

Nanotechnology is a showcase of demonstrating the ability of science to investigate beyond the molecular level with advantageous results. Nanotechnology is being applied for the development of new materials, systems, and devices at the nano-level. This technology includes a number of disciplines including engineering, biotechnology, electronics, bio-mechanics, coatings, construction materials, etc. [14,15]. In recent technological advancements in disease control, nanotechnology also plays a crucial role [16].

Nanocellulose (NC) is the most appealing topic in present-day forest products. Mukherjee and Woods isolated NC using sulfuric acid from ramie and cotton fibers [17]. Sulfuric acid hydrolysis of cellulose fibers formed a colloidal suspension [18,19]. Acid-hydrolysed crystalline cellulose is often referred to as nanocrystals, microcrystals, nanoparticles, nanofibers, microcrystallites, or whiskers. The properties of prepared NC depend on the sources and the treatment methods [2]. Lignocellulosic biomass is effectively fractionated into cellulose, lignin, and hemicellulose by organic acids [20,21]. The separated cellulose was subsequently converted to nanocrystalline cellulose (NCCs) by acid hydrolysis by dissolving the amorphous part of the cellulose chain and converted to biofuel [22]. Several commercial and pilot-scale plants for the extraction of cellulose nanofibers (CNFs) and NCCs from wood have been established in Europe and North America [23,24]. CelluForce is the first manufacturing facility of NCCs in the world. The plant was established in 2011 in Quebec, Canada with a production capacity of 3×10^5 Kg of NCCs per year [25].

Another important fraction of lignocellulosic is lignin, which is a polymeric material that constitutes about 13 – 35% of the total weight, which is an underutilized renewable component and it has the potential in developing green products. At present, lignin is used in generating energy in the pulp mill [26]. In recent years, interest is growing in nanolignin (NL) for potential biobased applications, but the commercialization of NL needs to be developed.

In this chapter, the extraction, and application of NC, NL, and nanohemicellulose in different processes from different sources are discussed. The methods for the production of NC and NL and their characterization and applications are also reviewed. A graphical representation of the of the work is shown in Fig. 2.

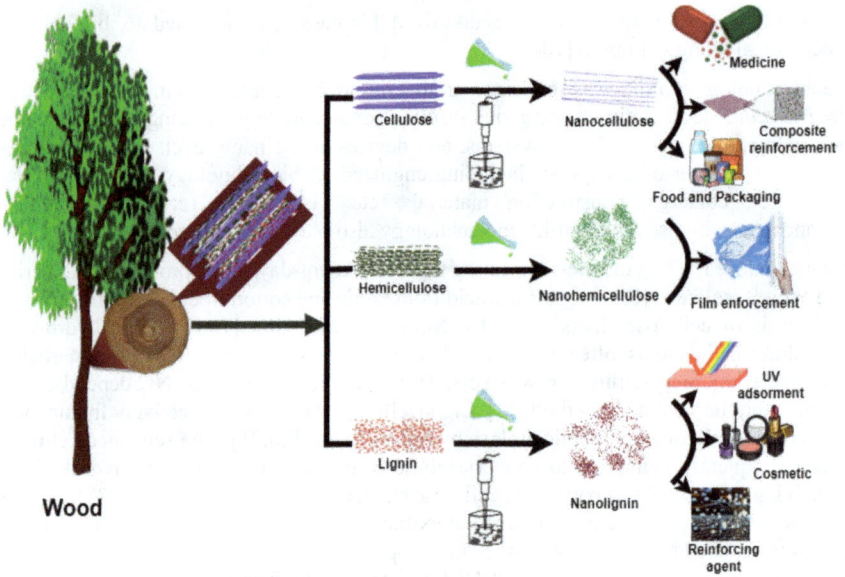

Fig. 2. Nanomaterials from forest resources.

2.   Nanocellulose

2.1   Types of nanocellulose

NC is isolated from fibers by mechanical, chemical means, or both together. The cellulose which has a width in the nanoscale (<100 nm) is termed NC. The size and properties of the NC highly depend on the methods of isolation [27]. NC is biodegradable, lightweight, is of low-density, and has exceptional strength. Specifically, NC has higher rigidity than Kevlar fiber, with an elastic modulus of up to 220 GPa. In addition, NC has an 8-times better specific strength than stainless steel as well as higher tensile strength than cast iron. NC is also transparent and has a lot of reactive surface hydroxyl functional groups which can be functionalized to various nanomaterials [28,29]. There are three main categories of NC are NCCs, CNFs and bacterial NC (BNCs). The chemical composition of all NCs is similar but differ in particle size, shapes, crystallinity, and morphology according to the sources and extraction methods [28,30]. The different types of NCs, their analogous name, sources, and methods of preparation are given in Table 2.

Table 2. Family of Nanocellulose Material [9]. (Reproduced with permission from John Wiley and Sons, Copyright © 2020)

Kinds of nano-cellulose	Analogous name	Sources	Preparation method/ size
Nanocellulose and microcellulose	Cellulose nanofiber, microfibrils	Lignocellulose (wood and Nonwoods)	Mechanical fibril width: From 5 to 60 nm Length: Few micrometers
Cellulose nanocrystal	Whisker nanocellulose, rod-like crystalinestals	Wood, nonwoods (agricultural residue, hemp, cotton etc) plant bark, microbial cellulose	Chemical Fibril width: 5–70 nm Length: From 100 to 250 nm
Microbial nanocellulose	Microbial cellulose, bio-cellulose	sugar and alcohols	Bacterial synthesis Fibril width: From 20 to 100 nm

2.1.1 Nanocrystalline cellulose

NCCs are rod-like-shaped nanofiber that is highly crystalline and has a large surface area. The morphology of NCC depends not only on the preparation methods but also on the source of cellulose. NCC is mostly produced by dissolving the amorphous portion of cellulose fiber [31]. The length and width of the NCCs varied between 200 and 500 nm, and 3 and 35 nm respectively [27].

2.1.2 Cellulose nanofibers

CNFs are different from NCCs, which are flexible nanofibers. It is literally elementary cellulose microfibrils consisting of alternating amorphous and crystalline regions. By means of mechanical force on cellulose fibers resulting the CNFs with a width of bellow100 nm and a length of several micrometers. The difference between NCCs and CNFs is that CNFs show a web-like structure whereas NCCs are rod-like nanomaterials. [1, 32]. CNFs have a high aspect ratio and surface area, and more widespread hydroxyl groups that can be used to modify their surfaces [33]. Applying mechanical force, CNFs can be produced by splitting cellulose fibrils according to the longitudinal axis [27].

2.1.3 Bacterial nanocellulose

BNC is also called microbial cellulose. It is produced from bacterial genera (e.g. Acetobacter, Rhizobium, etc.) [34] and cell-free systems [35]. The production of BNC need not require any further purification process because of the absence of lignin, pectin, and hemicellulose [36]. BNC differs from NCCs and CNFs since NCCs and CNFs are produced from lignocellulosic biomasses by a top-down process but BNC is produced by the bottom-up process where the low molecular weight glucose molecules join each other by bacteria during the biosynthesis process. The shape of BNC looks like twisting ribbons of microfibers with average diameters of 20–100 nm [36,37].

2.2 Preparation of nanocellulose

There are several methods such as mechanical, chemical or a combination of both have been applied for the isolation of highly purified NC from lignocellulosic biomasses. Pre-treatments like refining of pulp [38], high-pressure homogenizing [39], enzymatic treatment [40], ultrasonic technique [41,42], and cryocrushing [43,44], microuidization [44], steam explosion [45], etc. facilitated the extraction of NC. A graphical representation of nanocellulose production is shown in Fig. 3.

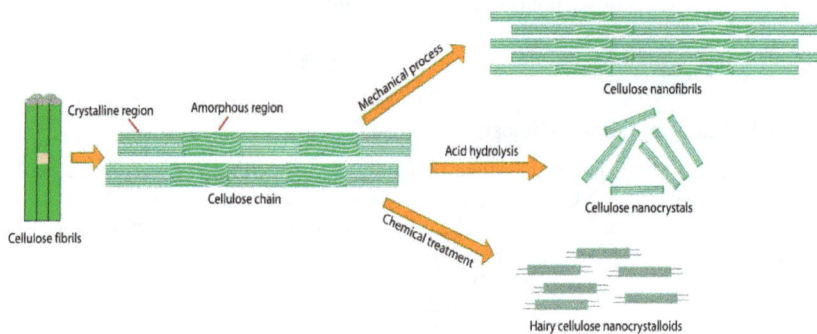

Fig.3. Schematic illustration of nanocellulose extraction pathways from forest resources. Chemically, cellulose nanocrystals can be produced through acid hydrolysis and cellulose nanofibrils can also be produced mechanically by cleaving the cellulose fibrils. Hairy cellulose nanocrystalloids can be produced by the chemical treatment where the amorphous regions are dissolved but few amorphous chains remain at both ends of crystalline regions.

The NCC is mostly produced by acid hydrolysis of cellulose in which the amorphous parts of cellulose are removed and crystalline parts remain intact [2,46,47]. NCCs are rod-shaped nanomaterials that have similar physical properties to the mother cellulose fibers. The most

Applications of Emerging Nanomaterials and Nanotechnology   Materials Research Forum LLC
Materials Research Foundations 148 (2023) 200-228   https://doi.org/10.21741/9781644902554-7

extensively used acids for the preparation of NCCs are sulfuric and hydrochloric acids [48-52], but phosphoric [53] and hydrobromic acid [54] were also used for the preparation of NCCs. NCCs produced by hydrochloric acid showed limited dispersibility and a tendency for flocculation in aqueous suspensions [49]. Different protocols for hydrolysis and separation processes have been developed based on cellulose sources. Wood [55,56] and nonwoods [21,22,57-60] are the most common sources of cellulose. The morphology of NCCs is presented in Fig. 4

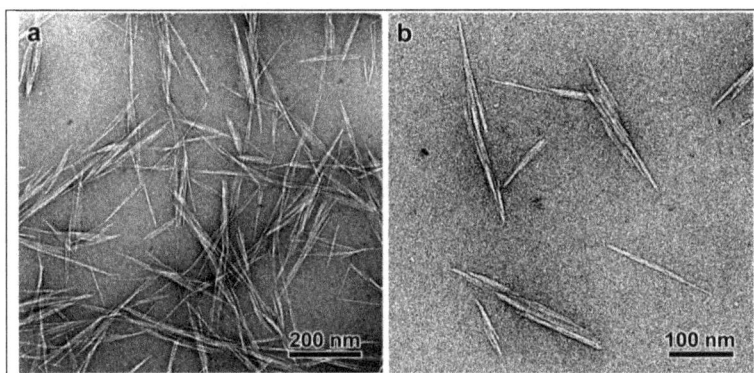

Fig.4. Cellulose nanocrystals under transmission electron microscopy (TEM) [48]. (Reproduced with permission from the Royal Society of Chemistry, Copyright © 2011)

CNFs are produced by only mechanical operation without the involvement of any chemical treatment. The mechanical process disintegrated the microfibrils from the interlinkage of the cellulose chain. The cellulose fibrils are isolated along their long axis [61]. Turbak et al. and Herrick et al. introduced CNFs as cellulosic nanomaterial for the first time from the soft-wood pulp by applying a high-pressure homogenization method [62,63]. Mechanical, chemical, enzymatic, or combined methods have been used for production of CNFs. Enzymatic treatments followed by chemical ones and in combination with mechanical treatments were applied to produce nanofiber [64]. A fine network of CNFs can be prepared from wood fiber by applying mechanical and/or enzymatic treatment [65]. Generally, several mechanical methods have been applied for the isolation of CNFs like refining, grinding, high-pressure homogenization, etc. In the chemi-mechanical processing of CNFs, chemical treatment reduces the energy consumption for mechanical defibrillation during the synthesis of CNFs from cellulose [66]. It is mostly employed in moderate climates to modify the hydroxyl groups into carboxyl derivatives. [67]. For example, individual CNFs can be produced by the oxidation with 2,2,6,6,-tetramethylpiperidine-1-oxyl (TEMPO) from wood under gentle stirring [68]. With this technique, a strong repulsive charge is created on the surface of CNFs and allows defibrillation with minimum energy input. [69,70]. To produce individual CNFs, both mechanical and chemical techniques can be applied. For instance, carboxymethylation followed by high-pressure homogenization can be used to prepare well-dispersed CNFs [62]. 19.7% CNFs with 1.7 nm diameter with several micrometers in length can be produced from rice straw by the TEMPO-mediated oxidation method [70] TEMPO-oxidation followed by ultrasound treatment was carried out to get NCCs with 6 nm diameter and 122 nm of length from the microcrystalline cellulose. The sono-assistance increased yield and carboxylate content compared to the non-assisted process [71].

BNCs can be produced from *Acetobacter* by using simple culturing methods [72]. Biosynthesized BNCs are partially transparent and resistant to mechanical force but show a property of elasticity and are soft in nature. BNC and plant cellulose are chemically identical, but the purity of bacterial cellulose is very high [73]. BNC is very much biocompatible and has numerous applications in medical science including drug delivery systems, wound healing, tympanic membranes and other. The high-purity BNCs are obtained by culturing microorganisms over the year [74,75]. Static and stirred culture methods are used to extract BNCs. Cowhide-like white BNCs are produced at the air-fluid interface of cultured bacteria in a static culture process. The important factors in selecting this method depend on the surface morphology, properties, and final applications of BNCs [76]. Moreover, several green methods such as deep eutectic solvents, solid acids, ionic liquids, etc. have been used for nanocellulose production [77].

Thus, chemical, mechanical, chemi-mechanical, and microbial processes can be applied for the production of NCCs, CNFs, and BNCs

2.3   Application of nanocellulose

NC, is a bio-based nanomaterial that exhibits impressive properties such as high crystallinity, high aspect ratio, large specific surface area, and superior mechanical properties. In addition to that NC is renewable, and biocompatible. Different types of nanomaterials with tunable properties can be produced chemically from NC due to the available hydroxyl groups. NC is a prospective candidate for use in industries like biomedicine, pharmaceuticals, electronics, supercapacitors, barrier films, membranes, nanocomposites, etc. [78-80]. As a sustainable raw material, NC has potential applications in the field of flexible electronics [81,82], sensors [83], energy storage materials [84], food and pharmaceuticals [85], biomedicine [86], packaging [87], and nanocomposite [88], etc. Due to its nanoscale dimensions, NC can easily fill the pores, which improves the superior barrier qualities of films. NC has the capacity to form hydrogen bonds, resulting in a strong network that makes it difficult for the molecules to pass through. The water vapor permeability of carboxymethyl cellulose (CMC) film containing nano-chitosan or nanocellulose at various concentrations improved with increasing nano-chitosan or nanocellulose content [89-91]. NCCs contain only crystalline region while CNFs consist of both crystalline and disordered regions which reflect better barrier properties [92]. CNF film showed excellent oxygen barrier properties with high air resistance capacity. A 21 μm thick film transmits only 17 ± 1 mL/m^2 oxygen per day. The oxygen barrier property of CNFs film is highly competitive with the best performed synthetic films. Acetylated CNFs films showed much lower oxygen permeability (<20 mL/m^2. per day), which can be used for packaging under atmospheric conditions [93]. Similarly to that, films made of carboxymethylated CNFs have extremely low oxygen permeabilities [94].

In the biomedical field, NC has many applications, including photodynamic and photothermal therapy of tumors, radical scavenging, healing from microbial infections, drug delivery, biosensors, isolation of different biomolecules, tissue engineering, advanced wound dressing, and engineering of blood vessels [95]. NC–carbon nanotube (CNT) composite can kill cancer cells in absence of anticancer drugs [96].

Neural interfaces are important for restoring the neural functions of paralysis patients. The most commonly used materials for conventional neural interfaces are platinum, titanium, gold, iridium oxide, and silicon. Recently, scientists have developed biocompatible materials based on NC as substrates for brain implants to reduce inflammatory response through softening neural interfaces [97-99]. These softening neural interfaces can be implanted into the brain. NC-incorporated scaffold showed improved mechanical properties [100]. In tissue engineering scaffolds, BNCs and NCCs are naturally suited [101]. NC based materials are successfully used for drug delivery, membranes technology, and as coating material for tablets [100,102,103]. Additionally, surface-modified NC-based materials provide superior stability and controlled drug release [104, 105].

NC-based materials have been used for fabricating flexible solar cells, sensors triboelectric nanogenerators (TENGs), and field-effect transistors (FET) [106]. The nanofibers ensure the improved mechanical performance of the devices. Good thermal stability and good wettability in various electrolytes along with a high aspect ratio of NC-based materials, is conducive to a wide potential window. These exceptional properties make it a highly desirable dielectric material for flexible organic light-emitting diodes (OLEDs), foldable batteries, organic thin-film transistors (OTFTs), and printed photovoltaic cells [74]. CNFs are fabricated by complex "bottom-up" nanostructures which make them perfect electrode components for piezoresistive sensors [107]. A conductive CNF/Ag nanowires-coated polyurethane (CA@PU) sensor was used to detect and record tiny sounds [108].

NC can be used to fabricate components of portable electrochemical energy storage (EES) devices. The NC-based EES has large absorptive bodies, hydrophilic surfaces, and hierarchical pores for charge storage which is good for the transportation and absorption of electrolytes [109]. NC reinforced multiwalled carbon nanotube forms a hybrid film with poly(3,4-ethylene dioxythiophene)-poly(styrene sulfonate) mixture in a ionic liquid of 1-butyl-3-methylimidazolium chloride by in situ polymerization process. The presence of NC provided higher flexibility and mechanical strength. The hybrid film was characterized as excellent supercapacitor with energy density and capacitance of 13.2 Wh kg^{-1} and 380 F g^{-1} respectively [110].

Thus, NC has versatile potential applications in energy storage devices, food, pharmaceuticals, biomedicine, flexible electronics, sensors, packaging, and nanocomposite materials, etc.

3. Nanolignin

3.1 Lignin

Lignin is the second most abundant natural polymer after cellulose in the flora world. It consists of 15-40% of mass fraction in the terrestrial plant and provides structural integrity [32]. It is a cross-linked biopolymer that consists of three phenylpropane units: guaiacyl (G), *p*-hydroxyphenyl (H), and syringyl (S). The quantitative presence of monomer units varied from plant species to species as well as their interlinkage attachment of monomers [111]. The lignin present in plant cells without any type of modification either chemical or mechanical is called native lignin. [112, 113]. On average about one-third of the plant biomass is made up of lignin. The forest product industry especially the pulp and paper industry and plant-based biorefinery industry produce sixty percent surplus lignin after meeting their need for energy production [114, 115]. High heterogeneity of chemical structure, molecular weight, and low free active group limit its downstream processing and valorization [116]. The lignin obtained as a byproduct from pulping and biorefineries by the chemical, mechanical, or biochemical process is recognized as technical lignin. Technical lignins are highly heterogeneous with regard to molecular structure, molecular weight, particle size depending on wood species, separation technique, growth rate, pulping method, and process intensity. Lignin has active hydroxyl, carbonyl, methoxyl, and carboxyl functional groups but their quantity and composition depend on the source and extraction processes [117]. The interlinkage of lignin is highly complex and there is no nondestructive method of isolation of native lignin from plants. Homogeneity of lignin fraction is crucial for its value-added applications. In recent years, interest is growing on green material development from technical lignin [118-120]. Lignin can be modified to nanoparticle as one of the high value applications of it.

3.1.1 Preparation of lignin nanoparticles

Lignins are soluble in alkaline aqueous medium, which limits their industrial applications, but the limitation could be overcome by dispersing lignin nanoparticles (LNPs) in water. Moreover, materials with nanostructures have many unique properties because of their large surface area. Various processes of LNPs production are described below.

Chemical process

Jiang et al. [121] used sulphate lignin to produce LNPs and natural rubber composite. The nanoparticle was co-precipitated with rubber. Lignin was dissolved in distilled water at the mass concentration of 0.5% and the pH was adjusted to 12 using sodium hydroxide. A 2% alkaline solution of poly(diallyldimethylammonium chloride) (PDAMAC) was prepared in water with pH 12. The two solutions were mixed with vigorous stirring to form a lignin-

PDAMAC (LPC). Finally, the colloidal solution was prepared by dropping the pH to 2 with sulfuric acid. It produced the LNPs of 90 – 100 nm.

Wang et al. [122] prepared LNPs by a solvent shifting method combined with sonication. A dilute solution (1 to 4 mg/mL) of acetylated lignin (AcL) in tetrahydrofuran (THF) was sonicated in an ultrasonic bath with various intensities (0, 100 kHz, and 200 W). LNPs were separated as distilled water was added drop by drop into the lignin/THF solution. The nanoparticles were separated immediately by centrifugation for 10 min at 10000 rpm under cool conditions. The lower the concentration of AcL and higher the sonication intensity, the lower was the particle size produced [122- 123]. Lievonen et al. prepared LNPs from the solution of softwood kraft lignin in THF (1-10 g/L) by precipitating in distilled water [124]. The THF solution of lignin in the dialysis bag of 6-8 kDA was placed in deionized water and kept for 24 h. This process produced colloidal LNPs of size between 200 and 500 nm. The particle size depends on the concentration of THF solution. The lower concentration provided smaller sized nanoparticle. Figueiredo et al. [125] produced LNPs of 221 ± 10 nm following the same method as described by Lievonen et al [124]. Xiong et al. [126] studied the LNPs form the organosolv lignin using the solvent shift method. However, the solution concentration was 10 – 100 g/L and the antisolvent ratio was 1:20 and 1:10. Lower lignin solution concentration and higher antisolvent ratio results in lower particle size. Lignin concentrations in THF of 10–20 g/L and antisolvent ration 1:20 produced an average nanoparticle size of 151 nm. Wang et al. [122] produced LNPs using a simple and green process. Initially, lignin was acetylated by microwave-assisted acetylation process in absence of any solvent or catalyst except the main reagent acetic anhydride. Acetic anhydride works as both reagent and solvent. Later, the acetylated lignin was converted to nanoparticles by applying solvent shifting and ultrasonication process.

Acetone-water solvent system was used to produce LNPs from commercial alkali lignin [127] and hardwood acidic dioxane lignin [128]. The lignin was dissolved in 9:1 (vol/vol) acetone: water mixture at a concentration of 10 g/L, the solution was added in a double volume of deionized water to separate lignin as nanoparticle. The average sizes of dioxane LNPs and alkali LNPs were 80 ± 27 nm and 82 ± 33 nm respectively [129]. Li et al. [130] prepared LNPs from the kraft lignin obtained from the pulp industry. A solution of kraft lignin in ethanol was subjected to the production of the nanoparticle. To precipitate lignin as nanoparticles, distilled water was added dropwise to the lignin solution in ethanol. It produced capsule-shaped nanoparticles in 10–100 nm range. Juikar and Vigneshwaran produced LNPs from coconut fiber through two different mechanical processes-homogenization and ultrasonication. The lignin was first extracted by digesting coconut fiber with soda at 170 °C. The soluble lignin was precipitated by reducing the pH to 2 by adding acid. In one process, the bulk lignin was diluted to a 7% mass fraction with distilled water. The aqueous suspension was homogenized at 10,000 rpm for 60 min. The mixture was allowed to settle down for 60 min. The large particles were precipitated and the nanoparticles remain suspended in the supernatant. In the ultrasonication process, the diluted lignin suspension was subjected to sonication in an ultrasonic water bath with 30 W power and 37 kHz frequency for 60 min. Later the mixture was allowed for sedimentation for 60 min. The supernatant contained the nanoparticle only. To avoid agglomeration, the supernatants were stored in an ice bath. The homogenization and sonication process produced particle sizes of 31.8 ± 11.2 and 58.6 ± 13.5 nm respectively [131].

Mechanical process

Nano-sized particles can be produced by various mechanical treatments; dry and wet milling is a widely used treatment. One of the obvious limitations is the uniformity of particle size. Many researchers produced LNPs by applying mechanical processes [132-135]. Nair et al. [136] prepared nano-sized lignin from softwood kraft lignin dispersed in water with a concentration of 5 g/L which was treated with a high shear homogenizer at 15000 rpm for up to 4 h. A shorter shearing time left larger particles. Approximately 75% of particles had diameters smaller than 100 nm after 2 h of shearing. However, after 4 h shearing, 100% of particles had diameters less the 100 nm.

Gilca et al. [137] reduced the wheat straw and Sarkanda grass macro sized lignin particle to nano size by applying the ultrasound technique. Lignin suspended in water at a concentration of 0.7% was sonicated with the intensity of 20000 kHz and 600W power. This technique successfully reduced the micro-sized lignin particle to 10–20 nm.
Ball milling is another well recognized mechanical treatment for reducing particle size. Shawn et al. [138] and Beisi et al. [139] produced LNPs of around 10 nm size by ball milling treatment at 18 °C.

Zimniewska et al. [140] prepared nanolignin from the kraft lignin by high intensity of ultrasound treatment. The LNPs were impregnated on the linen fabric structure lignin. The distribution of the nanoparticle was around 170 nm to 5 nm however, the particle size was mostly below 40 nm.

A water-miscible organic solvent is used to dissolve an organic compound. The organic compound is mixed with an excess of water, resulting in the formation of nanoparticles [141]. This process has been applied by many researchers to produce LNPs.

Microbial process

Juikar and Vigneshwaran [131] produced LNPs from the soda lignin of coconut fiber by a microbial process using Aspergillus sp. fungus. Aspergillus sp. was cultured on lignin as the only carbon source at 31 °C for 15 days with shaking at 110 rpm. The unhydrolyzed lignin was separated by filtration through a 1 μm filter. The filtrate contains the LNPs. The size of 25.8 LNPs was ± 8.9 nm.

Most of the LNPs were so far prepared by the solvent shifting system. The mechanical process was also widely applied to prepare LNPs. The size of the particles depends on processes, applied conditions, and sources.

3.1.2 Application of lignin nanoparticles

Nanomaterial has great importance for their target application efficiency, quality of product, and diversity of applicability. LNPs have been successfully applied for the applications in the fields of medicine, agriculture, and various engineering.

Bulk lignin and LNPs are incorporated into the phenol–formaldehyde resol resin for the production of biobased adhesive. The binding strength of the LNPs showed superior quality [142]. A nanocomposite from natural rubber and LNPs were also produced. The rubber-NL composite showed superior mechanical and thermal properties and oxidative stability to the rubber-carbon black composite. The nanolignin dispersed homogeneously in the nanocomposite matrix [121].

Lignin has antioxidant, UV absorbent, and antimicrobial properties and potential for use as a reinforcing agent [143]. A composite film was produced from polyvinyl alcohol, chitosan, and LNPs for biobased packaging. The film showed excellent thermal, optical, antioxidant, and antibacterial activities with high mechanical properties. The surface analysis showed a homogeneous distribution of nanoparticles which provided excellent mechanical properties. The film inhibited Gram-negative bacterial growth such as Erwinia carotovora subsp. Carotovora and Xanthomonas arboricola pv. pruni offers a novel way to control harmful bacterial plant/fruit pathogens [144].

Chitin nanofibrils-NL and chitin nanofibrils–NL–glycyrrhetinic acid were used to produce two composites for skin regeneration. The investigation showed that the complexes were cytocompatible with HaCaT cells and hMSCs. The complexes were capable to downregulate the expression of anti-inflammatory cytokines in human keratinocytes without altering their proliferative or osteogenic capabilities. In biomedical, personal care, and cosmetic applications, these materials are very promising. [145].

LNPs have been widely used for reinforcement in various polymer composites. Saz-Orozco et al. reinforced the phenolic foam with nano-sized lignin to improve the mechanical properties. The reinforced foam showed up to 128% and 174% higher compressive modulus and compressive strength in comparison to unreinforced foam [146]. LNPs were used to produce soy oil-based polyurethane biofoams. The foams showed higher thermal stability composite with improved mechanical properties in comparison to unreinforced foam. [147].

A transparent film was made through solvent casting process from LNPs and polyvinyl alcohol (LNS/PVA) for cutting the UV light penetration. The composite film showed excellent properties of UV light absorption with nice mechanical properties compared with lignin-PVA film. No phase separation occurred like macro particle rather evenly distributed nanoparticle transparent film was obtained. The UV radiation absorption and transparent property of the film makes it an excellent material for medicine bottle and food packaging [148].

Poly(L-lactide) based nanocomposite was produced with LNPs and various metal oxide nanoparticles (0.5% wt., Ag_2O, TiO_2, $_{WO3}$, Fe_2O_3 and $ZnFe_2O_4$). The synergistic effect of

LNPs and metal oxide was excellent in term of mechanical property, thermal stability, surface smoothness, UV absorption, antioxidant, and antibacterial properties. The LNPs played pivotal role in the mechanical property, surface wettability, and smoothness of the film as well as UV light absorption. LNPs show excellent protection against UV light while allowing visible light to pass. These wonderful properties make the composite film an attractive renewable additive for the fields of food, drug packaging, and biomedical applications [149].

LNPs can be used in the development of biomaterials, in biobased food packaging, in introducing antibacterial, antifungal, and UV light absorption properties of different consumer products, and in biomedical applications. NL helps in improving the mechanical properties of films/composites in all cases.

4.   Nanohemicellulose

Hemicellulose is mainly present in the primary and secondary walls of plant cells. It acts as a crosslinker between cellulose and lignin as well as cellulose microfibrils. Hemicellulose is composed of multiple sugar monomers with a short polymer chain and it contributes 20 to 30 % of dry mass of plants [150]. Composition of sugar units in the hemicelluloses depends on the type of plants. Hemicellulose extracted from rice straw is composed mainly of xylose and arabinose which contribute about 97% of total mass [151]. Although hemicelluloses are one of the important lignocellulosic biomasses, the research on their utilization is limited. Recently interest is growing in nanohemicellulose extraction. A biofilm produced from nanocellulose and reinforced with nanohemicellulose showed 18% higher tensile strength and 78% higher Young's modulus than the unreinforced biofilm [152]. In addition, the water vapor permeability (WVP) of the biofilm reduced remarkably. Louis et al [153] prepared starch nanocomposites using nanocellulose and nanohemicellulose, which improved mechanical properties and reduced WVP value. Hemicellulose from corn husk was extracted by sonication-assisted mild alkaline treatment followed by the treatment with deep eutectic solvent [154]. Nanohemicellulose was produced for the extracted hemicelluloses by mechanical homogenization. The properties of the marine calcium alginate film were improved using nanohemicellulose and nanocellulose. Small amounts of nanohemicellulose and nanocellulose synergistically improved the strength and gas barrier properties of the film.

The nanohemicellulose is the most recent topic for biomass chemists. Nanohemicellulose has rendered itself as a good reinforcing agent for improving the mechanical and barrier properties of biofilm.

Conclusion

This chapter provides an overview of nanomaterials from lignocellulosics (cellulose, hemicellulose, and lignin) and their applications. Cellulose fibers are composed of nano-sized fibrils which entangle together to form a cellulose fiber. Many chemical, mechanical, and microbial operations have been applied to separate the entangled fibrils to produce different types of NC. The aspect ratio of the NC is very high and applied to produce nanocomposites with special characteristics such as lightweight and high mechanical strength. Lignin and hemicellulose are nonfibrous macrostructure biopolymers. Various methods mostly chemical-mechanical methods are used to produce nanoparticles. LNPs have been used for producing lignin nanocomposites with improved special characteristics. LNPs enhance the strength and thermal stability of composite matrix materials, act as an antibacterial agent, flame retardant, and UV light adsorbent and exhibit a number of other properties. Lignin nanocomposites have many applications in engineering, agriculture, biomedical, drug delivery, and many others. Hemicellulose nanoparticles have been used for producing high strength biobased film materials. The research today is directed to the efficient synthesis of nanomaterials from lignocellulosic materials from natural forests with ease and exploitation of the advantageous features of the nanostructures for a range of applications.

Acknowledgements

MABHS acknowledges support from Bose Centre for Advanced Research in Natural Sciences and Semiconductor Technology Research Centre of the University of Dhaka, Bangladesh.

References

[1]   I. Siró, D. Plackett, Microfibrillated cellulose and new nanocomposite materials: a review, Cellulose 17(3) (2010) 459-494. https://doi.org/10.1007/s10570-010-9405-y

[2]   Y. Habibi, L. A. Lucia, O. J. Rojas, Cellulose nanocrystals: chemistry, self-assembly and applications, Chemical Reviews, 110(6) (2010) 3479-3500. https://doi.org/10.1021/cr900339w

[3]   H. Zhu, Z. Jia, Y. Chen, N. Weadock, J. Wan, O. Vaaland, Hu, L, Tin anode for sodium-ion batteries using natural wood fiber as a mechanical buffer and electrolyte reservoir, Nano Letters 13(7) (2013) 3093-3100. https://doi.org/10.1021/nl400998t

[4]   M.M. Haque, Y. Ni, A. J. U. Akon, M.A. Quaiyyum, M.S. Jahan, A review on Acacia auriculiformis: importance as pulpwood planted in social forestry, International Wood Products Journal 12(3) pp.194-205. https://doi.org/10.1080/20426445.2021.1949107

[5]   L. C. Malucelli, L. G. Lacerda, M. Dziedzic, da Silva, M. A. C. Filho, Preparation, properties and future perspectives of nanocrystals from agro-industrial residues: a review of recent research, Reviews in Environmental Science and Bio/Technology, 16 131-145. https://doi.org/10.1007/s11157-017-9423-4

[6]   M. S. Jahan, A. Al-Maruf, M. A. Quaiyyum, Comparative studies of pulping of jute fiber, jute cutting and jute caddis, Bangladesh Journal of Scientific and Industrial Research 42(4) 425-434. https://doi.org/10.3329/bjsir.v42i4.750

[7]   T. Ferdous, M. A. Quaiyyum, A. Salam, M. S. Jahan, Pulping of bagasse (Saccrarum officinarum), kash (Saccharum spontaneum) and corn stalks (Zea mays), Current Research in Green and Sustainable Chemistry, 3, 100017. https://doi.org/10.1016/j.crgsc.2020.100017

[8]   T. Ferdous, M.A. Quaiyyum, S. Bashar, and M.S. Jahan, Anatomical, morphological and chemical characteristics of kaun straw (Seetaria-ltalika), Nordic Pulp and Paper Research Journal, 35(2) pp.288-298. https://doi.org/10.1515/npprj-2019-0057

[9]   D. Klemm, F. Kramer, S. Moritz, T. Lindström, M. Ankerfors, D. Gray, A. Dorris, NCs: a new family of nature based materials, Angewandte Chemie International Edition, 50(24) (2011) 5438-5466. https://doi.org/10.1002/anie.201001273

[10]   A. Saeed, M. S. Jahan, H. Li, Z. Liu, Y. Ni, A. van Heiningen, Mass balances of components dissolved in the pre-hydrolysis liquor of kraft-based dissolving pulp production process from Canadian hardwoods, Biomass and Bioenergy 39 (2012) 14-19. https://doi.org/10.1016/j.biombioe.2010.08.039

[11]   L. Ahsan, M. S. Jahan, Y. Ni, Recovering/concentrating of hemicellulosic sugars and acetic acid by nanofiltration and reverse osmosis from prehydrolysis liquor of kraft based hardwood dissolving pulp process, Bioresource Technology, 155 (2014) 111-115. https://doi.org/10.1016/j.biortech.2013.12.096

[12]   H. Liu, H. Hu, M. S. Jahan, Y. Ni, Furfural formation from the pre-hydrolysis liquor of a hardwood kraft-based dissolving pulp production process, Bioresource Technology, 131 (2013) 315-320. https://doi.org/10.1016/j.biortech.2012.12.158

[13]   M. M. Rahman, R. S. Popy, J. Nayeem, K. M. Y. Arafat, M. S. Jahan, Dissolving pulp and furfural production from jute stick. Nordic Pulp and Paper Research Journal 37(4) (2022) 586-592. https://doi.org/10.1515/npprj-2022-0046

[14]   P. Balaguru, K. Chong, Nanotechnology and concrete: research opportunities. Proceedings of the ACI session on nanotechnology of concrete: recent developments and future perspectives, (2006) 15-28.

[15]   J. H. Thrall, Nanotechnology and medicine, Radiology, 230(2) (2004) 315-318. https://doi.org/10.1148/radiol.2302031698

[16]   S. M. Moghimi, A. C. Hunter, J. C. Murray, Nanomedicine: current status and future prospects, The FASEB Journal, 19(3) (2005) 311-330. https://doi.org/10.1096/fj.04-2747rev

[17]   S. M. Mukherjee, H. J. Woods X-ray and electron microscope studies of the degradation of cellulose by sulphuric acid, Biochimica et Biophysica Acta, 10 (1953) 499-511. https://doi.org/10.1016/0006-3002(53)90295-9

[18]   B. G. Ranby, Aqueous colloidal solutions of cellulose micelles, Acta Chemica Scandinavica 3(5) (1949) 649-650.

[19] B. G. Rånby, Fibrous macromolecular systems. Cellulose and muscle. The colloidal properties of cellulose micelles, Discussions of the Faraday Society 11 (1951) 158-164. http://doi.org/10.1039/DF9511100158

[20] M. S. Jahan, D. N. Chowdhury, M. K. Islam, Atmospheric formic acid pulping and TCF bleaching of dhaincha (Sesbaniaaculeata), kash (Saccharumspontaneum) and banana stem (Musa Cavendish), Industrial Crops and Products 26(3) (2007) 324-331. https://doi.org/10.1016/j.indcrop.2007.03.012

[21] M. Nuruddin, A. Chowdhury, S. A. Haque, M. Rahman, S. F. Farhad, M. S. Jahan, A. Quaiyyum, Extraction and characterization of cellulose microfibrils from agricultural wastes in an integrated biorefinery initiative, Biomaterials 3 (2011) 5-6.

[22] M. S. Jahan, A. Saeed, Z. He, Y. Ni, Jute as raw material for the preparation of microcrystalline cellulose, Cellulose 18(2) (2011) 451-459. https://doi.org/10.1007/s10570-010-9481-z

[23] H. Gu, R. Reiner, R. Bergman, A. Rudie, LCA study for pilot scale production of cellulose nano crystals (CNC) from wood pulp, Proceedings from the LCA XV Conference, A bright green future 6-8 October 2015 Vancouver, British Columbia, Canada P. 33-42.

[24] H. Wang, J. J. Zhu, Q. Ma, U. P. Agarwal, R. Gleisner, R. Reiner, J. Y. Zhu, Pilot-scale production of cellulosic nanowhiskers with similar morphology to cellulose nanocrystals, Frontiers in Bioengineering and Biotechnology 8(2020)565084. https://doi.org/10.3389/fbioe.2020.565084

[25] https://celluforce.com/about-celluforce/ accessed on 6 February 2023.

[26] A. J. Ragauskas, G. T. Beckham, M. J. Biddy, R. Chandra, F. Chen, M. F. Davis, C. E. Wyman, Lignin valorization: improving lignin processing in the biorefinery, Science, 344(6185) (2014)1246843. https://doi.org/10.1126/science.12468

[27] O. Nechyporchuk, M. N. Belgacem, J. Bras, Production of cellulose nanofibrils: A review of recent advances, Industrial Crops and Products 93(2016), 2-25. https://doi.org/10.1016/j.indcrop.2016.02.016

[28] R. J Moon, A. Martini, J. Nairn, J. Simonsen, J.Youngblood, Cellulose nanomaterials review: structure, properties and nanocomposites, Chemical Society Reviews 40(7) (2011) 3941-3994. https://doi.org/10.1039/C0CS00108B

[29] H. A. Khalil, A. H. Bhat, A. I. Yusra, Green composites from sustainable cellulose nanofibrils: A review, Carbohydrate Polymers 87(2) (2012) 963-979. https://doi.org/10.1016/j.carbpol.2011.08.078

[30] N. Lavoine, I. Desloges, A. Dufresne, J. Bras, Microfibrillated cellulose–Its barrier properties and applications in cellulosic materials: A review, Carbohydrate Polymers 90(2) (2012) 735-764. https://doi.org/10.1016/j.carbpol.2012.05.026

[31] D. Trache, M. H. Hussin, M. M. Haafiz, V. K. Thakur, Recent progress in cellulose nanocrystals: sources and production, Nanoscale 9(5) (2017) 1763-1786. https://doi.org/10.1039/C6NR09494E

[32]   G. Chauve, J. Bras, Industrial point of view of nanocellulose materials and their possible applications, In Handbook of Green Materials: 1 Bionanomaterials: separation processes characterization and properties, (2014) (pp.233-252). https://doi.org/10.1142/9789814566469_0014

[33]   N. Lavoine, I. Desloges, A. Dufresne, J. Bras, Microfibrillated cellulose–Its barrier properties and applications in cellulosic materials: A review, Carbohydrate Polymers 90(2) (2012) 735-764. https://doi.org/10.1016/j.carbpol.2012.05.026

[34]   M. W. Ullah, M. Ul-Islam, S. Khan, Y. Ki, J. K. Park, Structural and physico-mechanical characterization of bio-cellulose produced by a cell-free system, Carbohydrate Polymers 136 (2016) 908-916. https://doi.org/10.1016/j.carbpol.2015.10.010

[35]   M. W. Ullah, M. Ul-Islam, S. Khan, Y. Ki, J. K. Innovative production of bio-cellulose using a cell-free system derived from a single cell line, Carbohydrate Polymers 132 (2015) 286-294. https://doi.org/10.1016/j.carbpol.2015.06.037

[36]   S. P. Lin, I. LoiraCalvar, J. M. Catchmark, J. R. Liu, A. Demirci, K. C. Cheng, Biosynthesis, production and applications of bacterial cellulose, Cellulose 20(5) (2013) 2191-2219. https://doi.org/10.1007/s10570-013-9994-3

[37]   T. Abitbol, A. Rivkin, Y. Cao, Y. Nevo, E. Abraham T. Ben-Shalom, O. Shoseyov, nanocellulose, a tiny fiber with huge applications, Current Opinion in Biotechnology 39 (2016) 76-88. https://doi.org/10.1016/j.copbio.2016.01.002

[38]   A. N. Nakagait, H. Yano, The effect of morphological changes from pulp fiber towards nano-scale fibrillated cellulose on the mechanical properties of high-strength plant fiber based composites, Applied Physics A 78(4) (2004). 547-552. https://doi.org/10.1007/s00339-003-2453-5

[39]   P. Stenstad, M. Andresen, B. S. Tanem, P. Stenius, Chemical surface modifications of microfibrillated cellulose, Cellulose 15(1) (2008) 35-45. https://doi.org/10.1007/s10570-007-9143-y

[40]   F. Beltramino, M. B. Roncero, T. Vidal, C. Valls, A novel enzymatic approach to nanocrystalline cellulose preparation, Carbohydrate Polymers 189 (2018) 39-47. https://doi.org/10.1016/j.carbpol.2018.02.015

[41]   Q. Cheng, S. Wang, T. G. Rials, Poly (vinyl alcohol) nanocomposites reinforced with cellulose fibrils isolated by high intensity ultrasonication, Composites Part A: Applied Science and Manufacturing 40(2) (2009) 218-224. https://doi.org/10.1016/j.compositesa.2008.11.009

[42]   C. Salas, T. Nypelö, C. Rodriguez-Abreu, C. Carrillo, O. J. Rojas, Nanocellulose properties and applications in colloids and interfaces, Current Opinion in Colloid and Interface Science 19(5) (2014) 383-396. https://doi.org/10.1016/j.cocis.2014.10.003

[43]   A. Chakraborty, M. Sain, M. Kortschot, Cellulose microfibrils: a novel method of preparation using high shear refining and cryocrushing, Published by De Gruyte, (2005) 102-107. https://doi.org/10.1515/HF.2005.016

[44] H. Xie, H. Du, X. Yang, C. Si, Recent strategies in preparation of cellulose nanocrystals and cellulose nanofibrils derived from raw cellulose materials, International Journal of Polymer Science, 2018. https://doi.org/10.1155/2018/7923068

[45] E. Abraham, B. Deepa, L. A. Pothan, M. Jacob, S. Thomas, U. Cvelbar, R. Anandjiwala, Extraction of nanocellulose fibrils from lignocellulosicfibres: A novel approach, Carbohydrate Polymers, 86(4) (2011) 1468-1475. https://doi.org/10.1016/j.carbpol.2011.06.034

[46] M. N. Angles, A. Dufresne, Plasticized starch/tunicin whiskers nanocomposite materials. 2. Mechanical behavior, Macromolecules, 34(9) (2001) 2921-2931. https://doi.org/10.1021/ma001555h

[47] M. Matos Ruiz, J. Y. Cavaille, A. Dufresne, J. F. Gerard, C. Graillat, Processing and characterization of new thermoset nanocomposites based on cellulose whiskers, Composite Interfaces, 7(2) (2000) 117-131. https://doi.org/10.1163/156855400300184271

[48] W. P. F. Neto, J. L. Putaux, M. Mariano, Y. Ogawa, H. Otaguro, D. Pasquini, A. Dufresne, (2016). Comprehensive morphological and structural investigation of cellulose I and II nanocrystals prepared by sulphuric acid hydrolysis. RSC Advances, 6(79) 76017-76027.

[49] J. Araki, M. Wada, S. Kuga, T. Okano, Flow properties of microcrystalline cellulose suspension prepared by acid treatment of native cellulose, Colloids and Surfaces A: Physicochemical and Engineering Aspects 142(1) (1998) 75-82. https://doi.org/10.1016/S0927-7757(98)00404-X

[50] F. Beltramino, M. B. Roncero, A. L. Torres, T. Vidal, C. Valls, Optimization of sulfuric acid hydrolysis conditions for preparation of nanocrystalline cellulose from enzymatically pretreated fibers, Cellulose 23(3) (2016) 1777-1789. https://doi.org/10.1007/s10570-016-0897-y

[51] H. Y. Yu, Z. Y. Qin, L. Liu, X. G. Yang, Y. Zhou, J. M. Yao, Comparison of the reinforcing effects for cellulose nanocrystals obtained by sulfuric and hydrochloric acid hydrolysis on the mechanical and thermal properties of bacterial polyester, Composites Science and Technology 87(2013) 22-28. https://doi.org/10.1016/j.compscitech.2013.07.024

[52] H. Yu, Z. Qin, B. Liang, N. Liu, Z. Zhou, L. Chen, Facile extraction of thermally stable cellulose nanocrystals with a high yield of 93% through hydrochloric acid hydrolysis under hydrothermal conditions, Journal of Materials Chemistry A 1(12) (2013a) 3938-3944. https://doi.org/10.1039/C3TA01150J

[53] T. Koshizawa, Degradation of wood cellulose and cotton linters in phosphoric acid, Japan Tappi Journal 14(7) (1960) 455-458. https://doi.org/10.2524/jtappij.14.455

[54] H. Sadeghifar, I. Filpponen, S. P. Clarke, D. F. Brougham, D. S. Argyropoulos, Production of cellulose nanocrystals using hydrobromic acid and click reactions on

their surface, Journal of Materials Science 46(22) (2011) 7344-7355.
https://doi.org/10.1007/s10853-011-5696-0

[55]   L. Wang, X. Zhu, X. Chen, Y. Zhang, H. Yang, Q. Li, J. Jiang, Isolation and characteristics of nanocellulose from hardwood pulp via phytic acid pretreatment, Industrial Crops and Products 182 (2022) 114921. https://doi.org/10.1016/j.indcrop.2022.114921

[56]   A. Winter, B. Arminger, S. Veigel, C. Gusenbauer, W. Fischer, M. Mayr, W. Gindl-Altmutter, Nanocellulose from fractionated sulfite wood pulp, Cellulose, 27(16) (2020) 9325-9336. https://doi.org/10.1007/s10570-020-03428-8

[57]   V. A. Barbash O. V. Yashchenko, Preparation and application of nanocellulose from non-wood plants to improve the quality of paper and cardboard, Applied Nanoscience, 10(8) (2020) 2705-2716. https://doi.org/10.1007/s13204-019-01242-8

[58]   A. Dufresne, Nanocellulose: a new ageless bionanomaterial, Materials Today 16(6), (2013) 220-227. https://doi.org/10.1016/j.mattod.2013.06.004

[59]   J. Gröndahl, K. Karisalmi J. Vapaavuori, Micro-and NCs from non-wood waste sources; processes and use in industrial applications, Soft Matter, 17(43), (2021), 9842-9858. https://doi.org/10.1039/D1SM00958C

[60]   M. Rajinipriya, M. Nagalakshmaiah, M. Robert, S. Elkoun Importance of agricultural and industrial waste in the field of NC and recent industrial developments of wood based nanocellulose: a review, ACS Sustainable Chemistry and Engineering, 6(3) (2018) 2807-2828. https://doi.org/10.1021/acssuschemeng.7b03437

[61]   Y. Habibi, Key advances in the chemical modification of nanocelluloses, Chemical Society Reviews 43(5) (2014) 1519-1542. https://doi.org/10.1039/C3CS60204D

[62]   A. F. Turbak, F. W. Snyder K. R. Sandberg, Microfibrillated cellulose, a new cellulose product: properties, uses, and commercial potential, Journal of Applied Polymer Science. Applied Polymer Symposium 37(9) (1983) 815-827.

[63]   F. W. Herrick, R. L. Casebier, J. K. Hamilton, K. R. Sandberg, Microfibrillated cellulose: morphology and accessibility, Journal of Applied Polymer Science. Applied polymer Symposium 37(2) (1983) 797-813.

[64]   Y. Qing, R. Sabo, J. Y. Zhu, U. Agarwal, Z. Cai, Y. Wu, A comparative study of cellulose nanofibrils disintegrated via multiple processing approaches, Carbohydrate Polymers 97(1) (2013). 226-234. https://doi.org/10.1016/j.carbpol.2013.04.086

[65]   C. Miao, W. Y. Hamad, Cellulose reinforced polymer composites and nanocomposites: a critical review, Cellulose 20(5) (2013) 2221-2262. https://doi.org/10.1007/s10570-013-0007-3

[66]   T. Bhattacharjee, A Comprehensive Review on Important Preparation and Application of nanocelluloe, Medicon Medical Sciences, 4 (2023) 16-33. https://doi.org/10.55162/MCMS.04.091

[67]   H. L. Teo, R. A. Wahab, Towards an eco-friendly deconstruction of agro-industrial biomass and preparation of renewable cellulose nanomaterials: A review, International Journal of Biological Macromolecules 161(2020) 1414-1430. https://doi.org/10.1016/j.ijbiomac.2020.08.076

[68]   S. Fujisawa, Y. Okita, H. Fukuzumi, T. Saito, A. Isogai, Preparation and characterization of TEMPO-oxidized cellulose nanofibril films with free carboxyl groups, Carbohydrate Polymers 84(1) (2011) 579-583. https://doi.org/10.1016/j.carbpol.2010.12.029

[69]   T. Saito, S. Kimura, Y. Nishiyama, A. Isogai, Cellulose nanofibers prepared by TEMPO-mediated oxidation of native cellulose, Biomacromolecules 8(8) (2007) 2485-2491. https://doi.org/10.1021/bm0703970

[70]   F. Jiang, Y. L. Hsieh, Chemically and mechanically isolated nanocellulose and their self-assembled structures, Carbohydrate Polymers 95 (1) (2013) 32-40. https://doi.org/10.1016/j.carbpol.2013.02.022

[71]   R. Rohaizu, W. D. Wanrosli, Sono-assisted TEMPO oxidation of oil palm lignocellulosic biomass for isolation of nanocrystalline cellulose, Ultrasonics Sonochemistry 34(2017)631-639. https://doi.org/10.1016/j.ultsonch.2016.06.040

[72]   F. Barja, Bacterial nanocellulose production and biomedical applications Journal of Biomedical Research 35(4) (2021) 310. http://doi.org/10.7555/JBR.35.20210036

[73]   W. K.Czaja, D. J. Young, M. Kawecki, R. M. Brown, The future prospects of microbial cellulose in biomedical applications, Biomacromolecules, 8(1) (2007) 1-12. https://doi.org/10.1021/bm060620d

[74]   G. V. Sakovich, E. A. Skiba, V. V. Budaeva, E. K. Gladysheva L. A. Aleshina, Technological fundamentals of bacterial NC production from zero prime-cost feedstock, Doklady Biochemistry and Biophysics, 477(1) (2017) 357-359. https://doi.org/10.1134/S1607672917060047

[75]   C. H. Kuo, J. H. Chen, B. K. Liou, C. K. Lee, Utilization of acetate buffer to improve bacterial cellulose production by Gluconacetobacterxylinus, Food Hydrocolloids 53(2016) 98-103. https://doi.org/10.1016/j.foodhyd.2014.12.034

[76]   S. Mekhilef, R. Saidur, M. Kamalisarvestani, Effect of dust, humidity and air velocity on efficiency of photovoltaic cells, Renewable and Sustainable Energy Reviews 16(5) (2012) 2920-2925. https://doi.org/10.1016/j.rser.2012.02.012

[77]   J. Jiang, Y. Zhu, F. Jiang, Sustainable isolation of NC from cellulose and lignocellulosic feedstocks: Recent progress and perspectives, Carbohydrate Polymers 267 (2021) 118188. https://doi.org/10.1016/j.carbpol.2021.118188

[78]   M. B. Noremylia, M. Z. Hassan, Z. Ismail, Recent advancement in isolation, processing, characterization and applications of emerging nanocellulose: A review. International Journal of Biological Macromolecules. https://doi.org/10.1016/j.ijbiomac.2022.03.064

[79]   D. Trache, M. H. Hussin, M. M. Haafiz, V. K. Thakur, Recent progress in cellulose nanocrystals: sources and production, Nanoscale, 9(5) (2017) 1763-1786. https://doi.org/10.1039/C6NR09494E

[80]   C. Salas, T. Nypelö, C. Rodriguez-Abreu, C. Carrillo, O. J. Rojas, Nanocellulose properties and applications in colloids and interfaces, Current Opinion in Colloid and Interface Science, 19(5) (2014) 383-396. https://doi.org/10.1016/j.cocis.2014.10.003

[81]   T. Li, C. Chen, A. H. Brozena, J. Y. Zhu, L. Xu, C. Driemeier, L. Hu, Developing fibrillated cellulose as a sustainable technological material, Nature, 590(7844) (2021) 47-56. https://doi.org/10.1038/s41586-020-03167-7

[82]   D. Zhao, Y. Zhu, W. Cheng, W. Chen, Y. Wu, H. Yu, Cellulose based flexible functional materials for emerging intelligent electronics, Advanced Materials 33(28) (2021) 2000619. https://doi.org/10.1002/adma.202000619

[83]   Y. Ye, Y. Zhang, Y. Chen, X. Han, F. Jiang, Cellulose nanofibrils enhanced, strong, stretchable, freezing tolerant ionic conductive organohydrogel for multifunctional sensors, Advanced Functional Materials 30(35) (2020) 2003430. https://doi.org/10.1002/adfm.202003430

[84]   C. Vilela, A. J. Silvestre, F. M. Figueiredo C. S. Freire, Nanocellusoe-based materials as components of polymer electrolyte fuel cells, Journal of Materials Chemistry A 7(35) (2019) 20045-20074. https://doi.org/10.1039/C9TA07466J

[85]   S. A. Kedzior, V. A. Gabriel, M. A. Dubé, E. D. Cranston, Nanocellulose in emulsions and heterogeneous waterbased polymer systems: A review, Advanced Materials, 33(28), (2021), 2002404. https://doi.org/10.1002/adma.202002404

[86]   K. Heise, E. Kontturi, Y. Allahverdiyeva, T. Tammelin, M. B. Linder, O. Ikkala, Nanocellulose: recent fundamental advances and emerging biological and biomimicking applications, Advanced Materials, 33(3) (2021) 2004349. https://doi.org/10.1002/adma.202004349

[87]   M. N. F. Norrrahim, N. A. M. Kasim, V. F. Knight, F. A. Ujang, N. Janudin, M. A. I. A. Razak, W. M. Z. W. Yunus, Nancellulose: The next super versatile material for the military, Materials Advances, 2(5) (2021) 1485-1506. https://doi.org/10.1039/D0MA01011A

[88]   Q. F., Guan, H. B. Yang, Z. M. Han, Z. C. Ling, S. H. Yu, An all-natural bioinspired structural material for plastic replacement, Nature Communications 11(1) (2020) 1-7. https://doi.org/10.1038/s41467-020-19174-1

[89]   S. S. Nair, J. Y. Zhu, Y. Deng, A. J. Ragauskas, High performance green barriers based on nanocellulose, Sustainable Chemical Processes 2(1) (2014) 1-7. https://doi.org/10.1186/s40508-014-0023-0

[90]   T. A. Akter, J. Nayeem, A. H. Quadery, M. A. Razzaq, M. T. Uddin, M. S. Bashar, M. S. Jahan, Microcrystalline cellulose reinforced chitosan coating on kraft paper, Cellulose Chemistry and Technology 54(1–2) (2020) 95-102.

[91]   N. Jannatyha, S. Shojaee-Aliabadi, M. Moslehishad E. Moradi, Comparing mechanical, barrier and antimicrobial properties of NC/CMC and nanochitosan/CMC composite films, International Journal of Biological Macromolecules 164 (2020) 2323-2328. https://doi.org/10.1016/j.ijbiomac.2020.07.249

[92]   T. Saito, A. Isogai, TEMPO-mediated oxidation of native cellulose, The effect of oxidation conditions on chemical and crystal structures of the water-insoluble fractions, Biomacromolecules, 5(5), (2004), 1983-1989.

[93]   G. Rodionova, M. Lenes, O. Eriksen, O. Gregersen, Surface chemical modification of microfibrillated cellulose: improvement of barrier properties for packaging applications, Cellulose, 18 (2011) 127-134. http://doi.org/10.1007/s10570-010-9474-y

[94]   I. S. Bayer, D. Fragouli, A. Attanasio, B. Sorce, G. Bertoni, R. Brescia, A. Athanassiou, Water-repellent cellulose fiber networks with multifunctional properties, ACS Applied Materials and Interfaces, 3(10) (2011) 4024-4031. https://doi.org/10.1021/bm0497769

[95]   L. Bacakova, J. Pajorova, M. Tomkova, R. Matejka, A. Broz, J. Stepanovska P. Kallio, Applications of NC/nanocarbon composites: Focus on biotechnology and medicine, Nanomaterials, 10(2) (2020) 196. https://doi.org/10.3390/nano10020196

[96]   J. M. González-Domínguez, A. Ansón-Casaos, L. Grasa, L. Abenia, A. Salvador, E. Colom, W. K. Maser, Unique properties and behavior of nonmercerized type-II cellulose nanocrystals as carbon nanotube biocompatible dispersants, Biomacromolecules, 20(8) (2019) 3147-3160. https://doi.org/10.1021/acs.biomac.9b00722

[97]   N. G. Hatsopoulos J. P. Donoghue, The science of neural interface systems, Annual Review of Neuroscience, 32 (2009) 249.

[98]   J. R. Capadona, K. Shanmuganathan, D. J. Tyler, S. J. Rowan C. Weder, Stimuli-responsive polymer nanocomposites inspired by the sea cucumber dermis, Science, 319(5868) (2008) 1370-1374. http://doi.org/10.1126/science.115330

[99]   J. R. Capadona, K. Shanmuganathan, S. Trittschuh, S. Seidel, S. J. Rowan, C. Weder, Polymer nanocomposites with nanowhiskers isolated from microcrystalline cellulose, Biomacromolecules, 10(4) (2009) 712-716. https://doi.org/10.1021/bm8010903

[100] T. Abitbol, A. Rivkin, Y. Cao, Y. Nevo, E. Abraham, T. Ben-Shalom, O. Shoseyov, Nanocellulose, a tiny fiber with huge applications, Current Opinion in Biotechnology, 39 (2016) 76-88. https://doi.org/10.1016/j.copbio.2016.01.002

[101] R. M. Domingues, M. E. Gomes, R. L. Reis, The potential of cellulose nanocrystals in tissue engineering strategies, Biomacromolecules, 15(7) (2014) 2327-2346. https://doi.org/10.1021/bm500524s

[102] J. K. Jackson, K. Letchford, B. Z. Wasserman, L. Ye, W. Y. Hamad, H. M. Burt, The use of nanocrystalline cellulose for the binding and controlled release of drugs,

International Journal of Nanomedicine, 6 (2011) 321.
http://doi.org/10.2147/IJN.S16749

[103] X. Zhang, J. Huang, P. R. Chang, J. Li, Y. Chen, D. Wang, J. Chen, Structure and properties of polysaccharide nanocrystal-doped supramolecular hydrogels based on cyclodextrin inclusion, Polymer, 51(19) (2010) 4398-4407. https://doi.org/10.1016/j.polymer.2010.07.025

[104] D. O. Carlsson, K. Hua, J. Forsgren A. Mihranyan, Aspirin degradation in surface-charged TEMPO-oxidized mesoporous crystalline nanocellulose, International Journal of Pharmaceutics, 461(1-2), (2014), 74-81. https://doi.org/10.1016/j.ijpharm.2013.11.032

[105] N. Lin, A. Dufresne, Supramolecular hydrogels from in situ host–guest inclusion between chemically modified cellulose nanocrystals and cyclodextrin, Biomacromolecules, 14(3) (2013) 871-880. https://doi.org/10.1021/bm301955k

[106] S. Ling, W. Chen, Y. Fan, K. Zheng, K. Jin, H. Yu, D. L. Kaplan, Biopolymer nanofibrils: Structure, modeling, preparation, and applications, Progress in Polymer Science, 85 (2018) 1-56. https://doi.org/10.1016/j.progpolymsci.2018.06.004

[107] L. Zhu, X. Zhou, Y. Liu, Q. Fu, Highly sensitive, ultrastretchable strain sensors prepared by pumping hybrid fillers of carbon nanotubes/cellulose nanocrystal into electrospun polyurethane membranes, ACS Applied Materials and Interfaces, 11(13) (2019) 12968-12977. https://doi.org/10.1021/acsami.9b00136

[108] S. Zhang, H. Liu, S. Yang, X. Shi, D. Zhang, C. Shan, Z. Guo, Ultrasensitive and highly compressible piezoresistive sensor based on polyurethane sponge coated with a cracked cellulose nanofibril/silver nanowire layer, ACS Applied Materials and Interfaces, 11(11) (2019) 10922-10932. https://doi.org/10.1021/acsami.9b00900

[109] D. Zhao, Q. Zhang, W. Chen, X. Yi, S. Liu, Q. Wang, H. Yu, Highly flexible and conductive cellulose-mediated PEDOT: PSS/MWCNT composite films for supercapacitor electrodes, ACS Applied Materials and Interfaces, 9(15) (2017) 13213-13222.

[110] Y. Ko, M. Kwon, W. K. Bae, B. Lee, S. W. Lee, J. Cho, Flexible supercapacitor electrodes based on real metal-like cellulose papers, Nature Communications 8(1) (2017) 1-11. https://doi.org/10.1038/s41467-017-00550-3

[111] U. Vainio, N. Maximova, B. Hortling, J. Laine, P. Stenius, L.K. Simola, J. Gravitis, R. Serimaa, Morphology of dry lignins and size and shape of dissolved kraft lignin particles by X-ray scattering, Langmuir 20(22) (2004) 9736-9744. https://doi.org/10.1021/la048407v

[112] T. Higuchi, Lignin biochemistry: biosynthesis and biodegradation, Wood Science and Technology 24 (1990) 23-63. https://doi.org/10.1007/BF00225306

[113] C. Clemons, In: Wood -Polymer Composites. K. O. Niska, M. Sain, Eds. Woodhead Publ. Ltd.: Cambridge, U.K., 12, (2012)

[114] W. Boerjan, J. Ralph, M. Baucher, Lignin Biosynthesis, Annual Review of Plant Biology 54(2003) 519–546. http//doi.org/10.1146/annurev.arplant.54.031902.134938

[115] P. Sannigrahi, A.J. Ragauskas, Characterization of fermentation residues from the production of bio-ethanol from lignocellulosic feedstocks, Journal of Biobased Materials and Bioenergy 5 (2011) 514–519. https://doi.org/10.1166/jbmb.2011.1170

[116] C. Li, X. A. Zhao, Wang, G.W. Huber, T. Zhang, Catalytic transformation of lignin for the production of chemicals and fuels, Chemical Reviews 115 (2015) 11559–11624. https://doi.org/10.1021/acs.chemrev.5b00155

[117] R.J.A. Gosselink, A. H. AbächerliSemke, R. Malherbe, P. Käuper, A. Nadif, J.E.G. Van Dam, Analytical protocols for characterization of sulphur-free lignin, Industrial Crops and Products 19(3) (2004) 271-281. https://doi.org/10.1016/j.indcrop.2003.10.008

[118] R. S. Popy, Y. Ni, A. Salam, M. S. Jahan, Mild potassium hydroxide-based alkaline integrated biorefinery process of Kash (Saccharumspontaneum), Industrial Crops and Products 154, (2020), 112738. https://doi.org/10.1016/j.indcrop.2020.112738

[119] M. S. Jahan, M. M. Rahman, Y. Ni, Alternative initiatives for nonwood chemical pulping and integration with the biorefinery concept: A review, Biofuels, Bioproducts and Biorefining 15(1) (2021) 100-118. https://doi.org/10.1002/bbb.2143

[120] S. Sutradhar, K. M. Y. Arafat, J. Nayeem, M. S. Jahan, Organic acid lignin from rice straw in phenol-formaldehyde resin preparation for plywood, Cellulose Chemistry and Technology 54(5-6), (2020), 463-471.

[121] C. Jiang, H. He, H. Jiang, L. Ma, D.M. Jia, Nano-lignin filled natural rubber composites: Preparation and characterization, Express Polymer Letters, 7(5) (2013). https://doi.org/10.3144/expresspolymlett.2013.44

[122] B. Wang, D. Sun, H. M. Wang, T. Q. Yuan, and R. C. Sun, Green and facile preparation of regular lignin nanoparticles with high yield and their natural broad-spectrum sunscreens, ACS Sustainable Chemistry and Engineering 7(2) (2018) 2658-2666. https://doi.org/10.1021/acssuschemeng.8b05735

[123] Y. Qian, Y. Deng, X. Li, H. Qiu, D. Yang, Formation of uniform colloidal spheres from lignin, a renewable resource recovered from pulping spent liquor, Green Chemistry 16(4) (2014) 2156-2163. https://doi.org/10.1039/C3GC42131G

[124] M. Lievonen, J.J. Valle-Delgado, M.L. Mattinen, E.L. Hult, K. Lintinen, M.A. Kostiainen, A. Paananen, G.R. Szilvay, H. Setälä, and M. Österberg, A simple process for lignin nanoparticle preparation, Green Chemistry 18(5) (2016) 1416-1422. https://doi.org/10.1039/C5GC01436K

[125] P. Figueiredo, K., Lintinen, A. Kiriazis, V. Hynninen, Z. Liu, T. Bauleth-Ramos, A. Rahikkala, A. Correia, T. Kohout, B. Sarmento, In vitro evaluation of biodegradable lignin-based nanoparticles for drug delivery and enhanced

antiproliferation effect in cancer cells, Biomaterials, (2017) 121 97–108. https://doi.org/10.1016/j.biomaterials.2016.12.034

[126] K. Xiong, C. Jin, G. Liu, G. Wu, J. Chen, Z. Kong, Preparation and characterization of lignin nanoparticles with controllable size by nanoprecipitation method, Chemistry and Industry for Forest Products 2015 35 85–92

[127] S.R. Yearla, and K. Padmasree, Preparation and characterisation of lignin nanoparticles: Evaluation of their potential as antioxidants and UV protectants. Journal of Experimental Nanoscience 11 (2016) 289–302. https://doi.org/10.1080/17458080.2015.1055842

[128] R. S. Fukushima, R. D. Hatfield, Extraction and isolation of lignin for utilization as a standard to determine lignin concentration using the acetyl bromide spectrophotometric method, J. Agric. Food Chem. 49 (2001) 3133–3139. https://doi.org/10.1021/jf010449r

[129] A.P. Richter, B. Bharti, H. B. Armstrong, J.S. Brown, D. Plemmons, V.N. Paunov, S.D.Stoyanov, O.D. Velev, Synthesis and characterization of biodegradable lignin nanoparticles with tunable surface properties, Langmuir, 32 (2016) 6468–6477. https://doi.org/10.1021/acs.langmuir.6b01088

[130] H. Li, Y. Deng, B. Liu, Y. Ren, J. Liang, Y. Qian, X. Qiu, C. Li, D. Zheng, Preparation of nanocapsules via the self-assembly of kraft lignin: A totally green process with renewable resources, ACS Sustainable Chemistry and Engineering 4(4) (2016) 1946-1953. https://doi.org/10.1021/acssuschemeng.5b01066

[131] S.J. Juikar, and N. Vigneshwaran, Extraction of nanolignin from coconut fibers by controlled microbial hydrolysis, Industrial Crops and Products 109 2017 420-425. https://doi.org/10.1016/j.indcrop.(2017).08.067

[132] R. H. Müller, K. Peters, R. Becker, B. Kruss, Nanosuspensions for the iv administration of poorly soluble drugs-stability during sterilization and long-term storage, Proc. Int. Symp. Control Rel. Bioact. Mater. 22 (1995) 574–575.

[133] R. H. Müller, K. Peters, Nanosuspensions for the formulation of poorly soluble drugs, International Journal of Pharmaceuticals 160, (1998), 229–237. https://doi.org/10.1016/S0378-5173(97)00311-6

[134] E. Merisko-Liversidge, P. Sarpotdar, J. Bruno, S. Hajj, L. Wei, N. Peltier, J. Rake, J.M. Shaw, S. Pugh, L. Polin, Formulation and antitumor activity evaluation of nanocrystalline suspensions of poorly soluble anticancer drugs, Pharmaceutical Research 13 (1996) 272–278. https://doi.org/10.1023/A:1016051316815

[135] R. J. Malcolmson, J. K. Embleton, Dry powder formulations for pulmonary delivery, Pharmaceutical Science and Technology Today 1 (1998) 394–398. https://doi.org/10.1016/S1461-5347(98)00099-6

[136] S. S. Nair, S. Sharma, Y. Pu, Q. Sun, S. Pan, J. Y. Zhu, Y. Deng, A. J. Ragauskas, High shear homogenization of lignin to nanolignin and thermal stability of nanolignin-

polyvinyl alcohol blends, ChemSusChem 7 (2014) 3513–3520.
https://doi.org/10.1002/cssc.201402314

[137] I. A. Gilca, V. I. Popa, C. Crestini, Obtaining lignin nanoparticles by sonication. Ultrason, Sonochemistry 23 (2015) 369–375.
https://doi.org/10.1016/j.ultsonch.2014.08.021

[138] M.D. Shawn, K.N. Cicotte, D.R. Wheeler, D.A. Benko, Lignin Nanoparticle Synthesis, U.S. Patent 9,102,801, 11 (2015). https://doi.org/10.3390/ijms18061244

[139] S. Beisi, A. Miltner, A. Friedl, Lignin from Micro- to Nanosize: Production Methods, International Journal of Molecular Sciences 18(6), 1244 (2017)
https://doi.org/10.3390/ijms18061244

[140] M. Zimniewska, R. Kozłowski, J. Batog, Nanolignin modified linen fabric as a multifunctional product, Molecular Crystals and Liquid Crystals, 484(1) pp.43-409.
https://doi.org/10.1080/15421400801903395

[141] F.R. Wurm, C. K. Weiss, Nanoparticles from renewable polymers, Frontiers in Chemistry 2 (2008) p.49. https://doi.org/10.3389/fchem.2014.00049

[142] W. Yang, M. Rallini, M. Natali, J. Kenny, P. Ma, W. Dong, L. Torre, D. Puglia, Preparation and properties of adhesives based on phenolic resin containing lignin micro and nanoparticles: a comparative study, Materials and Design, 161 (2019) pp.55-63. https://doi.org/10.1016/j.matdes.2018.11.032

[143] D. Kai, M. J. Tan, P. L. Chee, Y. K. Chua, Y. L. Yap, X. J. Loh, Towards lignin-based functional materials in a sustainable world, Green Chemistry 18 (2016) 1175–1200. https://doi.org/10.1039/C5GC02616D

[144] W. Yang, J.S. Owczarek, E. Fortunati, M. Kozanecki, A. Mazzaglia, G. M. Balestra, J. M. Kenny, L. Torre, D. Puglia, Antioxidant and antibacterial lignin nanoparticles in polyvinyl alcohol/chitosan films for active packaging, Industrial Crops and Products 94 (2016) pp.800-811.
https://doi.org/10.1016/j.indcrop.2016.09.061

[145] S. Danti, L. Trombi, A. Fusco, B. Azimi, A. Lazzeri, P. Morganti, M.B. Coltelli, G. Donnarumma, Chitin nanofibrils and nanolignin as functional agents in skin regeneration, International Journal of Molecular Sciences 20(11) (2019) 2669.
https://doi.org/10.3390/ijms20112669

[146] B. Del Saz-Orozco, M. Oliet, M. V. Alonso, E. Rojo, F. Rodríguez, Formulation optimization of unreinforced and lignin nanoparticle-reinforced phenolic foams using an analysis of variance approach, Composite Science and Technology 72 (2012) 667−674. https://doi.org/10.1016/j.compscitech.2012.01.013

[147] X. Luo, A. Mohanty, M. Misra, Lignin as a reactive reinforcing filler for water-blown rigid biofoam composites from soy oil-based polyurethane, Industrial Crops and Products 47, (2013) 13−19. https://doi.org/10.1016/j.indcrop.2013.01.040

[148] Z. Chen, P. Li, Q. Ji, Y. Xing, X. Ma, Y. Xia, All-polysaccharide composite films based on calcium alginate reinforced synergistically by multidimensional cellulose and

hemicellulose fractionated from corn husks, Materials Today Communications 34105090. https://doi.org/10.1016/j.mtcomm.2022.105090

[149] M.M. Haque, Y. Ni, A. J .U. Akon, M. A. Quaiyyum, M. S. Jahan, , A review on Acacia auriculiformis: importance as pulpwood planted in social forestry International, Wood Products Journal 12(3) (2021) pp.194-205.

[150] L. C. Malucelli, L. G. Lacerda, M. Dziedzic, da Silva, M. A. C. Filho, Preparation, properties and future perspectives of nanocrystals from agro-industrial residues: a review of recent research, Reviews in Environmental Science and Bio/Technology 16 (2017) 131-145. https://doi.org/10.1007/s11157-017-9423-4

[151] M. S. Jahan, A. Al-Maruf, M. A. Quaiyyum, Comparative studies of pulping of jute fiber, jute cutting and jute caddis. Bangladesh Journal of Scientific and Industrial Research42(4), (2007) 425-434.

[152] T. Ferdous, M. A. Quaiyyum, A. Salam, M. S. Jahan, Pulping of bagasse (Saccrarum officinarum), kash (Saccharum spontaneum) and corn stalks (Zea mays), Current Research in Green and Sustainable Chemistry 3, (2020) 100017.

[153] A.C.F. Louis, S. Venkatachalam, S. Gupta, Innovative strategy for rice straw valorization into nanocellulose and nanohemicellulose and its application. Industrial Crops and Products, (2022),179, 114695 https://doi.org/10.1016/j.indcrop.2022.114695

[154] D. Klemm, F. Kramer, S. Moritz, T. Lindström, M. Ankerfors, D. Gray, A. Dorris, Nanocelluloses: a new family of nature based materials, Angewandte Chemie International Edition 50(24) (2011) 5438-5466. https://doi.org/10.1002/anie.201001273

Applications of Emerging Nanomaterials and Nanotechnology   Materials Research Forum LLC
Materials Research Foundations 148 (2023) 229-251   https://doi.org/10.21741/9781644902554-8

Chapter 8

Application of Nanoparticles in Soil and Water Treatment

Azizul Hakim[1*], Ferdouse Zaman Tanu[2], Hafiz Ashraful Haque[3], Md. Abu Bin Hasan Susan[4]

[1]Department of Soil Science, University of Chittagong, Chattogram 4331, Bangladesh

[2]Department of Soil and Environmental Sciences, University of Barishal, Barishal 8254, Bangladesh

[3]Department of Coastal Studies and Disaster Management, University of Barishal, Barishal 8254, Bangladesh

[4]Department of Chemistry, University of Dhaka, Dhaka 1000, Bangladesh

*ahakimsoil@cu.ac.bd

Abstract

Nanoparticles (NPs) exhibit size- and shape-dependent properties and may have distinctive colors that can be used in agriculture and biological applications as well as for physical and chemical studies. NPs have received significant attention in place of conventional bulk materials, and there has been an upsurge of interest in the exploitation of the unique features of NPs for soil and water treatment. In particular, nanoremediation is one of the most environmentally friendly techniques to use in soil, surface, and wastewater systems, especially with the expanding environmental pollution issues. The high surface-to-volume ratio of NPs results in a high absorption capacity for the remediation of contaminated soil and wastewater. This chapter focuses on the role of NPs in remediating polluted soil and water and the process of nanoremediation.

Keywords

Nanoparticles, Engineered Nanoparticles, Soil and Water, Wastewater, Heavy Metals, Contaminants

Contents

1.   Introduction

The "nanoparticle" is one of the most fascinating domains of modern research and since the last two decades it has been increasingly popular and practiced in the scientific world. Especially, eyes have been pointed on nanoparticles (NPs) to address issues regarding environmental contamination and pollution due to the rapid increase of industrialization. They have been found useful as novel tools for environmental protection, especially for soil and wastewater treatment [1]. The most recent applications of NPs have evolved nanotechnology, which have been key to the most successful achievements in the field of treatment of soil and water pollution. NPs used are nanoadsorbents, nanosensors, nanomembranes, and disinfectants, with size ranging from 1 to 100 nm in at least one-dimension. The NPs exhibit tremendous physical and chemical properties that significantly vary from those of the bulk materials. They typically have size-dependent properties, large specific surface area and consequently high catalytic activities, and high chemical reactivity, which give them a high degree of functionalization along with extraordinary optical, thermal, electromagnetic, and mechanical properties and structural and morphological features [2]. All these properties of NPs along with their wide availability [3-4] offer suitable features for wastewater treatment, water purification [5], and remediation processes of soil and sediments.

Lack of clean water is now a global concern, which is reflected in "The 2030 Agenda for Sustainable Development" and placed 6th entitled "Ensure availability and sustainable management of water and sanitation for all" among the 17 sustainable development goals (SDG), approved by all United Nations Member States in 2015. Sustainable soil remediation and water purification processes are required for a sustainable environment on the Earth. However, the goal is currently challenging due to the increasing human population growth, propagation of urbanization, and industrialization that disrupt natural ecosystems, deplete natural resources, cause a shortage of drinking water [6], and produce tremendous amounts of hazardous waste in the environment. Wastes from transportation, construction, manufacturing, mining, petroleum refining, etc. are discharged into the environment in various ways and contaminate the air, water, and soil. These pollutants have the potential to negatively impact human health [7-8] and endanger the quality of the environment since they enter the body by ingestion, inhalation, or absorption. The contaminants of soil and water may include organic materials such as insecticides,

Applications of Emerging Nanomaterials and Nanotechnology   Materials Research Forum LLC
Materials Research Foundations 148 (2023) 229-251   https://doi.org/10.21741/9781644902554-8

pesticides, hydrocarbons, phenols and heavy metals such as cadmium, lead, mercury, arsenic and pathogenic microorganisms. Persistent organic pollutants (POPs) [9-10] and heavy metals typically bioaccumulate through food chains [11] posing significant risks to humans and wildlife. Therefore, there is an urgent need to treat hazardous environmental pollutants. The growing interest in nanotechnology over conventional technologies [4] is due to threats of secondary pollution and high cost of conventional approaches for remediation such as chlorination and ozonation. Nanotechnology is providing promising, sustainable, cost-effective, as well as efficient approaches to materialize environmental remediation.

However, there are still certain restrictions on how NPs can be used to remediate soil and water. The ease of aggregation [4], rapid tendency for passivation, sensitivity to geochemical factors [12], and human health and environmental concerns of various NPs [13] are a few proven limitations on the application of NPs. Research to-date experienced many attempts to synthesize various nanomaterials, which *inter alia* include nanofibers, nanorods, nanowires, nanobelts, etc. using photocatalytic deposition (PD), deposition-precipitation (DP), chemical vapour decomposition (CVD), chemical solution decomposition (CSD), ultrasonic irradiation, wet chemical, sol-gel, thermal, and hydrothermal methods [14], etc. The production of NPs using microbes, microbially produced compounds, and plant extract-mediated agents expands the variety of NPs and is believed to be more affordable and environmentally benign than using conventional methods [15-16]. Novel materials may also be obtained by chemical and physical modification of the surfaces of NPs, and thus nanotechnology can overcome many complex challenges that hinder the application of NPs during remedial processes [17].

2.   History of nanoparticles and nanosciences

Richard R. Feynman, a Nobel Prize-winning physicist, coined the term "nanotechnology" in 1959 [18-19] after stating at Caltech that there was "plenty of room at the bottom" in his inspirational speech (USA). The quote was frequently used in publications devoted to nanotechnology to imply that there remained a plethora of tasks to be accomplished in terms of the fundamentals of nanoparticles and nanotechnology. Nanotechnology was defined by Taniguchi as a "production technology" with "ultrahigh accuracy and ultra-small sizes of approximately 1 nm" [20]. Eric Drexler later defined nanotechnology as the "management of atoms and molecules to build structures to sophisticated, atomic requirements" [21-22]. Atomic force microscopy (AFM) developed in 1986 and the invention of scanning tunnelling microscopy (STM) in 1983 were two pioneering innovations that launched the era of nanoparticle research [23-24].

NPs are either derived from natural sources such as viruses, bacteria, proteins, volcanic eruptions, nanostructured crystals and minerals, forest fires, etc. or may be artificially developed or be of human origin. Anthropogenic NPs can be subdivided into unintentionally and intentionally produced NPs [19]. The earlier class relates to NPs generated using specific manufacturing methods, while the later includes NPs formed via

Applications of Emerging Nanomaterials and Nanotechnology   Materials Research Forum LLC
Materials Research Foundations 148 (2023) 229-251   https://doi.org/10.21741/9781644902554-8

the combustion of diesel or gasoline, power plants, and incinerators. However, "nanotechnology" usually refers to deliberately produced artificial NPs called engineered nanoparticles. Additionally, according to the size of their three dimensions, nano-objects are categorized as NPs, and nanotubes or nanowires [19]. Various metals, including gold or silver, as well as metal oxides (e.g., titanium dioxide), can be used to produce NPs depending on the requirements of a particular application.

While clay was used in Cyprus to bleach wool and clothing as far back as 5000 BCE, it is customary to employ gold colloids as a colorant. [25]. Kaolin, an incredibly fine clay mineral, was used as the primary raw material in China around the ninth century CE to produce porcelains [25]. It is believed that the use of other metallic NPs dates back to the 14th and 13th centuries BCE. Gold NPs were used to make red-colored glasses known as "gold ruby glass" in Rome between the 4th and 12th centuries [19]. Silver NPs were also used for yellow stained glasses during the 14th centuries [26]. Metallic NPs were used to achieve optical effects in glasses and ceramics in addition to colored glasses and ceramics [27].

Early in the 1960s, the most significant studies on the distinctive characteristics of NPs and clusters were initiated [28]. Due to the growing variety of uses for NPs, several novel materials have been developed [13, 29]. Nowadays, there are numerous industrial processes including manufacture of several common goods and systems that use NPs. Large-scale applications include pharmacology, chemistry, photocatalysis, agriculture, packaging, shipping and navigation, construction, petroleum, photonic, textile products, electronics, healthcare, and other fields. Bao et al. [30] reported that 58,000 tonnes of NPs are now produced annually. In 2022, it has significantly increased [31].

Nano titanium dioxide (TiO_2), nano zinc oxide (ZnO), and nano silicon dioxide (SiO_2) are the most manufactured NPs in the world [32]. Iron and silver NPs have also dominated the global commercial market. Different synthesis techniques are used to prepare NPs with regulated shape, size, and composition.

3.   Nanoparticles in soil and water environment

The application of NPs in water and soil remediation techniques is constantly growing and developing new avenues. The most prevalent and, in some situations, unavoidable problem of the world today is environmental pollution. During the mining process, significant amounts of toxic heavy metals such as lead, arsenic, chromium, cadmium, iron, copper, zinc, and mercury are leached. This particular effluent is frequently released into moving rivers without proper treatments [33-34]. As a result, the solubilized heavy metals are easily transported into groundwater, sediments, and soils and pose a long-term threat as contaminants. In addition, the petroleum industry, manufacturing units, other industries, and underground storage tanks can all contribute to soil and groundwater pollution by releasing solvents and organics through leakage and irresponsible handling of synthetic chemicals.

Nanoscience enables researchers to study novel materials at the nanoscale, offering nanotechnology a potential and environmentally friendly answer to the problems that wastewater, solid waste, and contaminated sites pose to the environment [35]. The usage of nanotechnology in the realm of the environment involves the development of environmentally friendly products, the removal of dangerous contaminants from polluted materials, and the development of environmental agent sensors [36]. Due to the advantageous characteristics of NPs, the development of alternatives to the traditional remediation techniques, such as the use of NPs, has gained popularity. Because of their increased intrinsic reactivity, NPs can quickly convert and/or detoxify long-term contaminants and remove hazardous wastes. For instance, NPs like nanoscale zerovalent iron (nZVI), which outperforms typical iron species in terms of fast reactivity, limited interruption, non-toxicity, and cost-effectiveness, have assured significant uses in soil and water remediation [37, 38]. An example of *in situ* application of nanotechnology for treatment is the injection of nanosized (mobile) reactive or absorptive NPs in an aqueous system of contaminated regions. The occurrence and fate-transport of different NPs are presented in Figure 1.

In general, different NPs have been proved useful at cleaning up the environment, which are based on: metal, metal-oxide, carbon, silica, and polymer. Heavy metals and chlorinated organic contaminants can be removed from polluted water and soil using a variety of metals and NPs based on their oxides. TiO_2 can eliminate bacteria, odours, dirt, and other contaminants. TiO_2 can absorb UV photons, which aids in the production of oxidative radicals HO and O-/HO on the surface of TiO_2 semiconductors because of its photocatalytic activity [32]. These radicals have the power to oxidize organic contaminants and turn them into CO_2 and H_2O through mineralization. TiO_2 can kill bacteria and fungus due to the UV-photons. TiO_2 coated glass stays clear and clean because of the hydrophilic feature. TiO_2-based reactors are also being developed to treat air pollution and eliminate odors. The use of TiO_2 for water purification is currently the focus of the attention of the researchers around the globe.

Due to their great flexibility, adsorption, and catalysis in environmental remediation processes, silica-based NPs have a wide range of applications. The surface and pore walls of silica materials performed as good adsorbents and catalysts due to the presence of different functional groups. Thus, surfaces of silica materials with hydroxyl groups improved surface modification, wetting, and gas adsorption [40-41], whereas aluminosilicates with amine modifications took up CO_2 gases and carbonyl compounds [42]. Organic dyes, metal ions, and pollutants can be removed from surface of water bodies and wastewater by functionalizing mesoporous silica with COOH groups [43]. Functionalizing silica materials with amino, aminopropyl, and thiol groups [44, 45] is helpful in the remediation of metal ions like Cd^{2+}, Co^{2+}, Cu^{2+}, Zn^{2+}, Ni^{2+}, Al^{2+}, Cr^{3+}, Pb^{2+}, Hg^{2+}, and U^{6+}. Although all metal, silica, and carbon-based NPs have extremely large surface areas and great reactivity, their uses in environmental remediation are generally limited by their low stability and accumulation. The constraints have been somewhat removed by the use of polymer-based NPs, which serve as matrices for other nanomaterials

[40, 46]. The use of NPs in remediation has also been made possible by their catalytic, photocatalytic, and adsorption characteristics. Even at ambient temperature, a relatively modest concentration of gold NPs (diameter 5 nm) can catalyze the conversion of hazardous CO into CO_2 [33]. Masks and respirators of firefighters therefore contain gold NPs that oxidize CO. For the removal of arsenic from aquifers, iron oxide NPs are used. Both As^{3+} and As^{5+} can be very effectively absorbed by iron oxides. Arsenic water filters are now commercially available and utilize iron oxide-based resins [34].

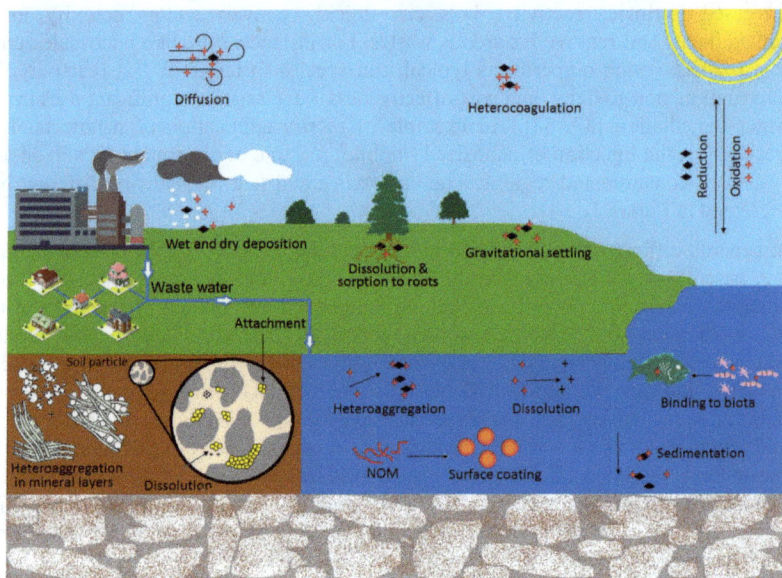

Fig. 1. Occurrence, distribution, and fate-transport of NPs in soil and water bodies (source: Abbas et al. [39], Environment International 138, reused under the Creative Commons Attribution License).

In the last few years, there have been many fascinating developments in the domains of agricultural, food, and fertilizer industries using NPs. For instance, smart tools are used to distribute fertilizers, nanoporous zeolites are utilized to increase fertilizer effectiveness, and nanosensors are used to gauge soil quality [47]. Other studies [47-48] are being conducted on the use of silver NPs and nanoclays to enhance water filtration as well as the use of nanosensors to find viruses in food. The detection of viral, bacterial, and fungal infections in soils and water has been made possible by the use of nanosensors [49-50].

During the manufacturing and application processes, engineered NPs will unavoidably be discharged into the environment contaminating the land, water, and food supply, which could endanger both human health and environmental quality. NPs also occur naturally in water and soil [51-52]. As a result, human exposure to environmental NPs has raised concerns about their adverse effects. The utilization of the modified NPs determines how they move to the environment.

According to Boxall et al. [53], NPs of TiO_2, ZnO, Fe_2O_3, Ag, and fullerene from personal care and cosmetic products can be found in surface water and wastewater, while the soil, surface water, ground water, and waste water may contain Fe, Pb, and polyurethane generated during water treatment and environmental remediation procedures. Porous SiO_2 can be found in the soil, surface water, and air, and which is employed in agrochemicals. Silver, nanoclay, and TiO_2 from the food packaging industry can also be found in soil and solid wastes, in addition to nanomedicines and carriers from the pharmaceutical and medical sectors. However, the use of pesticide is more likely to lead soil and surface water contamination, and direct discharges of NPs into surface and ground water are more likely to occur when NPs are used in water purification processes.

The degree of aggregation [54-55], size range of the aggregates, types of aggregation, and surface properties of the particles may influence how NPs behave and how they disintegrate in soil and water systems. Environmental variables including pH, ionic strength, and the amount of dissolved organic carbon are some examples of phenomena that may also have an impact on the aggregation and surface properties of natural and synthetic NPs [54-59].

NPs may sediment out in association with bed sediments after they have aggregated, according to some earlier investigations [60]. Some polymer-based particles may also degrade naturally [61]. NPs go to water bodies through runoff, leaching, and drain flow after possibly undergoing mechanisms of sorption, aggregation, and biotic and/or abiotic decomposition once released in water and soil environment. Surfactants can be used to change the mobility and transport of NPs depending on aggregation-dispersion behaviour of particles [56], which are affected by the environmental factors like pH and ion concentrations [54-55]. Figure 2 depicts the pH-dependent adsorption of heavy metal ions. To completely comprehend the complex interactions of aggregated NPs with sediments and floating solids like natural organic matter (NOM), as well as their persistence in the soil and aquatic environment, more research is nonetheless required.

The bioaccumulation and bioconcentration of NPs in aquatic and terrestrial organisms as well as in food is a current concern, as is the presence of NPs in the nature. Invertebrates, fish, algae, and bacteria are the most significant environmental species that are impacted by uptake of NPs [53, 63-64]. While concentrations of Ag, C60, and Al_2O_3 NPs in water were predicted to be in the ng/L range, sublethal effects on such species from engineered NPs like TiO_2, SiO_2, and ZnO were even seen in the µg/L range [65] in laboratory studies. However, under realistic environmental conditions, the impacts on microbial communities were different from those in the laboratory [66]. The types of synthesized NPs and variations in bioavailability are responsible for these discrepancies between laboratory and field studies [53].

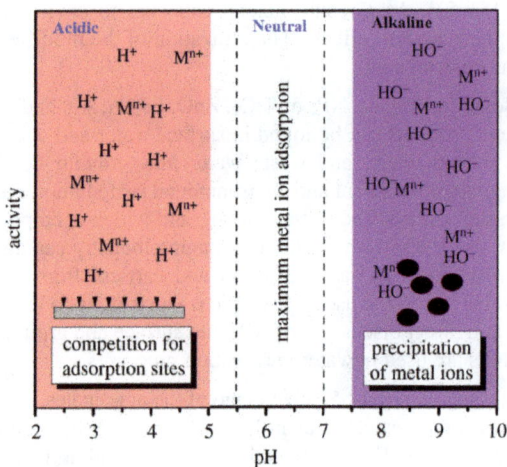

Fig. 2. Illustration of adsorption properties of heavy metal ion at various pH levels on inorganic nanoadsorbents (Source: Mensah et al. [62], Royal Soc. Open Sci. 8: 201485, reused under the Creative Commons Attribution License).

Even though previous data on the toxicity to aquatic organisms showed concentrations of currently-existing engineered NPs in the environment to be lower than those necessary to cause both lethal and sub-lethal effects, little research has been done on the potential health risks of exposure to engineered NPs in the environment [60]. Although some aquatic organisms have the capacity to absorb and retain particular NPs, it is still unclear how these particles may enter plants and possibly move up to the food chain. It is evident that when engineered NPs are used more often in daily life, there is a greater risk of environmental pollution and human exposure. Therefore, more research is required to determine the main exposure routes for humans as well as the current and future exposure routes.

4.   Application of nanoparticles for the treatment of soil

Heavy metal ions are the most common soil pollutants, but toxic levels rarely exist. Manufacturing, mining, landfills that take industrial wastes such as paint residues, batteries, electrical wastes, etc., and municipal or industrial sludge are the main sources of heavy metals in soil. These non-degradable metal substances stay in the contaminated sites for a long time. Microorganisms, however, have the ability to transform and evaporate the mercury and selenium. The best way to protect large areas of polluted soil is to use immobilization techniques to prevent heavy metal contamination or slow down the spread of heavy metals in soil [67]. The interaction of toxic heavy metals with soil constituents is mediated by sorption-desorption [68]. Ion exchange, surface precipitation, adsorption to mineral surfaces, and inducing the formation of stable complexes with organic ligands are some of the sorption processes [69]. In soil remediation, especially in heavy metal-contaminated soils, both mobilizing and immobilizing agents are used [70]. By increasing the mobility and bioavailability of heavy metals, mobilizing agents can better remove them through plant uptake and soil washing (phytoextraction). The immobilizing agents, on the other hand, decrease the bioavailability and mobility of heavy metals, preventing their leaching into groundwater and minimizing their transmission to the food chain (phytostabilization).

In order to immobilize heavy metals in soil, NPs have attracted a lot of attention. Two key requirements must be fulfilled while using NPs. One requirement is that they can be delivered to contaminated areas and should reside there even after the source is removed, while the other is that delivered NPs in contaminated areas act as an immobile sink for removing soluble metal ions. Due to their propensity to aggregate, NPs lose some of their high specific surface area. To solve this problem, stabilizers are typically incorporated on the NPs, such as starch [71] and carboxymethyl cellulose (CMC) [72]. Through steric and/or electrostatic stabilizing mechanisms, the encapsulated NPs with a higher specific surface area can prevent agglomeration and improve the stability of soil structure. For example, immobilization of arsenate As (V) and *in situ* enhanced sorption accomplished using magnetite NPs stabilized by starch [73] and a significant reduction in water-leachable As(V) could be observed.

In order to effectively immobilize heavy metals *in situ*, phosphate compounds including phosphoric acid, synthesized apatite, naturally occurring phosphate rocks, and even fishbone can be applied to soil. In contaminated soils, phosphate can also be applied to sequester lead [74]. As a result, apatite NPs accelerated the dissolution of phosphate and associated lead as pyromorphite, a stable lead phosphate compound, in soil [75]. The nZVI is commonly used to immobilize heavy metals in soil through *in situ* reduction. The use of nZVI has the disadvantage that it might aggregate or react rapidly with dissolved oxygen or water, decreasing its reactivity and solubility in soil. The influences of graphene oxide NPs (nGOx) and nZVI on the availability of different heavy metal ions in arsenic-metals-polluted soils (AM) are presented in Figure 3.

Due to their micron range, the agglomerated ZVI particles cannot be transported through soils or delivered there, and they are also not suitable for *in situ* treatments. A number of ZVI particle-stabilizing approaches are reported in the literature, which can minimize agglomeration. For the modification of nZVI [72], starch [77], polyvinylpyrrolidone [78], cetylpyridinium chloride [79], and sodium CMC can be employed for organic coatings. The high expense of chemical compounds comprising significant amounts of ferrous sulfate and ferrous chloride further restricts the use of iron-based NPs. Costs were reduced through the efficient removal of Cr(VI) from polluted soil using nZVI that had been stabilized by CMC [80].

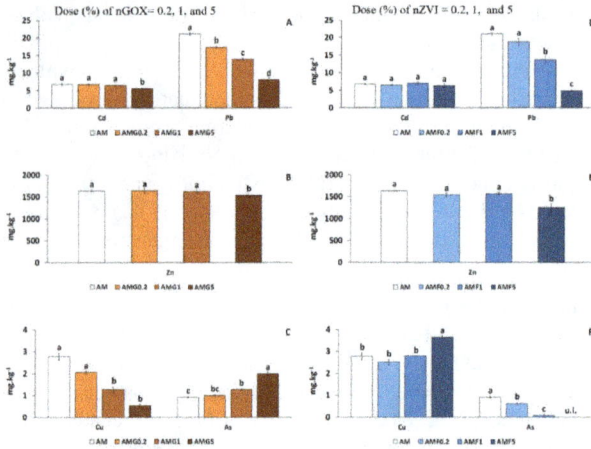

Fig. 3. Cd, Pb, Zn, Cu, and As levels in the arsenic-metals-polluted soil as a result of nGOX (A-C) and nZVI (D-F) effects (AM). Significant variations for each column are indicated by different letters in various samples (n = 3, ANOVA; P<0.05). (Source: Baragano et al. [76], Sci Rep 10, 1896, reused with the permission under the Creative Commons Attribution License).

There are numerous issues with the immobilization strategy for remediating polluted soil, including issues with cost and secondary contamination. For instance, the cost of materials and subsequent environmental pollution make it difficult to add considerable amounts of highly soluble phosphoric acid or phosphate salts to the subsoil. Since phosphate is highly soluble, an excessive intake of nutrients could contaminate groundwater and surface waters to cause eutrophication [81]. Furthermore, the CMC is prone to hydrolysis, and once it disintegrates, it loses its capacity to preserve stability of particles. The fine residual precipitates of the CMC are found in the soil phase [82]. Ecotoxicity studies on the chromium immobilized via CMC-stabilized nZVI revealed that such remediation had a deterrent effect on the plant growth. Fresh nZVI may increase iron uptake in plants by a number of different processes, one of which is that it may enter seeds and be taken up by seed embryos [83]. Endocytosis in root epidermal cells is another anticipated route for nZVI entry [84].

5.   Application of nanoparticles for the treatment of water

Nanotechnology provides incredibly effective, adaptable, and versatile methods to get around the substantial challenges brought on by the existing dearth of pure water [85]. How well nanoscale metal oxides perform as adsorbents and how they might be used to clean water are now under extensive investigation [86]. Because of their extraordinary magnetic properties, such as super paramagnetism and significant magnetic responsiveness under low applied magnetic fields [87], magnetic NPs have been suggested for use [88] among nanoscale metal oxides. However, several organic compounds, such as fulvic acid (FA), humic acid (HA), and hydroxylated and carboxylated fullerenes, among others, may have a considerable impact on copper (II) adsorption [89].

The membrane technique is a different practical way to clean water because of its better removal efficiency, simplicity of use, and lack of secondary pollution. The process is also feasible because it does not necessitate the use of chemicals or heat and does not require regeneration of the used medium [90, 91]. The membrane material naturally trades off membrane permeability for selectivity to determine the performance of the membrane material. Water purification membranes mostly use polymers such as polyamide (PA), polyacrylonitrile (PAN), and cellulose acetate (CA) [92]. The membrane method can be applied to a number of activities depending on the size of the pores and filtering application. Membrane processes are frequently used to remove suspended particles, bacteria, and protozoa, remove viruses and colloids, remove hardness, toxic metals, and dissolved organic matter, desalinate water, and purify it [90, 93]. The removal of bacteria, protozoa, and suspended particles from water is another common application for membrane processes.

Micron-sized particles may be effectively removed from aqueous solutions using polymer nanocomposites or composite nanofiber membranes, which have a high rejection rate and minimal fouling. They are adaptable and effective methods for constructing ultra-fine fibers from polymers, ceramics, or even metals [92, 94-95]. Higher specific surface area [91], controllable pore size [96], and high waterflux of nanocomposite membranes have increased their applicability as microfilters and ultrafilters [97-98]. The NPs may also be easily modified for a specific application. For example, they are utilized in filtration to remove toxic metals and organic chemicals and contaminants [96, 99]. Covalently bonded molecules to the surface of the nanomaterials cause them to become effective and functional. The chemical makeup of the substances is employed to functionalize the surface NPs and increase selectivity for heavy metal ions in solution as shown in Figure 4. The selectivity of Pd^{2+}, Pb^{2+}, Cd^{2+}, Zn^{2+}, Cu^{2+}, Co^{2+}, and Mn^{2+} was assessed by the covalent attachment of a dendrimer, L-cysteine methyl ester onto the surface of NPs [100], in order to improve the preferential adsorption of Hg^{2+} onto SiO_2-Al_2O_3-mixed oxide [62].

Membranes made of cellulose nanofibers can be made useful for the purification of albumin using Cibacron Blue [101]. Toxic metal ions can be removed through the processes of adsorption, chemisorption, and electrostatic attraction by manipulating polymer nanofiber membranes with ceramic NPs including iron oxide, alumina hydroxide, and hydrated alumina NPs [96]. In order to increase the affinity of the membrane for removing organic waste, cyclodextrin is added to a poly(methyl methacrylate) nanofiber membrane [102].

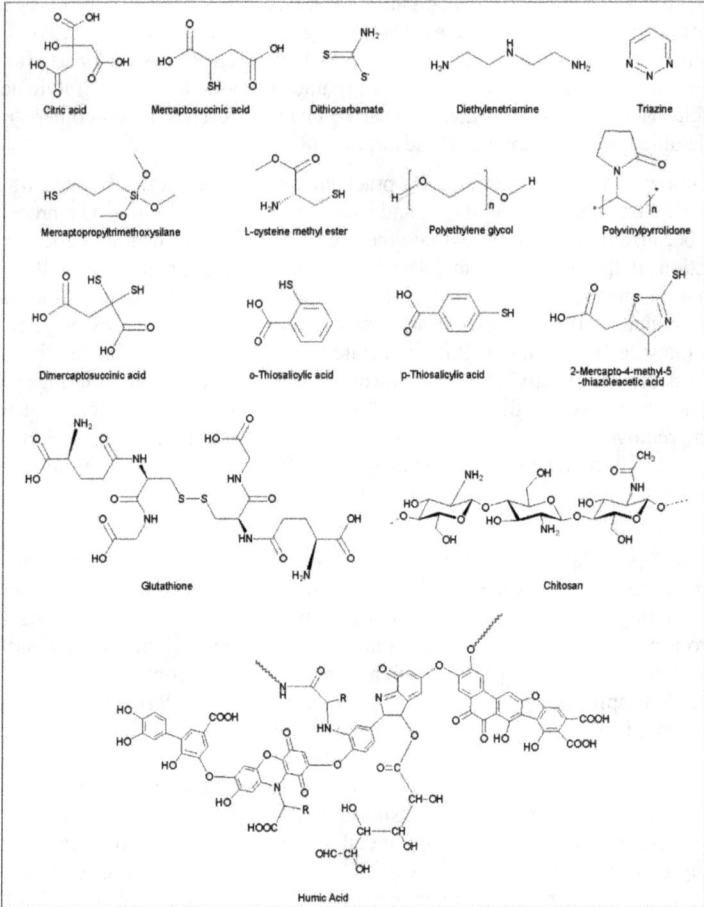

Fig. 4. Capping agents used for coating surface of NPs to improve the affinity and selectivity towards adsorption of mercury in water (Source: Mensah et al. [62], R. Soc. Open Sci. 8: 201485, reused with permission under the Creative Commons Attribution License).

The biggest problem with nanocomposite membranes is membrane fouling, which also increases use of energy, operational expenses, and complexity of the process. Biological and organic fouling are two categories for membrane fouling. The drop in flux in the membrane process is caused by both organic and biological fouling [103]. The quantity of natural organic matter in water is primarily responsible of organic fouling. Organic material that has been deposited and adsorbed on the surface of the membrane causes the pores to become blocked [103]. The attachment of bacteria to membrane surfaces results in the formation of a sticky biofilm of polysaccharides, organic compounds, and a complex community of microbial cells and cause biological fouling [104-105].

Numerous studies have been conducted on various modification techniques to enhance hydrophilicity and lower membrane fouling. These include coating, grafting [106], and blending with metal oxide NPs with good hydrophilicity and have been shown to be an efficient way to prepare nanocomposite membranes without a difficult manufacturing procedure [107]. According to Shen et al. [108], adding metal oxide NPs improves performance of the membrane and thermal stability without changing the structural integrity. In addition, in the treatment of consumable water the silver NPs have been used to stop the development of bacteria and biofilms, and inactivate viruses [107, 109] by coating, grafting and membrane fabrication processes [109]. The common biological contaminants in waterbodies, the *Escherichia coli* could be removed and deactivated by silver-nanocomposite membranes [109-110]. Whereas, the photocatalytic NPs, specifically TiO_2, have been used to improve thermal stability and the fouling resistance [111]. Membranes containing nZVI are used for the reductive dechlorination of contaminants. Generally, NPs are very effective for water purification and improving membrane process performance, with a few drawbacks in the secondary effects on nearby waterbodies.

Conclusions

In recent years, sustainable materials that could be used to clean up the environment have received a lot of attention. Nanoparticles are cost-effective and environmental friendly, and they are applicable in soil and water treatment. In this chapter, the potential applications of various NPs in soil and water treatment have been covered. There has now been a surge of interest in studying the toxicity of NPs for such applications to address the growing concern over the risks that NPs pose to human health and the environment. Definite improvements in nanotechnology and methods based on the use of NPs are needed to treat soil and water using metal-, metal oxide-, carbon-, silica-, and polymer-based NPs over conventional methods. Further research is required to exploit the full potential of various NPs by optimizing various parameters of the methods and by tuning the size and shape of the NPs for efficient treatment of soil and water at low cost.

Acknowledgements

MABHs acknowledges support from Bose Centre for Advanced Research in Natural Sciences and Semiconductor Technology Research Centre of the University of Dhaka, Bangladesh.

Reference

[1] G. Shan, R.Y. Surampalli, R.D. Tyagi, T.C. Zhang, Nanomaterials for environmental burden reduction, waste treatment, and nonpoint source pollution control: a review, Frontiers of Environmental Science & Engineering in China. 3(3) (2009) 249-264. https://doi.org/10.1007/s11783-009-0029-0

[2] M. Loos, Chapter 1—nanoscience and nanotechnology, in: M. Loos (Eds.), Carbon nanotube reinforced composites, William Andrew Publishing, Oxford, 2015, pp. 1–36. https://doi.org/10.1016/B978-1-4557- 3195-4.00001-1

[3] G. Fang, Y. Si, C. Tian, G. Zhang, D. Zhou, Degradation of 2,4-D in soils by Fe_3O_4 nanoparticles combined with stimulating indigenous microbes, Environmental Science and Pollution Research. 19(3) (2012), 784-793. https://doi.org/10.1007/s11356-011-0597-y

[4] G. Fan, L. Cang, W. Qin, C. Zhou, H. Gomes, D. Zhou, Surfactants-enhanced electrokinetic transport of xanthan gum stabilized nanoPd/Fe for the remediation of PCBs contaminated soils, Separation and Purification Technology. 114 (2013), 64-72. https://doi.org/10.1016/j.seppur.2013.04.030

[5] T.E. Cloete, M. De Kwaadsteniet, M. Botes, Nanotechnology in water treatment applications, Caister Academic Press, Poole, U. K. (2010). https://doi.org/10.21775/9781910190098

[6] C. Fishman, The big thirst: The decret life and turbulent future of water, Free Press, New York (2011).

[7] H. Fereidoun, M.S. Nourddin, N.A. Rreza, A. Mohsen, R. Ahmad, H. Pouria, The effect of long-term exposure to particulate pollution on the lung function of Teheranian and Zanjanian students, Pakistan, J Physiol. 3 (2007), 1–5.

[8] M. Kampa, E. Castanas, Human health effects of air pollution, Environ Pollut. 151 (2008), 362–367. https://doi.org/10.1016/j.envpol.2007.06.012

[9] M. Houde, D.C.G. Muir, K.A. Kidd, S. Guildford, K. Drouillard, M.S. Evans, X. Wang, D.M. Whittle, D. Haffner, H. Kling, Influence of lake characteristics on the biomagnification of persistent organic pollutants in lake trout food webs, Environ Toxicol Chem. 27 (2008), 2169 –2178. https://doi.org/10.1897/08-071.1

[10] B.C. Kelly, M.G. Ikonomou, J.D. Blair, A.E. Morin, F.A.P.C. Gobas, Food web-specific biomagnification of persistent organic pollutants, Science. 317 (2007), 236 –239. https://doi.org/10.1126/science.1138275

[11] B. Kumar, D. Mukherjee, S. Kumar, M. Mishra, D. Prakash, S. Singh, C. Sharma, Bioaccumulation of heavy metals in muscle tissue of fishes from selected aquaculture ponds in east Kolkata wetlands, Ann Biol Res. 2 (2011), 125–134.

[12] T. Shahwan, Ç. Üzum, A. Eroğlu, I. Lieberwirth, Synthesis and characterization of bentonite/iron nanoparticles and their application as adsorbent of cobalt ions, Applied Clay Science, 47 (2010), 257-262. https://doi.org/10.1016/j.clay.2009.10.019

[13] T. Ben-Moshe, I. Dror, B. Berkowitz, Transport of metal oxide nanoparticles in saturated porous media, Chemosphere, 81(2010), 387-393. https://doi.org/10.1016/j.chemosphere.2010.07.007

[14] M. Khajeh, S. Laurent, K. Dastafkan, Nanoadsorbents: classification, preparation, and applications (with emphasis on aqueous me- dia), Chem Rev. 113 (2013), 7728–7768. https://doi.org/doi.org/10.1021/cr400086v

[15] A. Husen, K.S. Siddiqi, Phytosynthesis of nanoparticles: concept, controversy and application, Nanoscale Res. Lett. 9 (2014), 229. https://doi.org/10.1186/1556-276X-9-229

[16] A. Husen, M. Iqbal, Nanomaterials and plant potential: an overview, in: A. Husen, M. Iqbal (Eds.), Nanomaterials and Plant Potential. Springer International Publishing AG, Cham, Switzerland, 2019. pp. 329.

[17] P.V. Kamat, D. Meisel, Nanoscience opportunities in environmental remediation, Comptes Rendus Chimie. 6:8-10 (2003), 999-1007. https://doi.org/10.1016/j.crci.2003.06.005

[18] R. Feynman, There's plenty of room at the bottom (reprint from speech given at annual meeting of the American Physical Society. Eng. Sci, 23 (1960), 22–36.

[19] D. Schaming, H. Remita, Nanotechnology: From the ancient time to nowadays. Found. Chem, 17 (2015), 187-205. https://doi.org/10.1007/s10698-015-9235-y

[20] N. Taniguchi, On the basic concept of ''nano-technology'', in: Proceedings of International Conference on Production Engineering, Tokyo, Part II, Japan Society of Precision Engineering (1974).

[21] J.C. Glenn, Nanotechnology: Future military environmental health considerations, Technol. Forecast, Soc. Chang. 73 (2006), 128–137. https://doi.org/10.1016/j.techfore.2005.06.010

[22] E.K. Drexler, Engines of Creation: The Coming Era of Nanotechnology; Anchor Books, Doubleday: New York, NY, USA, (1986), ISBN 0-385-19973-2.

[23] G. Binnig, H. Rohrer, Scanning tunneling microscopy, Surf. Sci. 126 (1983), 236–244. https://doi.org/10.1016/0039-6028(83)90716-1

[24] G. Binnig, C.F. Quate, C. Gerber, Atomic Force Microscope, Phys. Rev. Lett. 56 (1986), 930–933. https://doi.org/10.1103/PhysRevLett.56.930

[25] G. Rytwo, Clay minerals as an ancient nanotechnology: historical uses of clay organic interactions, and future possible perspectives. Macla. 9 (2008), 15–17. https://doi.org/10.13140/2.1.4481.0884

[26] J. Delgado, M. Vilarigues, A. Ruivo, V. Corregidor, R.C. da Silva, L.C. Alves, Characterization of medieval yellow silver stained glass from Convento de Cristo in Tomar, Portugal, Nucl. Instrum. Methods B. 269 (2011), 2383–2388. https://doi.org/10.1016/j.nimb.2011.02.059

[27] P. Colomban, The use of metal nanoparticles to produce yellow, red and iridescent colour, from Bronze age to present times in lustre pottery any glass: solid state chemistry, spectroscopy and nanostructure, J. Nano Res. 8 (2009), 109–132. https://doi.org/10.4028/www.scientific.net/JNanoR.8.109

[28] J. Belloni, The role of silver clusters in photography, C.R. Phys. 3 (2002), 381–390. https://doi.org/10.1016/S1631-0705(02)01321-X

[29] J. Fatisson, S. Ghoshal, N. Tufenkji, Deposition of carboxymethylcellulose-coated zero-valent iron nanoparticles onto Silica: Roles of Solution Chemistry and Organic Molecules, Langmuir, 26(15), (2010), 12832-12840. https://doi.org/10.1021/la1006633

[30] Y. Bao, J. He, K. Song, J. Guo, X. Zhou, S. Liu, Plant-extract-mediated synthesis of metal nanoparticles, J. Chem. (2021), 6562687. https://doi.org/10.1155/2021/6562687

[31] R. Kuhn, I. M. Bryant, R. Jensch, J. Böllmann, Applications of environmental nanotechnologies in remediation, wastewater treatment, drinking water treatment, and Agriculture, Appl. Nano. *3* (2022), 54–90. https://doi.org/10.3390/applnano3010005

[32] M. Naghdi, S. Metahni, Y. Ouarda, S.K. Brar, R.K. Das, M. Cledon, Instrumental approach toward understanding nano-pollutants, Nanotechnol, Environ. Eng. 2:3 (2017). https://doi.org/10.1007/s41204-017-0015-x

[33] T. Ma, Y. Sheng, Y. Meng, J. Sun, Multistage remediation of heavy metal contaminated river sediments in a mining region based on particle size, Chemosphere. 225 (2019), 83–92. https://doi.org/10.1016/j.chemosphere.2019.03.018

[34] Y. Lin, F. Meng, Y. Du, Y. Tan, Distribution, speciation, and ecological risk assessment of heavy metals in surface sediments of Jiaozhou Bay, China, Hum. Ecol. Risk Assess. Int. J. 22 (2016), 1253–1267. https://doi.org/10.1080/10807039.2016.1159503

[35] S. Das, B. Sen, N. Debnath, Recent trends in nanomaterials applications in envi ronmental monitoring and remediation, Environ, Sci. Pollut. Res. Int. 22 (2015). 18333-18344. https://doi.org/10.1007/s11356-015-5491-6

[36] P.G. Tratnyek, R.L. Johnson, Nanotechnologies for environmental clean-up, Nano Today 1 (2006), 44-48. https://doi.org/10.1016/S1748-0132(06)70048-2

[37] A. Mondal, B.K. Dubey, M. Arora, K. Mumford, Porous media transport of iron nanoparticles for site remediation application: A review of lab scale column study, transport modelling and field-scale application, J. Hazard. Mater. 403 (2021), 123443. https://doi.org/10.1016/j.jhazmat.2020.123443

[38] X. Zhao, W. Liu, Z. Cai, B. Han, T. Qian, D. Zhao, An overview of preparation and applications of stabilized zero-valent iron nanoparticles for soil and groundwater remediation, Water Res. 100 (2016), 245–266. https://doi.org/10.1016/j.watres.2016.05.019

[39] Q. Abbas, B. Yousaf, A. Muhammad, U. Ali, M. Ahmed, M. Munir, A. El-Naggar, J. Rinklebe, M. Naushad, Transformation pathways and fate of engineered nanoparticles (ENPs) in distinct interactive environmental compartments: A review, Environmental International 138, (2020), 105646. https://doi.org/10.1016/j.envint.2020.105646

[40] F. Guerra, M. Attia, D. Whitehead, F. Alexis, Nanotechnology for environmental remediation: materials and applications, Molecules 23(7), (2018), 1760. https://doi.org/10.3390/molecules23071760

[41] H.Y. Huang, R.T. Yang, D. Chinn, C.L. Munson, Amine-grafted MCM-48 and silica xerogel as superior sorbents for acidic gas removal from natural gas, Ind. Eng. Chem. Res. 42 (2003), 2427-2433. https://doi.org/10.1021/ie020440u

[42] A. Nomura, C.W. Jones, Amine-functionalized porous silicas as adsorbents for aldehyde abatement, ACS App. Mat. Interf. 5 (2013), 5569-5577. https://doi.org/10.1021/am400810s

[43] C.H. Tsai, W.C. Chang, D. Saikia, C.E. Wu, H.M. Kao, Functionalization of cubic mesoporous silica SBA-16 with carboxylic acid via one-pot synthesis route for effective removal of cationic dyes, J. Hazard. Mat. 309 (2016), 236-248. https://doi.org/10.1016/j.jhazmat.2015.08.051

[44] S. Wang, K. Wang, C. Dai, H. Shi, J. Li, Adsorption of Pb^{2+} on amino-functionalized coreshell magnetic mesoporous SBA-15 silica composite, Chem. Eng. J. 262 (2015), 897-903. https://doi.org/10.1016/j.cej.2014.10.035

[45] D. Lei, Q. Zheng, Y. Wang, H. Wang, Preparation and evaluation of aminopropyl-functionalized manganese-loaded SBA-15 for copper removal from aqueous solution, J. Environ. Sci. 28 (2015), 118-127. https://doi.org/10.1016/j.jes.2014.06.045

[46] C.H. Deng, J.L. Gong, P. Zhang, G.M. Zeng, B. Song, H.Y. Liu, Preparation of melamine sponge decorated with silver nanoparticles-modified graphene for water

disinfection, J. Colloids Interf. Sci. 488 (2017), 26-38.
https://doi.org/10.1016/j.jcis.2016.10.078

[47] C.R. Chinnamuthu, P.M. Boopathi, Nanotechnology and agro-ecosystem. Madras Agric. J. 96 (2009), 17-31.

[48] R. Prasad, V. Kumar, K.S. Prasad, Nanotechnology in sustainable agriculture: present concerns and future aspects. Afr. J. Biotechnol. 13 (2014), 705-713.

[49] K.S. Yao, S. Li, K. Tzeng, T.C. Cheng, C.Y. Chang, C. Chiu et al., Fluorescence silica nanoprobe as a biomarker for rapid detection of plant pathogens. Adv. Mat. Res. 79-82 (2009), 513-516. https://doi.org/10.4028/www.scientific.net/AMR.79-82.513

[50] N. Chartuprayoon, Y. Rheem, W. Chen, N.V. Myung, Detection of plant pathogen using LPNE grown single conducting polymer nanoribbon, in: Proceedings of the 218th ECS Meeting, Las Vegas, NV, October 10-15, (2010), pp. 22-78.

[51] M.S. Diallo, C.J. Glinka, W.A. Goddard, J.H. Jhonson, Characterization of nanoparticles and colloids in aquatic systems 1. Small angle neutron scattering investigations of Suwannee River fulvic acid aggregates in aqueous solutions, J. Nanoparticle Res. 7(4-5), (2005), 435-448. https://doi.org/10.1007/s11051-005-7524-4

[52] N.M. Nagy, J. Konya, M. Beszeda et al., Physical and Chemical formation of lead contaminants in clay and sediment, J. Colloid Interface Sci. 263 (1), (2003), 13-22. https://doi.org/10.1016/S0021-9797(03)00284-4

[53] A.B.A. Boxall, K. Tiede, Q. Chaudhry, Engineered nanomaterials in soils and water: how do they behave and could they pose a risk to human health?, Nanomedicine. 2(6), (2007), 919-927. https://doi.org/10.2217/17435889.2.6.919

[54] A. Hakim, T. Suzuki, M. Kobayashi, Strength of humic acid aggregates: Effects of divalent cations and solution pH, ACS Omega. 4(5), (2019), 8559–8567. https://doi.org/10.1021/acsomega.9b00124

[55] A. Hakim, M. Kobayashi, Charging, aggregation, and aggregate strength of humic substances in the presence of cationic surfactants: Effects of humic substances hydrophobicity and surfactant tail length, Colloids and Surfaces A: Physicochemical and Engineering Aspects. 577 (2019), 175-184. https://doi.org/10.1016/j.colsurfa.2019.05.071

[56] A. Hakim, M. Kobayashi, Aggregation and charge reversal of humic substances in the presence of hydrophobic monovalent counter-ions: effect of hydrophobicity of humic substances, Colloids and Surfaces A: Physicochemical and Engineering Aspects. 540 (2018), 1-10. https://doi.org/10.1016/j.colsurfa.2017.12.065

[57] W.K.W.A. Khodir, A. Hakim, M. Kobayashi, Strength of flocs formed by the complexation of lysozyme with leonardite humic acid, Polymers. 12(8):1770 (2020). https://doi.org/10.3390/polym12081770

[58] A. Hakim, M. Kobayashi, Aggregation and aggregate strength of microscale plastic particles: effects of ionic valance, J. Polymers Environ. 29 (2021), 1921-1929. https://doi.org/10.1007/s10924-020-01985-4

[59] K. Omija, A. Hakim, K. Masuda, A. Yamaguchi, M. Kobayashi, Effect of counter ion valence and pH on the aggregation and charging of oxidized carbon nanohorn (CNHox) in aqueous solution, Colloids and Surfaces A: Physicochemical and Engineering Aspects. 619 (2021), 126552. https://doi.org/10.1016/j.colsurfa.2021.126552

[60] J.A. Brant, J. Labille, C. Ogilvie Robichaud, M. Wiesner, Fullerol cluster formation in aqueous solutions: implications for environmental release, J. Colloid Interface Sci. 314(1), (2007), 281-288. https://doi.org/10.1016/j.jcis.2007.05.020

[61] B. Liu, S.Q. Jian, W.D. Zhang, F. Ye, Y.H. Wang, J. Wu, D.Y. Zhang, Novel biodegradable HSAM nanoparticle for drug delivery, Oncol. Rep. 15(4), (2006), 957-961. https://doi.org/10.3892/or.15.4.957

[62] M.B. Mensah, D.J. Lewis, N.O. Boadi, J.A.M. Awudza, Heavy metal pollution and the role of inorganic nanomaterials in environmental remediation. Royal society open science 8(10), (2021), 201485. https://doi.org/10.1098/rsos.201485

[63] S.B. Lovern, R. Klaper, Daphnia magna mortality when exposed to titanium dioxide and fullerene (C60) nanoparticles, Environ. Toxicol. Chem. 25(4), (2006), 1132-1137.

[64] N. Lubick, Nanosilver toxicity: ions, nanoparticles—or both? Environ. Sci. Technol. 2008, 42, 23, 8617. https://doi.org/10.1021/es8026314

[65] J. Fang, D.Y. Lyon, M.R. Wiesner, J. Dong, P.J.J. Alvarez, Effect of a fullerene water suspension on bacterial phospholipids and membrane phase behaviour, Environ. Sci. Technol. 41:7 (2007), 2636-2642. https://doi.org/10.1021/es062181w

[66] Z. Tong, M. Bischoff, L. Nies, B. Applegate, R.F. Turco, Impact of fullerene (C60) on a soil microbial community, Environ. Sci. Technol. 41:8 (2007), 2985-2991. https://doi.org/10.1021/es0619531

[67] Q.Y. Ma, S.J. Traina, T.J. Logan, J.A. Ryan, In situ Pb immobilization by apatite, Environ Sci Technol. 27:9 (1993), 1803-1810. https://doi.org/10.1021/es00046a007

[68] S.P. Singh, L.Q. Ma, W.G. Harris, Heavy metal interactions with phosphatic clay: sorption and desorption behaviour. J Environ Qual. 30 (2001), 1961-1968. https://doi.org/10.2134/jeq2001.1961

[69] J. Kumpiene, A. Lagerkvist, C. Maurice, Stabilization of As, Cr, Cu, Pb and Zn in soil using amendments—a review, Waste Manag. 28 (2008), 215–225. https://doi.org/10.1016/j.wasman.2006.12.012

[70] B.H. Robinson, G. Bañuelos, H.M. Conesa, M.W.H. Evangelou, R. Schulin, The phytomanagement of trace elements in soil, Crit. Rev. Plant Sci. 28 (2009), 240–266. https://doi.org/10.1080/07352680903035424

[71] F. He, D. Zhao, Preparation and characterization of a new class of starch-stabilized bimetallic nanoparticles for degradation of chlorinated hydrocarbons in water, Environ. Sci. Technol. 39 (2005), 3314–3320. https://doi.org/10.1021/es048743y

[72] F. He, D. Zhao, Manipulating the size and dispersibility of zerovalent iron nanoparticles by use of carboxymethyl cellulose stabilizers, Environ. Sci. Technol. 41 (2007), 6216–6221. https://doi.org/10.1021/es0705543

[73] Q. Liang, D. Zhao, Immobilization of arsenate in a sandy loam soil using starch-stabilized magnetite nanoparticles, J. Hazard Mater. 271 (2014), 16–23. https://doi.org/10.1016/j.jhazmat.2014.01.055

[74] J. Yang, D.E. Mosby, S.W. Casteel, R.W. Blanchar, Lead immobilization using phosphoric acid in a smelter-contaminated urban soil, Environ. Sci. Technol. 35 (2001), 3553–3559. https://doi.org/10.1021/es001770d

[75] R. Liu, D. Zhao, Synthesis and characterization of a new class of stabilized apatite nanoparticles and applying the particles to in situ Pb immobilization in a fire-range soil, Chemosphere, 91 (2013), 594–601. https://doi.org/10.1016/j.chemosphere.2012.12.034

[76] D. Baragano, R. Forjan, L. Welte, J.L.R. Gallego, Nanoremediation of As and metals polluted soils by means of graphene oxide nanoparticles, *Sci Rep* 10, 1896 (2020). https://doi.org/10.1038/s41598-020-58852-4

[77] L.A. Reyhanitabar, A. Khataee, S. Oustan, Application of stabilized Fe0 nanoparticles for remediation of Cr (VI)-spiked soil, Eur J Soil Sci. 63 (2012), 724–732. https://doi.org/10.1111/j.1365-2389.2012.01447.x

[78] Z. Fang, X. Qiu, J. Chen, X. Qiu, Debromination of polybrominated diphenyl ethers by Ni/Fe bimetallic nanoparticles: influencing fac- tors, kinetics, and mechanism, J Hazard Mater. 185 (2011), 958–969. https://doi.org/10.1016/j.jhazmat.2010.09.113

[79] S.S. Chen, H.D. Hsu, C.W. Li, A new method to produce nanoscale iron for nitrate removal, J Nanoparticle Res. 6 (2004), 639–647. https://doi.org/10.1007/s11051-004-6672-2

[80] Y. Wang, Z. Fang, B. Liang, E.P. Tsang, Remediation of hexavalent chromium contaminated soil by stabilized nanoscale zero-valent iron prepared from steel pickling waste liquor, Chem Eng J. 247 (2014), 283–290. https://doi.org/10.1016/j.cej.2014.03.011

[81] J.H. Park, N. Bolan, M. Megharaj, R. Naidu, Comparative value of phosphate sources on the immobilization of lead, and leaching of lead and phosphorus in lead contaminated soils, Sci Total Environ. 409:4 (2011), 853-860. https://doi.org/10.1016/j.scitotenv.2010.11.003

[82] Y. Xu, D. Zhao, Reductive immobilization of chromate in water and soil using stabilized iron nanoparticles, Water Res. 41 (2007), 2101–2108. https://doi.org/10.1016/j.watres.2007.02.037

[83] Y. Wang, Z. Fang, Y. Kang, E.P. Tsang, Immobilization and phyto- toxicity of chromium in contaminated soil remediated by CMC- stabilized nZVI, J Hazard Mater. 275 (2014), 230-237. https://doi.org/10.1016/j.jhazmat.2014.04. 056

[84] D.L. Slomberg, M.H. Schoenfisch, Silica nanoparticle phytotoxicity to Arabidopsis thaliana, Environ Sci Technol. 46 (2012), 10247–10254. https://doi.org/10.1021/es300949f

[85] X. Qu, J. Brame, Q. Li, P.J.J. Alvarez, Nanotechnology for a safe and sustainable water supply: enabling integrated water treatment and reuse, Acc Chem Res. 46 (2013), 834–843. https://doi.org/10.1021/ar300029v

[86] K. Engates, H. Shipley, Adsorption of Pb, Cd, Cu, Zn, and Ni to titanium dioxide nanoparticles: effect of particle size, solid concentration, and exhaustion, Environ Sci Pollut Res. 18 (2011), 386–395. https://doi.org/10. 1007/s11356-010-0382-3

[87] M. Kilianová, R. Prucek, J. Filip, J. Kolarik, L. Kvitek, A. Panacek, J. Tucek, R. Zboril, Remarkable efficiency of ultrafine superparamagnetic iron(III) oxide nanoparticles toward arsenate removal from aqueous environment, Chemosphere. 93 (2013), 2690–2697. https://doi.org/10.1016/j.chemosphere.2013.08.071

[88] X. Xin, Q. Wei, J. Yang, L. Yan, R. Feng, G. Chen, B. Du, H. Li, Highly efficient removal of heavy metal ions by amine functionalized mesoporous Fe_3O_4 nanoparticles, Chem Eng J. 184 (2012), 132–140. https://doi.org/10.1016/j.cej.2012.01.016

[89] J. Wang, Z. Li, S. Li, W. Qi, P. Liu, F. Liu, Y. Ye, L. Wu, L. Wang, W. Wu, Adsorption of Cu(II) on oxidized multi-walled carbon nanotubes in the presence of hydroxylated and carboxylated fullerenes, PLoS ONE. 8:8 (2013), e72475. https://doi.org/10.1371/journal.pone. 0072475

[90] R. Balamurugan, S. Sundarrajan, S. Ramakrishna, Recent trends in nanofibrous membranes and their suitability for air and water filtrations, Membranes. 1 (2011), 232–248. https://doi.org/10.3390/membranes1030232

[91] M.G. Buonomenna, Membrane processes for a sustainable industrial growth, RSC Advances 3 (2013), 5694 -5740. https://doi.org/10.1039/ C2RA22580H

[92] H.L. Yang, J.C.T. Lin, C. Huang, Application of nanosilver surface modification to RO membrane and spacer for mitigating biofouling in seawater desalination, Water Res. 43 (2009), 3777–3786. https://doi.org/10.1016/j. watres.2009.06.002

[93] P. Bernardo, E. Drioli, G. Golemme, Membrane gas separation: a review/state of the art, Ind Eng Chem Res. 48 (2009), 4638–4663. https://doi.org/10.1021/ie8019032

[94] TE Cloete, M de Kwaadsteniet, M Botes and JM Lopez-Romero, Nanotechnology in water treatment applications. Caister Academic Press, New York (2010).

[95] X. Li, C. Zhang, R. Zhao, X. Lu, X. Xu, X. Jia, C. Wang, L. Li, Efficient adsorption of gold ions from aqueous systems with thioamide-group chelating nanofiber membranes, Chem Eng J. 229 (2013), 420–428. https://doi.org/10.1016/j.cej.2013.06.022

[96] S. Ramakrishna, K. Fujihara, W.E. Teo, T. Yong, Z. Ma, R. Ramaseshan, Electrospun nanofibers: solving global issues, Mater Today. 9 (2006), 40–50. https://doi.org/10.1016/S1369-7021(06)71389-X

[97] R. Gopal, S. Kaur, Z. Ma, C. Chan, S. Ramakrishna, T. Matsuura, Electrospun nanofibrous filtration membrane, J Membr Sci. 281 (2006), 581–586. https://doi.org/10.1016/j.memsci.2006.04.026

[98] R. Gopal, S. Kaur, C.Y. Feng, C. Chan, S. Ramakrishna, S. Tabe, T. Matsuura, Electrospun nanofibrous polysulfone membranes as prefilters: particulate removal, J Membr Sci. 289 (2007), 210–219. https://doi.org/10.1016/j.memsci.2006.11.056

[99] X. Qu, P.J.J. Alvarez, Q. Li, Applications of nanotechnology in water and wastewater treatment, Water Res. 47 (2013), 3931–3946. https://doi.org/10.1016/j.watres.2012.09.058

[100] M. Arshadi, H. Firouzabadi, A. Abbaspourrad, Adsorption of mercury ions from wastewater by a hyperbranched and multi-functionalized dendrimer modified mixed-oxides nanoparticles, J. Colloid Interface Sci. 505, (2017), 293-306. (https://doi.org/10.1016/j.jcis.2017.05.052).

[101] Z. Ma, M. Kotaki, S. Ramakrishna, Electrospun cellulose nanofiber as affinity membrane, J Membr Sci. 265 (2005), 115–123. https://doi.org/10.1016/j.memsci.2005.04.044

[102] S. Kaur, M. Kotaki, Z. Ma, R. Gopal, S. Ramakrishna, S.C. NG, Oligosaccharide functionalized nanofibrous membrane, Int J Nanosci. 5:1 (2006), 1–11. https://doi.org/10.1142/S0219581X06004206

[103] F. Meng, S.R. Chae, A. Drews, M. Kraume, H.S. Shin, F. Yang, Recent advances in membrane bioreactors (MBRs): membrane fouling and membrane material, Water Res. 43 (2009), 1489–1512. https://doi.org/10.1016/ j.watres.2008.12.044

[104] S. Ciston, R.M. Lueptow, K.A. Gray, Controlling biofilm growth using reactive ceramic ultrafiltration membranes, J Membr Sci. 342 (2009), 263–268. https://doi.org/10.1016/j.memsci.2009.06.049

[105] I. Sawada, R. Fachrul, T. Ito, Y. Ohmukai, T. Maruyama, H. Matsuyama, Development of a hydrophilic polymer membrane containing silver nanoparticles with both organic antifouling and antibacterial properties, J Membr Sci. 387–388 (2012), 1–6. https://doi.org/10.1016/j.memsci. 2011.06.020

[106] A. Rahimpour, UV photo-grafting of hydrophilic monomers onto the surface of nano-porous PES membranes for improving surface properties, Desalination. 265 (2011), 93–101. https://doi.org/10.1016/j.desal.2010.07. 037

[107] R.K. Ibrahim, M. Hayyan, M.A. AlSaadi, A. Hayyan, S. Ibrahim, Environmental Application of nanotechnology: Air, soil and water, Environ Sci Pollut Res. 23 (2016), 13754–13788. https://doi.org/10.1007/s11356-016-6457-z

[108] J.N. Shen, H.M. Ruan, L.G. Wu, C.J. Gao, Preparation and characterization of PES–SiO$_2$ organic–inorganic composite ultrafiltration membrane for raw water pretreatment, Chem Eng J. 168 (2011), 1272– 1278. https://doi.org/10.1016/j.cej.2011.02.039

[109] K. Zodrow, L. Brunet, S. Mahendra, D. Li, A. Zhang, Q. Li, P.J.J. Alvarez, Polysulfone ultrafiltration membranes impregnated with sil- ver nanoparticles show improved biofouling resistance and virus removal, Water Res. 43 (2009), 715–723. https://doi.org/10.1016/j.watres.2008.11.014

[110] L. Obalová, M. Reli, J. Lang, V. Matějka, J. Kukutschová, Z. Lacný, K. Kočí, Photocatalytic decomposition of nitrous oxide using TiO2 and Ag-TiO2 nanocomposite thin films, Catal Today. 209 (2013), 170– 175. https://doi.org/10.1016/j.cattod.2012.11.012

[111] G. Wu, S. Gan, L. Cui, Y. Xu, Preparation and characterization of PES/TiO2 composite membranes, Appl Surf Sci. 254 (2008), 7080–7086. https://doi.org/10.1016/j.apsusc.2008.05.221

Applications of Emerging Nanomaterials and Nanotechnology   Materials Research Forum LLC
Materials Research Foundations 148 (2023) 252-275   https://doi.org/10.21741/9781644902554-9

Chapter 9

Nanotechnology in Agricultural Practices: Prospects and Potential

N.G. Giri[1], N.S. Abbas[2] and Saroj Kumar Shukla[2]*

[1]Department of Chemistry, Shivaji College, University of Delhi, Delhi-110027, India

[2]Bhaskaracharya College of Applied Sciences, University of Delhi, Delhi-110075, India

*sarojshukla2003@yahoo.co.in

Abstract

The novel features of nano confined materials properties are widely explored in several agricultural practices *i.e.* seed coating, soil conditioning, preservatives and packaging of agro based products. The evolution of materials properties due to nano confinements like controlled degradability, solubility, responsiveness, gas barrier and porosity has added several suitable advantageous features in agriculture practices to help the farmers, traders, consumers and policy makers. In the context of the above development, the present chapter describes the basic features of nano materials to use in agricultural practices along with their impact and effectiveness. Further, the suitable scheme and illustration are used on the basis of available literatures to make the subject lucid and effective along with existing challenges.

Keywords

Nanomaterials, Properties, Soil Nutrient, Protection and Monitoring

Contents

Applications of Emerging Nanomaterials and Nanotechnology   Materials Research Forum LLC
Materials Research Foundations 148 (2023) 252-275   https://doi.org/10.21741/9781644902554-9

1. Introduction

Recent scientific study reveals that the nanotechnology has efficacy to positively impact on the field of agricultural practices to minimize the adverse problems of agricultural practices on the environment, economy, and human health. Food productivity, soil management, storage and safety [1]. The potential of nano technology and materials promises to irradicate the hunger and maintain the food security for world community along with monitoring soil condition and food quality [2]. Pertaining to this different novel nanomaterials are used in agriculture are polymeric nano composite, nano hydrogel, lipid nanoparticles and inorganic and organic nano materials developed by using different nano fabrication techniques *i.e.* top down and bottom up [1]. One of the other important use of nanoparticles is in recycling and valorization of agricultural wastes with objective control pollution and their judicious conversion into adsorbent, fertilizer and other organic compound after using different thermo-chemicals, mechano-chemical, physio-chemicals and biochemical treatments [3]. Although, the recycling of crop wastes is a natural process and practiced by farmer since ancient time but the current development in nano technology improves the efficiency of these methods along with better projection of these project into commercial scales [4-5]. The importance of these areas yields exponential increasing trends of publications *i.e.* research publications, review and patents, which is depicted in Fig. 1.

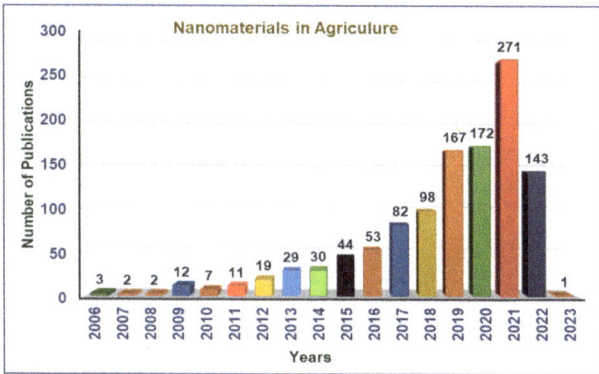

Figure 1. Publication frequency trend on nanomaterials in agriculture (www.scopus.com)

This exponential increase in publication frequency is confirming the growth of field and need of more and more research studies for commercialization of nano technologies in agriculture practices for economic growth and commercialization. In this context, several reviews and book chapter are published but its update in the form of book chapter is still demand for educating researchers and scientists. In the context of above development, present book chapter describe the updates about the nano technology, nano materials and their applications in agriculture practices. The advantageous features of nanotechnology influenced agricultural practice are discussed with the help of scheme and illustrations.

2.   Nano technology in agriculture

In general, the nanotechnology deals the physics, chemistry, fabrication and engineering of materials in the scale of 1 to 100 nm. This materials dimension so important to create a materials transition due to properties confinements' related Bohr radius, and it generates the unique surface reactivity, sustained degradability, functionality and responsiveness to use in different aspects of agriculture and agricultural practices. The use and practice of nanomaterials and nanotechnology have been created second revolution in agriculture practices and industries with the significant   benefit to each stake holders [6-7]. Although, the concept of nano technology has been frequently used by natures as phytochemical biology, biochemistry and properties but its commercialization is a newer development for improved production and processing of agricultural products. The basic need of this technique is with evolving problems in conventional agriculture like limit of conventional farming and nutrient management, and it has invited the scientist to explore the use on nano technology for sustainable advances in agriculture and practices. In this context nano chemist has designed and developed several pristine and hybrid nano materials with tunable chemical and physical properties. Some of the significant nano structured materials used in agriculture practices along with their brief properties and applications are listed in Table 1.

Table 1. Different nanostructured materials, properties and applications

SN	Nano materials	Properties	Applications
1	Hydrogel	High water retaining capacity with excellent mechanical flexibility	Alternate source of water in soil
2	Nano coating	Improved mechanical strength with gas barrier.	Seed coating and active packaging
3	Nano particles	Large surface area and reactivity	Fertilizer and soil nutrients
4	Nano polymer composite	Tunable biodegradability, reactivity and conductivity	Sustained released and monitoring of soil nutrients.

Applications of Emerging Nanomaterials and Nanotechnology   Materials Research Forum LLC
Materials Research Foundations 148 (2023) 252-275   https://doi.org/10.21741/9781644902554-9

These nano materials prepared after using different bottom up and top down methods with their own inherited advantages and disadvantages. The comparison of both methods in the term of synthesis used for preparation of nano particles has been illustrated in Fig. 2.

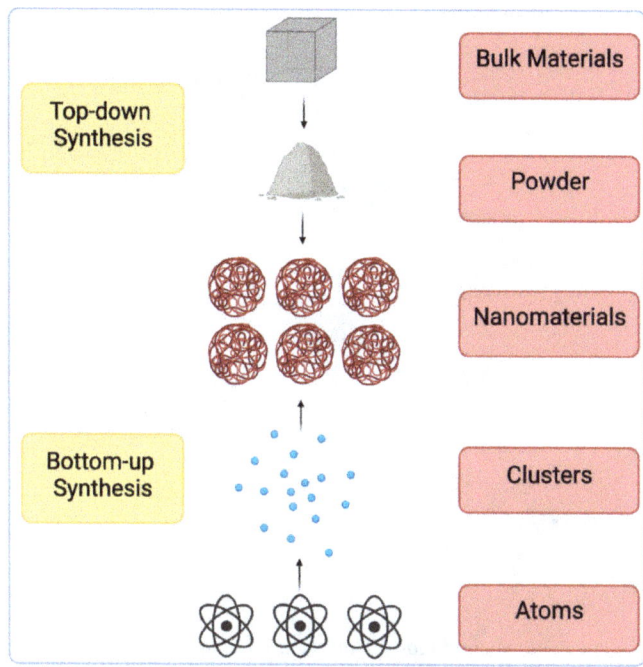

Figure 2. The comparison of preparative methods of nanoparticles

In general, the top down method deals with physical methods like evaporation, sputtering, ablation and ball milling, while the bottom up method includes precipitation, solgel, hydrothermal, LB and reverses micelles. Further, to avoid the agglomeration both hard and soft templates are used with their own inherited merit and demerits. Furthermore, during the preparation of hybrid form of different nano materials, integration of both the techniques are exercised after optimizing the processing condition. In an example, the different templates both hard and soft are used in order to control the aggregation of additives along with influenced dispersion of dispersion phase into dispersion matrix [8]. Thus, a dispersed additive in polymer matrix develops several innovative synergistic features for applications in agriculture for coating, controlled release of nutrient, sensing agro-chemicals and preservatives. Thus, different applications of nano technology and nano materials in agriculture are shown in Fig. 3. The basic properties makes nanomaterials superior than conventional materials are high surface area, reactivity, responsiveness and retaining capacity for use in different applications like genetic modification, targeted delivery, crop protection and nano fertilizer [9].

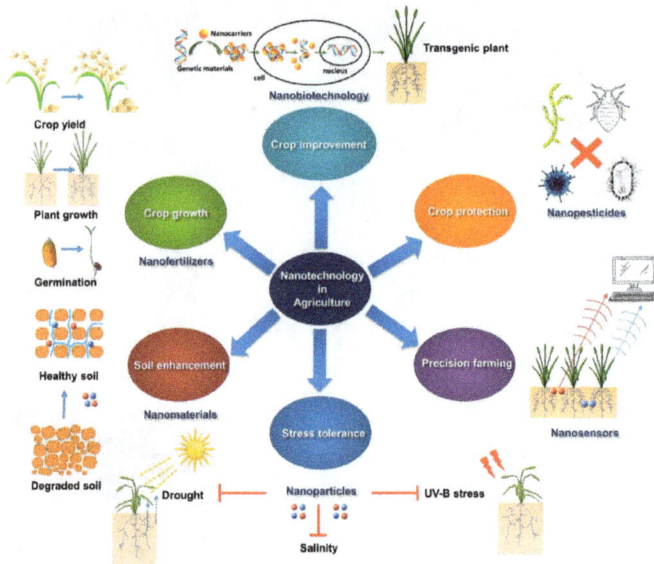

Figure 3. Schematic presentation for application of nano materials in agriculture

3. Applications

3.1 Valorization of agricultural wastes and management

Term "valorization" refers the industrial processing, recycling, and reuse of used agricultural products like seed husk, cake, straw and peels. In general the natural uses of agro waste are vermi compositing to convert into fertilizer, fodder and microbial degradation into several value added products [10]. However, it current time after realization the corporate social responsibility, it is very important to increase the consciousness towards conversion of huge quantities of agricultural food and agro wastes into different value added product not only to meet the environmental problem but also increase the financial gain of all stake holders. The reduction in food wastage would not only circumvent the burden on availability of natural resources but also help to raise food production for insuring food availability [11-12]. The basic routes and importance of reuse and valorizations of agro wastes are shown in Fig. 4.

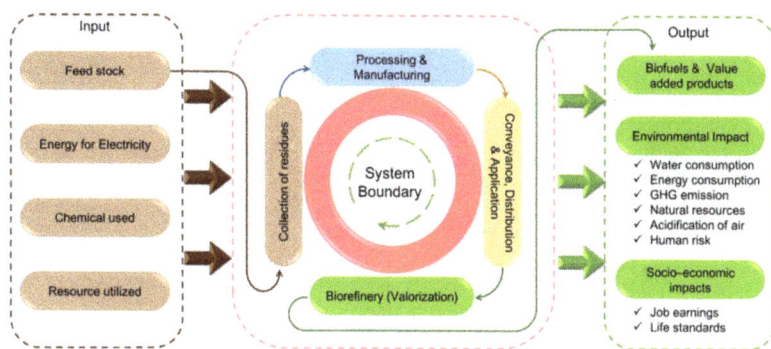

Figure 4. Illustrative importance and methods for recycling of agro wastes [10]

In this process the significance nanomaterials are to provide catalysts, facile conversion process, better yields and to control the environmental issue, economy and society [13-14]. For example, the ligno cellulose biomass present in agro wastes are explored for production of biofuels, organic acids, protein enriched feed, bioactive secondary metabolites, aroma compounds after using different nano materials. [15-16] The physio chemical treatments of agro waste also yield several nano materials to use as fertilizers, pesticides, filters, sensors, and photocatalysts with improved properties [17-18]. The potential of agriculture waste valorization has attracted the attention of technocrat to design several biochemical reactors to valorize agro wastes like rice husk into aromatics by thermolysis in the presence nano catalysts like zeolites [19-20]. The proposed experimental set up for effective conversion of rice husk into aromatics is shown in Fig. 5. The diagram is containing three reactors R1 CH4 decomposition reactor; R2 thermal conversion reactor and R3: for upgrading the pyrolysis vapor.

Similarly the presence huge biopolymer like proteins, starch, cellulose and carbohydrate are explored for preparation of bionano composites, biochar, and activated carbon and carbon nanostructure with significant economic valorization [21-23]. Further the nanotechnology induced transformation of organic agro waste cotton fibers and cellulose are good option to convert into biofuel like ethanol due to global increase into price of maize [24-26]. Here, the nano engineering tool increases the performance of the enzymes (α-L-arabinofuranosidase, α-fucosidase, α-galactosidase, β-glucosidase, β-xylosidases, endo-galactanase, β-galactosidase and β-mannosidase during conversion of cellulose into ethanol [27].The technique has been explored by major countries for production of bio-ethanol and a comparative picture is given in Figure 6. Further, nano technology has been also used to prepare different nano particle like cellulose nanocrystals from different agro wastes like corn husk, rice straw and bamboo pulp using nano technology induced acid hydrolysis [28-30]. The significant properties of nanocellulose are transparency, tensile strength, low coefficient of thermal expansion, which make it suitable to use as reinforcement for polymers, pharmaceuticals, biomedical, fibers and textiles, antimicrobial films, supercapacitors [31]. The conversion processes use some hazardous chemical like chlorinate solvent for bleaching and sulfuric acid in hydrolysis, hence the efforts are made to integrate the principle of green chemistry and green engineering during valorization and recycling [32]. In this regards, Shukla et al has reported the use of hydrogen peroxide a greener solvent to prepare nano sized cellulose to use a reinforcing agent. Thus, obtained cellulose was used to prepare polyvinyl alcohol film to replace non-biodegradable plastic film with heat seaing behaviour [33].

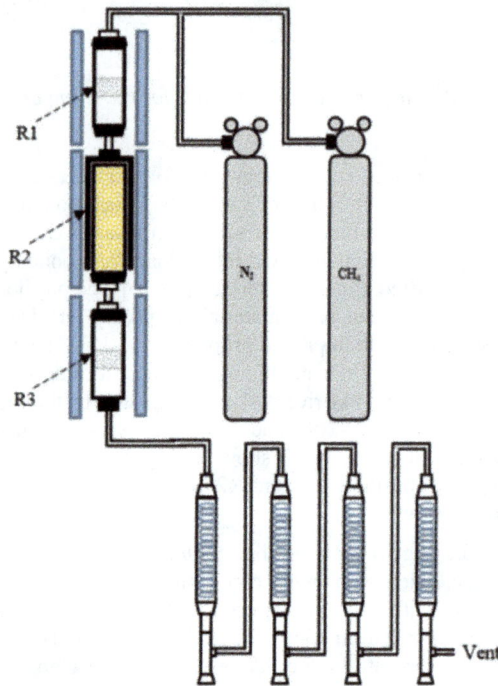

Figure 5. Schematic diagram of the experimental setup to valorize rice husk [19]

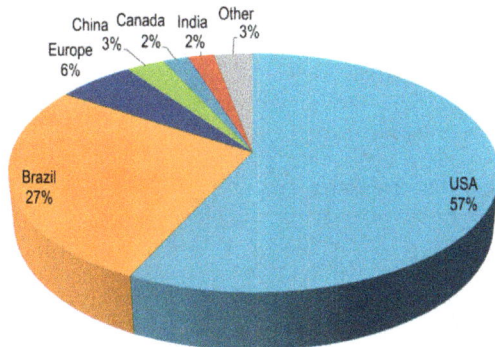

Figure 6. Bioethanol production by major countries

3.2 Soil nutrient and Fertilizers

The nano confine materials used to increase the fertility of soil, high crop production, and minimum pollution are refereed as nano fertilizer. The use nano fertilizer improves the efficiency of fertilizer along with lesser requirement and bioaccumulation into the food chain. Some of the commercially available fertilizer are nano urea, the use nano urea increases the nitrogen use efficiency, since the use of conventional urea looses their 70 % nitrogen contents during surface leaching [34]. In this regard, the improved solubility of urea further, improves at nano scale, thus for sustainable release several modifications are reported. For example, the urea was coated with bio-based epoxy comprised of nano-SiO_2 and organosilicon for improvement in properties like hydrophobicity. Water absorption, ammonium adsorption, and water contact angles. Thus, modified epoxy coated urea was enhanced urea release for longer time for 56 days with potential to use in modern agricultural practices [35].

The plants also need several micronutrients like zinc, iron, magnesium and born, however their normal salts either loss their retention capacity or absorption into plants body. These metal-based nanomaterials also work as anti-bacterial, antifungal, antiviral apart from micronutrients [36-37]. However, the novelty of materials at nano scale improves its both negative feature and several nano sized compound like iron oxide, zinc oxide, aluminum oxide, magnesium oxide and copper oxide are used a suitable micro nutrients [38-39]. The silver in nano sized are also in wheat crops to reduce phytotoxic nature of ozone as abiotic stress to improve the yield of crop. In general, the presence tropo-spheric ozone causes several damage including leaf injuries, which reduces the crop production. However, the anti-ageing feature of silver nano particles control negative feature of ozone exposure as well as improve the crop production [40-42]. Similarly, the presence of Zinc, in trace amounts is important and essential for optimum metabolic activities of plants and its deficiency reduces the production rate of crop [41-42]. In general the zinc manages the reactive oxygen species towards protection of plant cells against oxidative stresses as well as formation of chlorpophyll and carbohydrate. It is reported that wide range of soil are having deficiency, therefore the use nano sized zinc oxide as nano fertilizer improves the health of plants as well as productivity of crop. Magnesium oxide (MgO) is another important inorganic metal oxide with significant applications in agriculture chemical with antifungal and anti-microbial in nature. Further, the non-toxic nature of magnesium for human, frequently used in agriculture practices to inhibits pathogens and different microbes. Its working mechanism varies like coordination between with cell wall, enhanced formation of reactive oxygen species [43-46]. The other important point of nano fertilizer is formulation to ensure its quality availability for longer time. Currently, the blending of micronutrient and fertilizers are also used at nano scale to improve the functionality and effectiveness of nano technology. In this context, Kundu et al prepared iron oxide and urea nano composite as a novel fertilizer with improved nutrient utility and reduced environmental pollutions. The composite was used for seedlings of Oryza sativa L and checked for 14 days under hydroponic conditions and found significant up-regulation in photosynthetic efficiency as well as nitrogen metabolism due to better availability of both of nitrogen and iron [47]. In the above line the preparative mechanism for zinc and urea composite was shown in Figure 7 an along with its working impact on a wheat plant. The observation reveals the better growth of plants along with better agronomic performances [48].

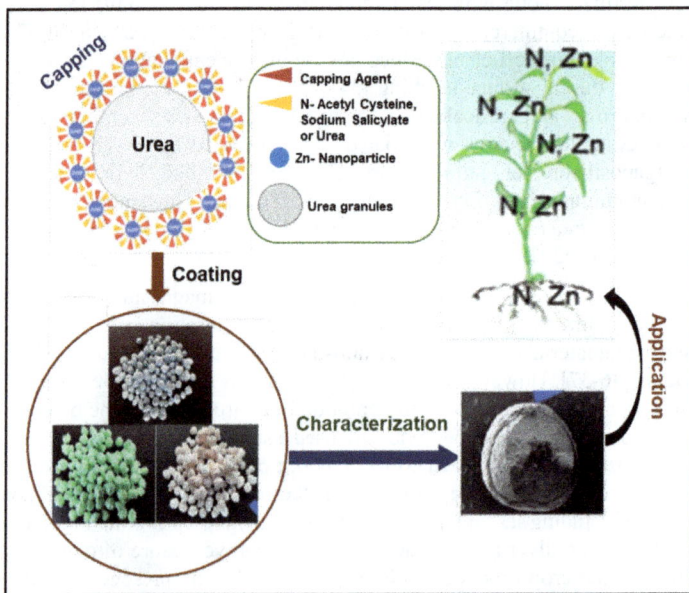

Figure 7. Preparation and use of zinc urea nanocomposite nano fertilizer on wheat plant [48].

Applications of Emerging Nanomaterials and Nanotechnology   Materials Research Forum LLC
Materials Research Foundations 148 (2023) 252-275   https://doi.org/10.21741/9781644902554-9

3.3   Sensing and monitoring

Different nanomaterials-based sensors are used in agricultural practice for monitoring of pH of soil, water contents, volatile ammonia and organic vapors. These parameters are important to grade soil nutrition index as well as improving the soil quality in the term of supplying additional fertilizer and nutrients. In this regards electrical, optical and gravimetric sensors are used for monitoring agrochemical and soil nutrients with precise sensing parameters [49-50]. In an example, Shukla et al has reported chemi-resistive sensors for monitoring of volatile ammonia from soil as nutrition index. The scheme of sensor along with basic components and parameters is shown below in Figure 8.

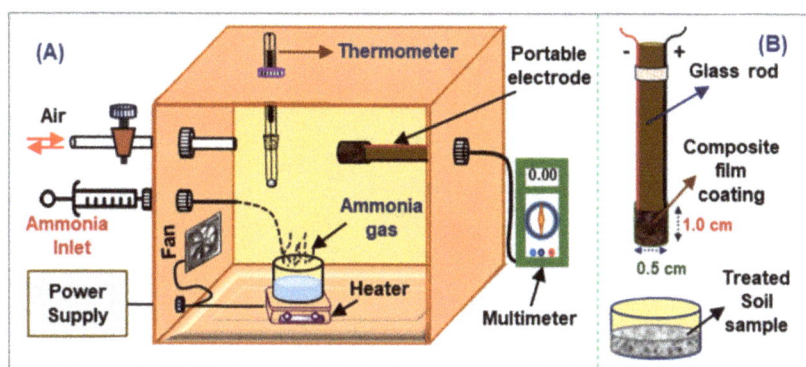

Figure 8. The nanocomposite based electrochemical sensor for volatile ammonia [51]

The basic principle and importance of ammonia sensing is based on adsorption of volatile ammonia after hydrolysis of urea from the adsorbed ammonia which changes the resistance of sensing layer and serve as a basic principle in ammonia sensing along effective sensing parameters i.e. response time, recovery time and sensitivity of sensing [51]. Similarly, pH monitoring is other significant parameters of soil to decide the fertility and selection of the crop for cultivation. In general, pH is sensed after measuring the induced color blue to red from acidic pH to basic pH or transition in resistance and current. Here, the use of nano particle controls the effective reactivity towards the induced changes because of interacting sites and surface area. Further, the integration pH responsiveness with remote control like Arduino helps the agriculture scientist to monitor the soil condition for policy making and mapping of soil pH Chajanovsky et al. have prepared a thin film of carbon nanotube and polyaniline using inverse polymerization using sonication. The obtained film was found suitable for electrochemical sensing of pH with high capacity due to synergistic effect of high surface area of CNTs and high electrical conductance of polyaniline. Thus, the developed property of composite was also found suitable for sensing of organic compounds like aminophenol present in soil with high sensitivity [53]. Humidity is another important component of soil to insure the cultivation of crop. Several sensors are used for monitoring humidity using different nano materials like metal oxides, carbon nano structure and their polymer composites [54]. The nanostructure in humidity sensitive materials in the term of hydrophilicity and interactivity has improved the humidity sensitive nature, in this context several nano structured hydrogel, xerogel and composites are used efficient monitoring of precise humidity. In general the monitoring of humidity from soil is consist of different parts i.e. soil moisture metrics, in-situ soil moisture sensing methods, remote sensing methods and proximal in-field soil moisture monitoring. Thus, highly hydrophilic nano composite and hydrogels are most promising materials for humidity sensing with competitive sensing parameters. The commercially available temperature sensor is shown in Figure 9, which is an integral effect of heating probe and humidity sensors with sensitivity of 0.632 ^{0}C per 1% in the range of 5 to 41 %.

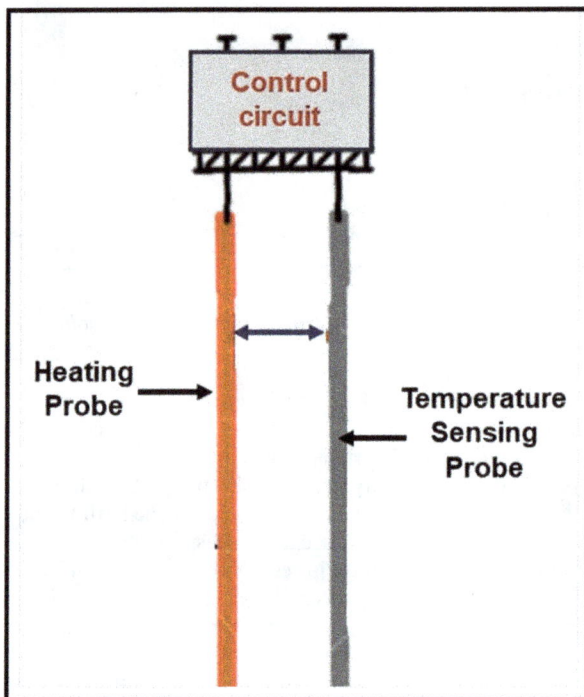

Figure 9. The schematic of dual probe-based soil moisture sensor [55].

Applications of Emerging Nanomaterials and Nanotechnology   Materials Research Forum LLC
Materials Research Foundations 148 (2023) 252-275   https://doi.org/10.21741/9781644902554-9

The use of nanomaterials enhances the sensing parameters in the term of sensing range, sensitivity and response time [55]. In an example the use of reduced graphene oxide as sensing film has improved the sensing response of 60 -75 s along with stability for 5 months. The sensing principle of this kind of sensor is based on change in resistance due to after dissociative adsorption with water molecule, which is shown in Figure 10. It reveals that the first layer of H_2O adsorption is non ionizing due to covalent nature, while the subsequent layer of adsorption in ionizing nature and it yield hydronium and hydroxide ions. The generation of both ions increases conductance and decrease in resistance [56].

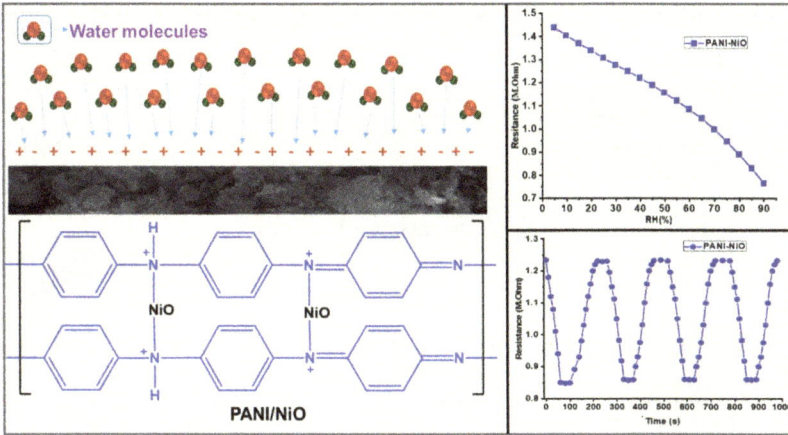

Figure 10. Chemi-resistive sensing mechanism for humidity over NiO and PANI nano composite [52]

Some more examples of nano technology based sensors along with brief properties for agriculture are listed in Table 2.

Table 2. Nanotechnology based sensors and their properties

S. No.	Materials	Type of Sensors	Analytes and sensing parameters	Reference
1	Zinc oxide encapsulated polypyrrole	Electrochemical	Soil volatile ammonia, in the range of 1–100 ppm with sensitivity 0.4947 $k\Omega$-ppm^{-1} for 90 days.	51
2	ZnS quantum dots	Fluorescence	Urea, with dynamic limit 4×10^{-9} M to 4×10^{-3} M and low detection limit ($< 4 \times 10^{-9}$ M)	57
3	Graphene quantum dots	Electro-chemical low-cost sensors	Soil moisture in the range of water content 0% to 32% and response time 180 s	58
4	One-dimensional nanomaterial polyaniline		Ecotoxicity to earthworms, the soil enzyme activities i.e sucrase, phosphatase, and protease.	59
5	Multiwall carbon nanotubes	Electrochemical	Chlorpyrifos in the concentrationof 0.5 pM to 500 nMand detection limit of 0.16 pM	60
6	Graphitic carbon nitride	Square wave voltammetric technique	Amino-triazole and linuron in range of 3.0×10^{-7} to 4.5×10^{-5} M, and 1.2×10^{-7} to 3.0×10^{-4} M respectively	61
7	Polyaniline/nickel oxide	Electrocatalytic detection	epinephrine Very good performances achieved at a molecular level	62
8	Sodium alginate grafted polyaniline	Potentiometric	Mercury from 1.0 nM to 1000 μM with sensitivity of 0.25 mV μM^{-1} cm^{-2}.	63

3.4   Crop protection and storage

The several nanomaterials are used as pesticides, insecticides, herbicides and preservatives to secure crops from microbial attack and environmental ageing. In this context different nanoparticles like metal oxides, metals, carbon nanostructure and polymer nano structure are used as agrochemicals like improved herbicides along with reduced negative effect [64]. In general, these agrochemical are two types one is exclusive in nano size, while other is hybrid of nano size.In an example Chitosan/tripolyphosphate nano particles were used to encapsulate nonselective widely used herbicides i.e paraquat. Thus, obtained nano formulation was found less toxic with conserve effectiveness for herbicidal treatment of

Applications of Emerging Nanomaterials and Nanotechnology   Materials Research Forum LLC
Materials Research Foundations 148 (2023) 252-275   https://doi.org/10.21741/9781644902554-9

different plants [65]. Furthermore, the use of pristine pesticides in nano size improves the chemical efficiency of pesticides in term of reactivity and toxicity. In general, the use of traditional pesticides bears larger size, poor dispersibility, low stability and low biological activity along with low target utilization i.e. less than 30%. However the nano sized pesticide shows smaller size with high dispersibility and high target utilization, therefore the nano technology reduces the required amount of pesticides along with controls of the negative effect of pesticides towards environments. However, the improved dispersibility and scattering effects of nano sized pesticides also enters into the body of non-targeted organism and may cause unpredictable consequence as illustrated in Figure 11.

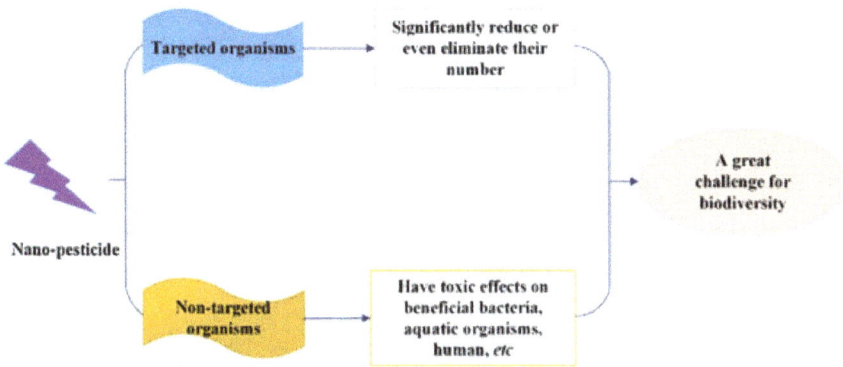

Figure 11. Conclusive impact of nanosized pesticides

Therefore, huge precaution is needed before commercializing a nano scale agrochemical like pesticides. Some representative nano sized agrochemicals along with their applications are listed in Table 3.

Table 3. Some different nano-sized agrochemicals and their applications in promoting agricultural sustainability.

S.No.	Nano material as fertilizer	Nutrient	Applications	Reference
1	Hydroxyapatite and urea	Nitrogen	35 % slower release of nitrogen for better supplements of N and K for leaves	66
2	Hybrid nanocomposite of Urea, hydroxyapatite and montmorillonite	Nitrogen	Sustainable release of urea for better crop yield	67
3	Nano sized zeolites	Phosphorous and Potassium	High increased release rate with better accumulation of P and K at optimum pH and moisture	68
4	Chitosan nanoparticle	N,P, and K	Better harvest and crop index with increase in root and shoot length	69
5	Nanosized calcium phosphate	N,P, and K	Better fertilization efficiency for précised agriculture with increased nitrogen efficiency	70, 71
6	Calcium nanoparticles	Calcium	Increased food nutrition quality for essential nutrition plant food	72
7	Zinc and Copper nanoparticles	Zn and Cu	Increase chlorophyll and carotenoid in the leaves Improved phenolic and flavonoid contents with better quality and quantity of bassil	73
8	Nano sized zeolite	N,P, and K	Better vegetative growth, photosynthesis with increase in fresh and dry weight with phenols, tannins, flavonoids, and micro-elements	74

Nanopesticides				
S. No.	Nanomaterial	Nature	Target	Reference
9	Nanostructured alumina	Nanoinsecticide	- Sitophilus oryzae - Oryzaephilussurinamensis - Ceratitis capitate - Leaf-cutting ants	75,76
10	Nano silica	Nanoinsecticide	-Sitophilus oryzae -Rhizoperthadominica -Triboliumcastaneum -Orizaephilussurinamenisis	77
11	Silver nanoparticles	Nano bactericides	- Xanthomonas axonopodis	78
12	Cu and Ag nanoparticles	Fungicides	Effective against Rhizoctonia solani, Phytophthora cactorum, Fistulina hepatica and Grifolafrondosa	79
13	Polycaprolactone nano capsules with atrazine	Herbicides	Amaranthus viridis and Bidens pilosa	80
14	Essential oil Nanoencapsulation of savory (Satureja hortensis L.) essential oil	Herbicides	Effective against Lycopersicon esculentum Mill and Amaranthus retroflexus L.	81

Conclusions and future prospects

The basic and novelty of nano particles has been described along with their applications in different agriculture practices. The adopted methods for synthesis are discussed along with secondary used tools for confinement effect and size particularly for geometry, physical and chemical properties. Furthermore, the role of different nano particles in valorization of agriculture wastes, nutrients, sensing and agrochemical are discussed along with effective illustration and schemes.

References

[1] A.S.Prabha, J.A. Thangakani, N.R. Devi, R. Dorothy, T.A. Nguyen, S.S. Kumaran, S. Rajendran, Nanotechnology and sustainable agriculture. In Nanosensors for Smart Agriculture, Elsevier, (2022)25-39. https://doi.org/10.1016/B978-0-12-824554-5.00016-1

[2] U. Muhammad, F.Muhammad, W.Abdul, N. Ahmad, A.C. Sardar, R.U.Hafeez , A. Imran, S. Muhammad, Nanotechnology in agriculture: Current status, challenges, and future opportunities. Sci. Total Environ., 721(2020) 137778. https://doi.org/10.1016/j.scitotenv.2020.137778

[3] L. Muthukrishnan, An overview on the nanotechnological expansion, toxicity assessment and remediating approaches in Agriculture and Food industry, Environ. Technol. Innov., 25,(2022)102136, https://doi.org/10.1016/j.eti.2021.102136

[4] P. Zhang, Z. Guo, S. Ullah, G. Melagraki, A. Afantitis, I. Lynch, Nanotechnology and artificial intelligence to enable sustainable and precision agriculture. Nat. Plants, 7(7), (2021)864-876. https://doi.org/10.1038/s41477-021-00946-6

[5] H. Singh, A. Sharma, S.K. Bhardwaj, S.K.Arya, N. Bhardwaj, M. Khatri, Recent advances in the applications of nano-agrochemicals for sustainable agricultural development. Environ. Sci.: Process. Impacts, 23(2), (2021)213-239. https://doi.org/10.1039/D0EM00404A

[6] A.D. Tiple, V.J. Badwaik, S.V. Padwad, R.G. Chaudhary, N.B. Singh, A Review on Nanotoxicology: Aquatic Environment and Biological System, Mater. Today: Proc, 29 (4),(2020) 1246-1250. https://doi.org/10.1016/j.matpr.2020.05.755

[7] N. Baig, I. Kammakakam, W. Falath, Nanomaterials: A review of synthesis methods, properties, recent progress, and challenges. Mater. Adv. 2(6),(2021)1821-1871. https://doi.org/10.1039/D0MA00807A

[8] C. Bartolucci, V. Scognamiglio, A. Antonacci, L.F. Fraceto, What makes nanotechnologies applied to agriculture green? Nano Today, 43(2022)101389. https://doi.org/10.1016/j.nantod.2022.101389

[9] Y. Shang, M. Hasan, G.J. Ahammed, M. Li, H. Yin, J. Zhou, Applications of nanotechnology in plant growth and crop protection: a review. Molecules, 24(14),(2019) 2558. https://doi.org/10.3390/molecules24142558

[10] G. Ginni, S. Kavitha, Y. Kannah, S.K. Bhatia, A. Kumar, M. Rajkumar, N.T.L. Chi, Valorization of agricultural residues: Different biorefinery routes. J. Environ Chem Eng, 9(4),(2021) 105435.

[11] Food wastage Footprint-Impacts on natural resources: Summary Report. Food and Agriculture Organization of the United Nations, Europe: Springer; (2003).

[12] O. Tepe, A.Y. Dursun, Exo-pectinase production by Bacillus pumilus using different agricultural wastes and optimizing of medium components using response surface methodology. Environ Sci Pollut Res In.21(16),(2014) 9911-9920. https://doi.org/10.1007/s11356-014-2833-8

[13] E. Capanoglu, E. Nemli, F. Tomas-Barberan, Novel Approaches in the Valorization of Agricultural Wastes and Their Applications. J Agric Food Chem 70, (23)(2022) 6787-6804. https://doi.org/10.1021/acs.jafc.1c07104

[14] R.G. Chaudhary, G.S. Bhusari, A.D. Tiple, A.R. Rai, Metal/Metal Oxide Nanoparticles: Toxicity, Applications, and Future Prospects, Curr Pharm Des, 25,(2019) 4013-4029. https://doi.org/10.2174/1381612825666191111091326

[15] I. De Luca, F. Di Cristo, A. Valentino, G. Peluso, A. Di Salle, A. Calarco, Food-Derived Bioactive Molecules from Mediterranean Diet: Nanotechnological Approaches and Waste Valorization as Strategies to Improve Human Wellness. Polymers, 14(9)(2022)1726. https://doi.org/10.3390/polym14091726

[16] P. Nigam, Production of bioactive secondary metabolites. Biotechnology for agro-industrial residues utilisation (2009)129-145. https://doi.org/10.1007/978-1-4020-9942-7_7

[17] A.V. Shekdar, Sustainable solid waste management: an integrated approach for Asian countries. Waste Manag.29,(2009)1438-1448. https://doi.org/10.1016/j.wasman.2008.08.025

[18] T. Robinson, Membrane bioreactors: Nanotechnology improves landfill leachate quality. Filtr. Sep.44,(2007) 38-49. https://doi.org/10.1016/S0015-1882(07)70288-4

[19] S. Moogi, J. Lee, J. Jae, C. Sonne, J. Rinklebe, D.H. Kim, Y.K. Park, Valorization of rice husk to aromatics via thermocatalytic conversion in the presence of decomposed methane. Chem Eng J, 17,(2021) 129264 and15. https://doi.org/10.1016/j.cej.2021.129264

[20] B. Hu, K. Wang, L. Wu, S.H. Yu, M. Antonietti, M.M. Titirici, Engineering carbon materials from the hydrothermal carbonization process of biomass. Adv Mater 22,(2020) 813-828. https://doi.org/10.1002/adma.200902812

[21] W.J. Liu, H. Jiang, H.Q. Yu, Development of biochar-based functional materials: toward a sustainable platform carbon material. Chem Rev, 115,(2015)12251-12285. https://doi.org/10.1021/acs.chemrev.5b00195

[22] M.S. Umekar, A.K. Potbhare, G.S. Bhusari, M.F. Desimone, R.G. Chaudhary, Bioinspired reduced graphene oxide based nanohybrids for photocatalysis and antibacterial applications. Curr Pharm Biotechnol, 22,(2021) 1759 - 1781. https://doi.org/10.2174/1389201022666012311115826

[23] N. Liu, K. Huo, M.T. McDowell, J. Zhao, Y. Cui, Rice husks as a sustainable source of nanostructured silicon for high performance Li-ion battery anodes. Sci Rep 3,(2013)1919. https://doi.org/10.1038/srep01919

[24] S. Ramakrishna, K. Fujihara, W.E. Teo, T. Yong, Z. Ma, R. Ramaseshan, Electrospun nanofibers: solving global issues. Mater Today 9,(2006)40-50. https://doi.org/10.1016/S1369-7021(06)71389-X

[25] S. Meraz-Dávila, C.E. Pérez-García, Feregrino-Perez, A. Ana, Challenges and advantages of electrospun nanofibers in agriculture: a review. Mater. Res. Express 8,(2021)042001. https://doi.org/10.1088/2053-1591/abee55

[26] M. Aman Mohammadi, S.M. Hosseini, M. Yousefi, Application of electrospinning technique in development of intelligent food packaging: A short review of recent trends, Food Sci. Nutr.8(2020) 4656-4665. https://doi.org/10.1002/fsn3.1781

[27]., A. Arora, P. Nandal, J. Singh, M.L. Verma, Nanobiotechnological advancements in lignocellulosic biomass pretreatment, Mater Sci Technol,3,(2020) 308-318. https://doi.org/10.1016/j.mset.2019.12.003

[28] A. Musa, M. Ahmad, M.Z. Hussein, S.M. Izham, Acid hydrolysis-mediated preparation of nanocrystalline cellulose from rice straw, Int j nanomater.nanotechnol nanomedicine,3(2),(2017) 51-56.

[29] P. Kampeerapappun, Extraction and characterization of cellulosenanocrystalsfromcornstover, J Met Mater Miner,25(1),(2015)19-26.

[30] B. Hong, F. Chen, G. Xue, Preparation and characterization of cellulose nanocrystals from bamboo pulp, Cellul Chem Technol, 50(2),(2016) 225-231.

[31] R.J. Moon, A. Martini, J. Nairn, J. Simonsen, J. Youngblood, Cellulose nanomaterials review: structure, properties and nanocomposites. Chem Soc Rev. 40(7),(2011) 3941-3994. https://doi.org/10.1039/c0cs00108b

[32] L. Castro, M.L. Blázquez, J.A. Muñoz, F. González, C. García-Balboa, A. Ballester, Biosynthesis of gold nanowires using sugar beet pulp. Process Biochem., 46(5),(2011)1076-1082. https://doi.org/10.1016/j.procbio.2011.01.025

[33] D. Sharma, M.K. Varshney, S. Prasad, Bhawana, S. K. Shukla, Preparation and characterization of rice husk derived cellulose and polyvinyl alcohol blended heat sealable packaging film, Indian J Chem Technol, 28,(2021)453-459.

[34] B. Beig, M.B.K. Niazi, F. Sher, Z. Jahan, U.S. Malik, M.D. Khan, D.V. N. Vo, Nanotechnology-based controlled release of sustainable fertilizers. A review. Environ Chem Lett, 20, (2022)2709-2726. https://doi.org/10.1007/s10311-022-01409-w

[35] A. Mondal, M.S. Umekar, G.S. Bhusari, S. Mondal, R.G. Chaudhary, M. Sami, Biogenic synthesis of metal/metal oxide nanostructured materials, Curr Pharm Biotechnol, 22,(2021)1782 - 1793. https://doi.org/10.2174/1389201022666210111122911

[36] K. Shankramma, S. Yallappa, M.B. Shivanna, J. Manjanna, Fe2O3 magnetic nanoparticles to enhance S. lycopersicum (tomato) plant growth and their biomineralization. Appl Nanosci 6,(2016) 983-990. https://doi.org/10.1007/s13204-015-0510-y

[37] D. Lin, B. Xing Phytotoxicity of nanoparticles: Inhibition of seed germination and root growth. Environ Pollut 150(2),(2007) 243-250. https://doi.org/10.1016/j.envpol.2007.01.016

[38] N.M. Salem, L.S. Albanna, A. Abdeen, O.Q. Ibrahim, A.I. Awwad, Sulfur nanoparticles improves root and shoot growth of tomato. J Agric Sci 8(4),(2016)179-185. https://doi.org/10.5539/jas.v8n4p179

[39] A.Y. Ghidan, T.M. Al-Antary, A.M. Awwad, O.Y. Ghidan, S.E. Al Araj, M.A. Ateyyat, Comparison of different green synthesized nanomaterials on green peach aphid as aphicidal potential. Fresenius Environ Bull 27(10),(2018) 7009-7016.

[40] R. Kannaujia, P. Singh, V. Prasad, V. Pandey, Evaluating impacts of biogenic silver nanoparticles and ethylenediurea on wheat (Triticum aestivum L.) against ozone-induced damages. Environ Res, 203,(2022)11857. https://doi.org/10.1016/j.envres.2021.111857

[41] V. Krishna ,U. Hina , S. Arshdeep, R. Manisha, Influence of Zinc Application In Plant Growth: An Overview, Eur J Mol Clin Med, 7(7),(2020)2321-2326.

[42] R. Komal, P. Vishal, P. Kalavati, The Importance Of Zinc In Plant Growth- A Review, Int res j nat sci, 5(2),(2018)38-48.

[43] S. Lidvin Daisy, A. Christy Catherine Mary, Devi Kasthuri, S. Santhana, S. Rajendran, Prabha, S. SyedZahirullah, S. Magnesium oxide nanoparticles- Synthesis and Charecterisation, Int j nano corros sci eng 2(5),(2015) 64-69.

[44] S. Abinaya, P. Kavitha Helen, M. Prakash, A. Muthukrishnaraj, Green synthesis of magnesium oxide nanoparticles and its applications: A review, Sustain Chem Pharm 19(7),(2021)100368-100378. https://doi.org/10.1016/j.scp.2020.100368

[45] S.K. Moorthy, C.H. Ashok, K.V. Rao, C. Viswanathana, Synthesis and characterization of MgO nanoparticles by neem leaves through green method. Mater Today: Proc 2, (2015) 4360-4368. https://doi.org/10.1016/j.matpr.2015.10.027

[46] A. Awwad, A. Ahmad, Biosynthesis, characterization, and optical properties of magnesium hydroxide and oxide nanoflakes using Citrus limon leaf extract. Arab J Chem 1(2014)65-70.

[47] T. Guha, G. Gopal, A. Mukherjee, R. Kundu, Fe3O4-urea nanocomposites as a novel nitrogen fertilizer for improving nutrient utilization efficiency and reducing environmental pollution. Environ Pollut, 292(2022)118301. https://doi.org/10.1016/j.envpol.2021.118301

[48] C.O. Dimkpa, M.G. Campos, J. Fugice, K. Glass, A. Ozcan, Z. Huang, S. Santra, Synthesis and characterization of novel dual-capped Zn-urea nanofertilizers and application in nutrient delivery in wheat. Environmental Science: Advances, 1(1), (2022)47-58. https://doi.org/10.1039/D1VA00016K

[49] H. Yin, Y. Cao, B. Marelli, X. Zeng, A.J. Mason, C. Cao, Soil sensors and plant wearables for smart and precision agriculture. Adv Mater, 33(20),(2021) 2007764. https://doi.org/10.1002/adma.202007764

[50] M. Nadporozhskaya, N. Kovsh, R. Paolesse, L. Lvova, Recent Advances in Chemical Sensors for Soil Analysis: A Review. Chemosensors, 10(1),(2022) 35. https://doi.org/10.3390/chemosensors10010035

[51] P. Singh, C.S. Kushwaha, V.K. Singh, G.C. Dubey, S.K. Shukla, Chemiresitive sensing of volatile ammonia over zinc oxide encapsulated polypyrrole based nanocomposite, Sens Actuators B Chem, 342(10),(2021) 130042 https://doi.org/10.1016/j.snb.2021.130042

[52] M. Setka, J. Drbohlavova, J. Hubalek, Nanostructured Polypyrrole-Based Ammonia and Volatile Organic Compound Sensors, Sensors,17,(2017) 562. https://doi.org/10.3390/s17030562

[53] I. Chajanovsky, S. Cohen, G. Shtenberg, R.Y. Suckeveriene, Development and Characterization of Integrated Nano-Sensors for Organic Residues and pH Field Detection. Sensors, 21(17),(2021)5842. https://doi.org/10.3390/s21175842

[54] N.S. Abbas, S.K. Shukla, Contemporary Advances in humidity sensing materials method and performances, Adv Mater Lett, 12,(2021) 21061634. https://doi.org/10.5185/amlett.2021.061634

[55] B. Kashyap, R. Kumar, Sensing methodologies in agriculture for soil moisture and nutrient monitoring. IEEE Access, 9,(2021) 14095-14121. https://doi.org/10.1109/ACCESS.2021.3052478

[56] P. Singh, C.S. Kushwaha, S.K. Shukla, G.C. Dubey Synthesis and Humidity Sensing Properties of NiO Intercalated Polyaniline Nanocomposite, . Polym Plast Technol Eng, 58(2),(2018) 1-9. https://doi.org/10.1080/03602559.2018.1466170

[57] E. Safitri, L.Y. Heng, M. Ahmad, T.L. Ling, Fluorescence bioanalytical method for urea determination based on water soluble ZnS quantum dots. Sens Actuators B Chem, 240,(2017)763-769. https://doi.org/10.1016/j.snb.2016.08.129

[58] H. Kalita, V.S. Palaparthy, M.S. Baghini, M. Aslam, Electrochemical synthesis of graphene quantum dots from graphene oxide at room temperature and its soil moisture sensing properties. Carbon, 165,(2020) 9-17. https://doi.org/10.1016/j.carbon.2020.04.021

[59] W. Shu, Z. Yang, Z. Xu, T. Zhu, X. Tian, Y. Yang, Effects of one-dimensional nanomaterial polyaniline nanorods on earthworm biomarkers and soil enzymes. Environ Sci Pollut Res, 29 (2022), 35217-35229. https://doi.org/10.1007/s11356-021-18260-1

[60] H. Ehzari, M. Safari, M. Samimi, M. Shamsipur, M.B. Gholivand, A highly sensitive electrochemical biosensor for chlorpyrifos pesticide detection using the adsorbent nanomatrix contain the human serum albumin and the Pd: CdTe quantum dots. Microchem J, 179(2022)107424. https://doi.org/10.1016/j.microc.2022.107424

[61] D. Ilager, N.P. Shetti, K.R. Reddy, S.M. Tuwar, T.M. Aminabhavi, Nanostructured graphitic carbon nitride (g-C3N4)-CTAB modified electrode for the highly sensitive detection of amino-triazole and linuron herbicides. Environ Res, 204,(2022)111856. https://doi.org/10.1016/j.envres.2021.111856

[62] G.E. Uwaya, Y. Wen, K. Bisetty, A combined experimental-computational approach for electrocatalytic detection of epinephrine using nanocomposite sensor based on polyaniline/nickel oxide. J Electroanal Chem, 911(2022) 116204. https://doi.org/10.1016/j.jelechem.2022.116204

[63] C.S. Kushwaha, V.K. Singh, S.K. Shukla, Electrochemically triggered sensing and recovery of mercury over sodium alginate grafted polyaniline. New J Chem, 45(24),(2021) 10626-10635. https://doi.org/10.1039/D1NJ01103K

[64] N. Yadav, A.K. Singh, Talha Bin Emran, R.G. Chaudhary, R. Sharma, S. Sharma, K. Barman, Salicylic acid treatment reduces lipid peroxidation, chlorophyll degradation and preserves quality attributes of pointed gourd fruit, J Food Qual, (2022) 2090562 https://doi.org/10.1155/2022/2090562

[65] R. Grillo, A.E.S. Pereira, C.S. Nishisaka, R. De Lima, K. Oehlke, R. Greiner, L.F. Fraceto, Chitosan/tripolyphosphate nanoparticles loaded with paraquat herbicide: An environmentally safer alternative for weed control. J. Hazard. Mater. 278,(2014) 163-171. https://doi.org/10.1016/j.jhazmat.2014.05.079

[66] N. Kottegoda, C. Sandaruwan, G. Priyadarshana, A. Siriwardhana, U.A. Rathnayak, D.M. BerugodaArachchige, A.R. Kumarasinghe, D. Dahanayake, V. Karunaratne, G.A. Amaratunga, Urea-hydroxyapatite nanohybrids for slow release of nitrogen, ACS Nano, 11,(2017)1214-1221. https://doi.org/10.1021/acsnano.6b07781

[67] N. Madusank, C. Sandaruwan, N. Kottegoda, D. Sirisena, I. Munaweera, A. De Alwis, V. Karunaratne, G.A. Amaratunga, Urea-hydroxyapatite-montmorillonite nanohybrid composites as slow release nitrogen compositions, Appl. Clay Sci., 150,(2017) 303-308. https://doi.org/10.1016/j.clay.2017.09.039

[68] A.A. Rajonee, S. Zaman, S.M.I. Huq, Preparation, characterization and evaluation of efficacy of phosphorus and potassium incorporated nano fertilizer, Adv. Nanopart., 6,(2017)62. https://doi.org/10.4236/anp.2017.62006

[69] H.M.A. Aziz, M.N. Hasaneen, A.M. Omer, Nano chitosan-NPK fertilizer enhances the growthand productivity of wheat plants grown in sandy soil, Span. J. Agric. Res., 14,(2016)17. https://doi.org/10.5424/sjar/2016141-8205

[70] G.B. Ram'ırez-Rodr'ıguez, C. Miguel-Rojas, G.S. Montanha, F.J. Carmona, G.D. Sasso, J.C. Sillero, J.S. Pedersen, N. Masciocchi, A. Guagliardi, P'erez-de-Luque, A. Reducing Nitrogen Dosage in Triticum durum Plants with Urea-Doped Nanofertilizers, Nanomaterials, 10,(2020)1043. https://doi.org/10.3390/nano10061043

[71] G.B. Ram'ırez-Rodr'ıgue, G. DalSasso, F.J. Carmona, C. Miguel-Rojas, A. P'erez-de-Luque, N. Masciocchi, A. Guagliardi, J.M. Delgado-L'opez, EngineeringBiomimetic Calcium Phosphate Nanoparticles: A Green Synthesis of Slow-Release Multinutrient(NPK) Nanofertilizers, ACS Appl. Bio Mater., 3,(2020) 1344-1353. https://doi.org/10.1021/acsabm.9b00937

[72] L. Azeez, A.L. Adejumo, O.M. Simiat, A. Lateef, Infuence of calcium nanoparticles (CaNPs) on nutritional qualities, radical scavenging attributes of Moringa oleifera and risk assessments on human health, J. Food Meas. Charact., 14, (2020)2185-2195. https://doi.org/10.1007/s11694-020-00465-6

[73] A. Abbasifar, F. Shahrabadi, B. ValizadehKaji, Effectsof green synthesized zinc and copper nano-fertilizers on the morphological and biochemical attributes of basil plant, J. Plant Nutr.,43,(2020)1104-1118. https://doi.org/10.1080/01904167.2020.1724305

[74] M.A. Mahmoud, H.M. Swaefy, Comparison between effect of commercial and nano NPK in presence of nano zeolite onsage plant yield and components under drought stress, J. Agric. Res., 47,(2020) 435-457. https://doi.org/10.21608/zjar.2020.94486

[75] T. Stadler, M. Buteler, S.R. Valdez, J.G. Gitto, Particulate nanoinsecticides: a new concept in insect pest management, Insecticides: Agriculture and Toxicology, 2018, 83. https://doi.org/10.5772/intechopen.72448

[76] M. Butele, G. Lopez Garcia, T. Stadler, Potential of nanostructured alumina for leaf-cutting ants Acromyrmexlobicornis (Hymenoptera: Formicidae) management, Austral Entomol., 57,(2018) 292-296. https://doi.org/10.1111/aen.12277

[77] M.E. El-Naggar, N.R. Abdelsalam, M.M. Fouda, M.I. Mackled, M.A. Al-Jaddadi, H.M. Ali, M.H. Siddiqui, E.E. Kandil, Soil Application of Nano Silica on Maize Yield and Its Insecticidal Activity Against Some Stored InsectsAfterthePost-Harvest, Nanomaterials, 10,(2020)739. https://doi.org/10.3390/nano10040739

[78] A. Sherkhane, H. Suryawanshi, P. undada, B. Shinde, Control of bacterial blight disease of pomegranate using silver nanoparticles, J. Nanosci. Nanotechnol., 9,(2018) 1-5.

[79] S. Banik, A.P. Luque, In vitro effects of copper nanoparticles on plant pathogens, benefcialmicrobes and crop plants, Span. J. Agric. Res., 15(2),(2017) 1-15. https://doi.org/10.5424/sjar/2017152-10305

[80] G.F. Sousa, D.G. Gomes, E.V. Campos, J.L. Oliveira, L.F. Fraceto, R. Stolf-Moreira, H.C. Oliveira, Post- emergence herbicidal activity of nanoatrazine against susceptible weeds, Front. Environ. Sci., 6(12),(2018) 1-6. https://doi.org/10.3389/fenvs.2018.00012

[81] A. Taban, M.J. Saharkhiz, M. Khorram, Formulation and assessment of nano-encapsulated bioherbicides based on biopolymers and essential oil, Ind. Crops Prod., 149,(2020) 112348. https://doi.org/10.1016/j.indcrop.2020.112348

Applications of Emerging Nanomaterials and Nanotechnology   Materials Research Forum LLC
Materials Research Foundations 148 (2023) 276-303   https://doi.org/10.21741/9781644902554-10

Chapter 10

Overview of Nanotechnology in Food Sciences

Ritika Arora, Rizwana and Saroj Kumar Shukla*

Bhaskaracharya College of Applied Sciences, University of Delhi, Delhi-110075, India

*sarojshukla2003@yahoo.co.in

Abstract

Innovations in materials science at nanoscale has produced the wide range for pristine and hybrid materials with high chemical reactivity, surface area and tunable physical properties to use as food preservatives, packaging materials, monitoring of adulterant and contaminants as well as reuse of waste foods. The present chapter presents the overview of advances in nanotechnology and its application for preparation of different types of nano materials with controlled composition, morphology, crystallinity and structural alignment. Further, the applications of these nanomaterials in different food industries, consumers and service providers are discussed with suitable scheme and illustrations as preservatives, nutraceuticals, packaging and degradation of food wastes.

Keyword

Nanotechnology, Nanomaterials, Food Preservatives, Nutraceuticals, Packaging

Contents

Applications of Emerging Nanomaterials and Nanotechnology   Materials Research Forum LLC
Materials Research Foundations 148 (2023) 276-303   https://doi.org/10.21741/9781644902554-10

1.   Introduction

Innovative features of materials at nanoscale have been a significant driving force for expanding the horizon of food products and related industries in the term of quality improvement, availability, shelf life, nutritional values and reusability. The size confined properties of both edible and nonedible items like chemical reactivity, stability, physical properties have played significant role as additives, preservation, degradation, processed foods and nutraceuticals [1]. However, the improved mechanical strength, surface interactivity and gas permeability of processable nano materials like polymer composite have significantly improved the scope of packaging industries as active, passive and intelligent packaging applications [2]. Furthermore, the integration of chemical responsive nature and mechanical properties of nano materials due to improved surface responsiveness has been explored for remote sensing and monitoring of food quality, microbial growth, and adulteration [3]. Thus, the wider use and impact of nano materials has leads to exponential growth in continuous publications and patents [Fig. 1]. This continuous increasing in publication has confirmed the significance of area for study and has also motivated to compile the developments in field and use of nano materials in different food sectors like processing, nutraceuticals, sensing and packaging applications. The concept of nano confinements, properties and applications of nano materials have been also explained with the help of scheme and illustrations from recent publications.

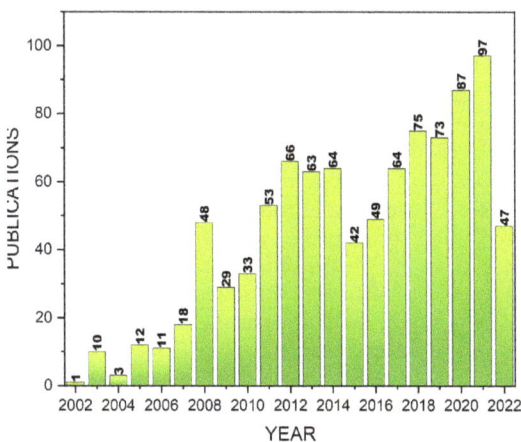

Figure 1. Trends of publication frequency (www.scopus.com)

Applications of Emerging Nanomaterials and Nanotechnology   Materials Research Forum LLC
Materials Research Foundations 148 (2023) 276-303   https://doi.org/10.21741/9781644902554-10

2.   Overview of nanotechnology

In general, nanotechnology deals with manipulation of materials, processing and science in terms of chemicals compositions, physical states and molecular interaction to evolve the novel features of materials in the scale of 1 to 100 nm. The simple comparative illustration of different sized materials including nano materials are shown in Fig. 2.

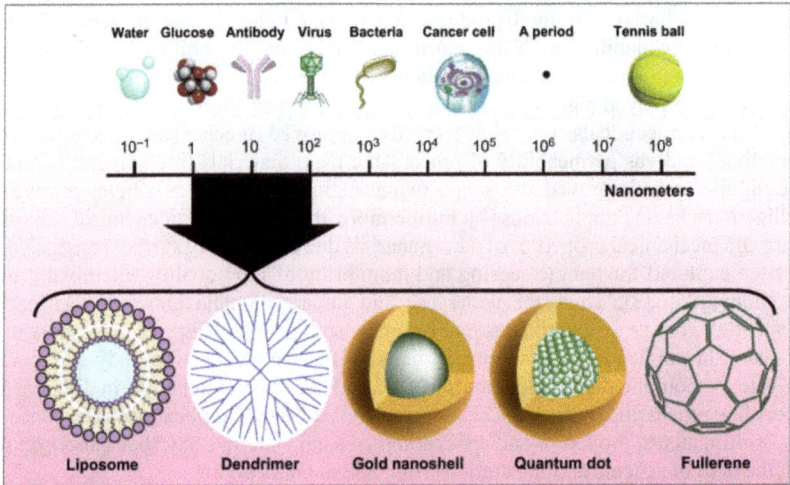

Figure 2. Comparative illustration of different sized materials. [4]

The manifestation of particles at nano scale yields several novel features like high surface area, improved reactivity, dispersibility, and physical strength than the its bulk materials. Thus, the nano materials have drastically improved the working and yields of most of the industries including food industries such as food processing, packaging and storage [5]. In general, nature has used a wide range of nano technology in growth of plants, flowering, fruit production, preservation of nutrients and safety of plants to yield better crops, long life, flowers, fruits and seeds. These natural inspirations have encouraged scientist to explore the natural practice in laboratory for improvements in practice of different food sectors [6] The pictorial representation about the use of nano technology in different applications in food sectors is shown in Fig. 3.

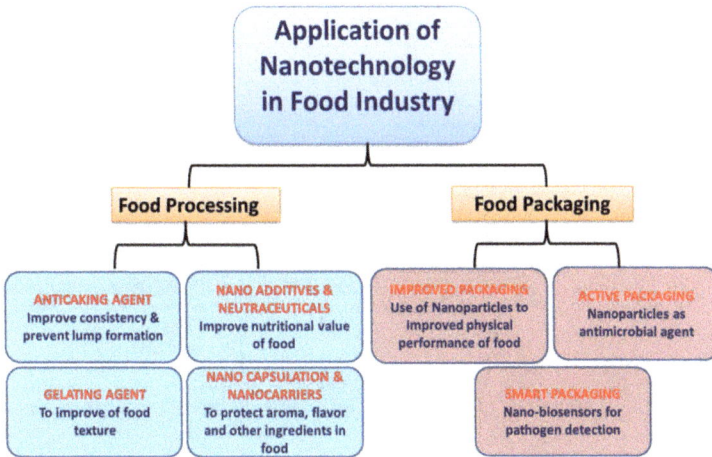

Figure 3. Different applications of nanotechnology in food sectors[6]

It reveals the broad application of nano materials are food processing and packaging after exploring their different applications. The food processing means the manipulation of edible contents and composition for improvement in their stability and nutritional indexing. In general, nature stores the most of nutrients in fruits and vegetable in nanoscale to maintain their properties, compositional balance and properties. However, the post harvesting practices has created several physical and chemical transitions in edible items, which are responsible for several negative changes and loss of nutritional indexing [6]. In examples, the loss of water in fruits and vegetable deteriorates the antioxidation activities and it leads to chemical and microbials degradation along these have been eaten safely for generations. Thus, several integrated steps are practiced in food processing industries to convert the agricultural crop into food with several advantageous features including better nutritional values. The concise impact and applications of nano technology in food processing industries are shown in Fig. 4.

The basic applications of nanotechnologies are in encapsulation to enhance the nutritional values and improved flavor in food items. However, the other class of application of nanomaterials in restructuring of matrix and the texture after using as emulsifier and surfactants. Further, structural modification and composite modification has also explored in nano technologies to develop different nano materials with specific properties for use in food industries. The brief properties of different nano materials in food processing are given in Table 1 along with brief descriptions and applications [7].

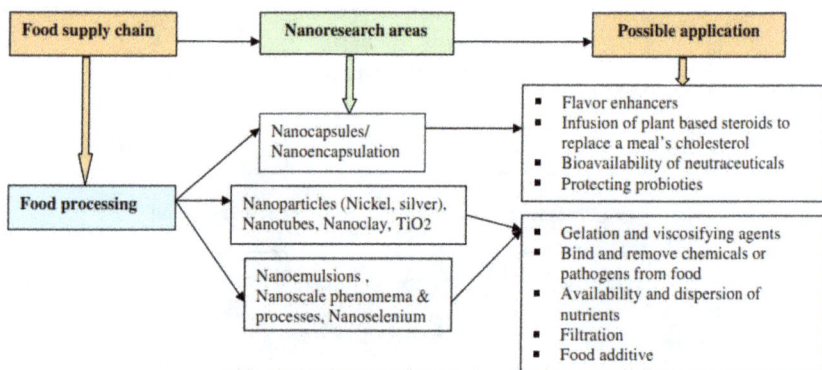

Figure 4. Applications of nanotechnologies in food processing industries [5]

Gelation is another important tool for improving the processing of food products after making gel and hydrogel. The use of nano additives improves the gelation capacity during processing of ingredients along with better properties of gel. For, example the hydrogel prepared from different edible polymer like polysaccharides, protein and lipid has open several doors for their applications like edible coating, pharmaceutical and drug delivery [8]. In this context the structure of polysaccharides has played significant role in hydrogel formation in example the linear polysaccharide exhibits stiff and rigid properties and yield membrane, sheet and coating, while the polysaccharides with globular morphology produce films type hydrogels and till date several polysaccharides based hydrogel are used as edible coating [9]. Currently, the chemical and physical treatments like surface plasmonic wielding of nano materials are also explored to develop the advance features in materials to use specifically in packaging. The integration of nano fillers in process able polymer generates high mechanical strength with high surface resistance for use as a better packaging material. In this, the important fillers used are nanoclay, nanosized carbon structure and metal oxides nanoparticles. The presence of nano fillers also develops additional features like UV absorption capacity and gas barrier nature for advancing the features of composite to use in packaging applications. Similarly, the grafting of two polymers with nano confinement develops responsive features against external perturbation like hydronium ions and scavenging nature. These types of nanocomposites reveal its suitability to use as active packaging materials for fruits and vegetable [10]. Thus, the developed nano materials with effective degradability, gas permeability and strength has been used to develop the sachets and edible coating for spectrum of applications like packaging of banana with scavenging features for residual oxygen and ethylene.

Table 1. Various nano techniques for encapsulation and release of functional components [6].

Nano technique	Characteristic	Examples
Nanoemulsions	Increased droplet agglomeration and gravitational separation stability, increased optical clarity, improved oral bioavailbility	β-carotenebased nanoemlusion β-carotene based nanoemulsion
Edibe coatings	Helps maintain the freshness of food throughout a prolonged duration of storage	edible coatings made of gelatin and cellulose nanocrystals chitosan/ nanosilica coatings chitosan film with nano-SO_2 alginate/lysozyme nanolaminate coatings
Hydrogels	Simple to incorporate in capsules, shields drugs from harsh conditions and allows them to be delivered in reaction to atmospheric factors like pH and temperature	Protein hydrogels
Liposomes	Liposomes can be employed as carriers for hydrophobic moieties as they enclose an aqueous solution inside a hydrophobic membrane (hold within the bilayer)	Cationic lipid incorporated liposomes modified with an acid labile polymer hyperbranched poly (glycidol) (HPG)
Inorganic NPs	Their rigid surfaces offer regulated functionalization, and they have high encapsulating capabilities.	Mesoporous silica nanoparticles
Polymeric micelles	High solubility, low toxicity, solubilize water-insoluble molecules in the hydrophobic region	PEO-b-PCL [poly(ethylene glycol)block-poly (caprolactone)] Methoxy poly (ethylene glycol) palmitate polymeric miceles)

3.   Applications

The basic size confined and surface synergized properties of nano materials are explored for use in different applications in food processing and processing industries. The applications of nano particles in food sectors are based on selective advanced features and the simple illustration for properties induced application in food industries is shown in Fig. 5 and same are discussed in followed sections.

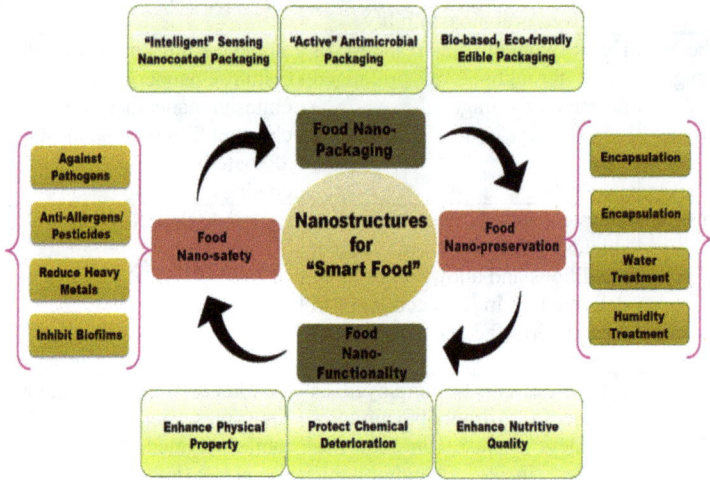

Figure 5 Application of nano particles in food industries [10].

3.1   Food processing

In general the food processing is an integrated thermal, chemical and physical transition to prevent contaminants and increase shelf life during storage and transportation. In this prospect the nanomaterials are used as preservatives, colorants, flavor, and emulsifier. The use of nano additives is during processing of crop and crop produces during conversion of agriculture products into processed products like fruit juices, fruit pulps, fruit cake and preservation of fruits and vegetables. The smaller size particles of colorant improve the color imparting capacity after use of small amount colorants and it avoid the excess use of metallic pigments like titanium oxide for reduced toxicity. In example the nano sized amorphous silica prepared by different method has been used as nano additives in different food products as antistatic agent as well as creamers. However, the use silica in nano size requires lesser amount for its effective applications, which control the silica toxicity due to use in higher than prescribed limit i.e. 250 ppm [11]. Similarly, the nano sized titanium oxide has been also used as an additive in gum, candy and pudding under controlled amount. Further, the use of nano particles again helps to induced toxicity due to use in limited amount i.e. 100=250 ppm in food grade products [12]. Zinc oxide is another important oxide with multiple functionalities and negligible toxicity to use in different prospects even as micronutrients [13]. Furthermore, the ZnO also exhibits antimicrobial, antifungal and antibacterial nature without human toxicity. Hence, these properties have supported the use of nano size ZnO in food processing industries i.e. medicinal formulation and preservatives of processed foods. The presence of optimum zinc concentration is essential for maintaining the effective activities of enzyme for regulation of metabolic body activities helpful [14].

The chemical and physical properties nano particles are also used for food preservation against denaturation and degradation of food from microbial and chemical interaction. The basic strategy for food preservation is emulsification, solvent extraction and homogenizations to insure the stable of edible bioactive food contents. The use of polymer nano size like celluloses has been used to supports the dispersion of food gradients in preserved foods. The nanostructure used in such applications is nano liposomes, colloidosome, chelates and nano carriers [15]. Emulsification has been explored for protection of different bioactive food items like vitamins, antioxidants, proteins, and lipids to use as functional food. The preservation induced interactive nature of nano polymers has been explored of delivery of active compounds like curcumin. The gelation properties of chitosan nano particle are used to prepare ionic gel of curcumin due to positively charged CHIT for nano carrier in different applications. The scheme of interaction is shown in Fig. 6 along with interacting structure and SEM pictures [15].

The concept has been for several polymeric and non-polymeric nano particles to ensure secure and effective availability of different active components. In this context, the different natural and synthetic nano materials are used as preservative of foods [14]. Chitosan is a natural copolymer-based polysaccharide consists of β--linked D-glucosamine and N-acetyl-D-glucosamine with several unique features i.e. Non-toxicity, antimicrobial nature and biocompatibility for use as food preservatives in processed foods. The important applications of nanoparticle-based food additives are listed in Table 2 along with properties and applications. The nano structure hybrid particles with chemical functionalities increasingly used as functional ingredients to food and beverage products. However, another important area is effect of processed food containing nano particle for their effects on gastrointestinal fate due to their large surface area. The area is still less studied but important to facilitate the safe design of edible items for commercialisation and consumption [23].

Materials Research Forum LLC
https://doi.org/10.21741/9781644902554-10

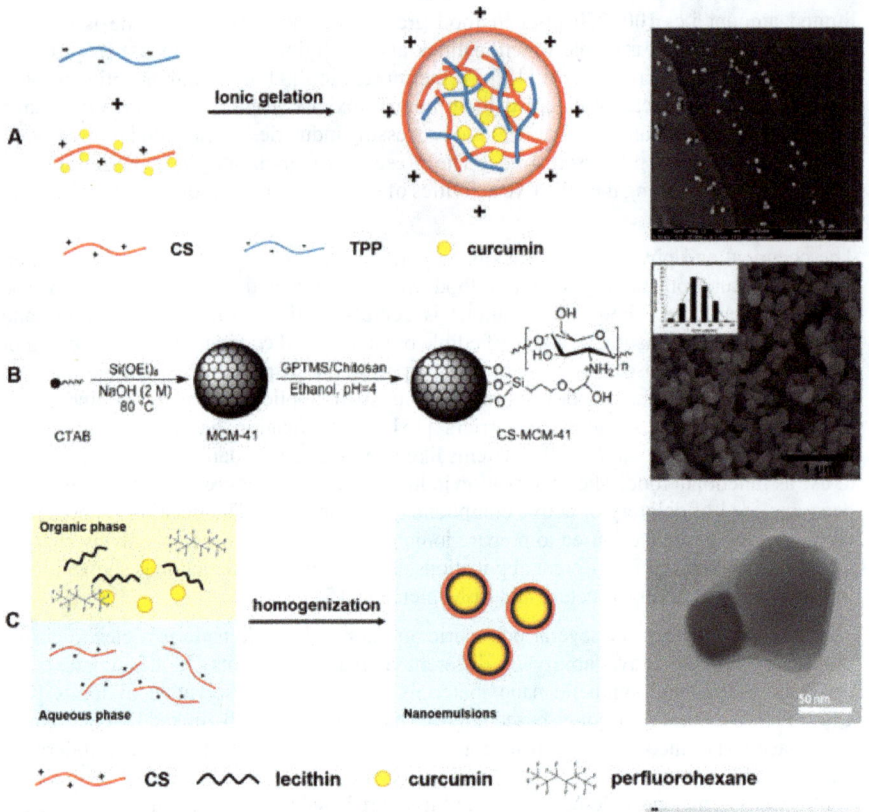

Figure 6. Scheme for preparation chitosan based curcumin carrier [15]

Table 2. Nano particles based food processing

S N	Nanoparticle (NP)	Properties	Applications	Ref.
1.	Chitosan NPs	Antimicrobial; Antioxidant Anticancer; Mucoadhesive	Cryprotectant in grey mullet surimi	[17]
2.	ZnO-NPs	Antimicrobial; Antifungal Immunity ; Enzyme activity Cell function; Heat stable	Zn-O-NPs in canned sea foods	[18]
3.	PLA-ZnO NCs	Antibacterial activity	Fused filament fabrication	[19]
4.	Curcumin-loaded Poloxamer$_{188}$	Stability	Improved antioxidant, anti-degradation, water solubility and stability of curcumin	[20]
5.	TiO$_2$ (titanium dioxide) NPs	Food pigment Confers white color Increases opacity	Natural whiteness and opacity to foods such as icing on cakes	[21]
6.	SiO$_2$ (silicon dioxide) NP	Anticaking agent	Incorporated in powdered foods such as salts, dried milk, and icing sugar	[22]

3.2 Value addition and nutraceuticals

The addition of values in edible items is currently used for cure of patient other than nutrition to survive, grow and reproduce. The process is consisting of extraction of nutrients and then encapsulation into a polymer matrix with objective to preserve for selective and controlled delivery using different polymeric as well as non-polymeric nano materials. Thus, the resulted hybrid nano materials enhanced the value of nutrient after covering therapy to remedies without using hazardous synthetic chemicals. Currently, this is the group of nano structure hybrid edible materials with properties other than nutrition i.e. medical and health benefits to prevent the infection or diseases. For this purpose, the bioactive compound is initially extracted from food matrix and then encapsulated in external matrix for preservative delivery cum availability. The nanoencapsulation technique has several advantages: improved stability and integrity, protection against oxidation and rancidity, preservation of volatile compounds, flavor masking, pH control, and moisture barrier, regulated release, several active substances delivered in a row, modification in flavor, long-lasting sensory attributes, and improved bioavailability and effectiveness. Singh et al has published an exhaustive review on the use of nano technology in applications and prospects of nutraceuticals. The fundamental properties of nano materials used in this regard are nano confined high surface to volume ratio, strength, solubility, chemical functionality and diffusivity. The brief application of nano materials in nutraceuticals are shown in Fig. 7 along with brief examples.

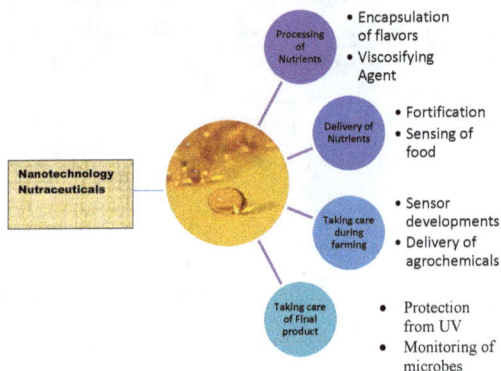

Figure 7. Basic role of nutraceuticals

Further, integration of different properties of different components of nano structure evolves the responsiveness and encapsulation of commercialization of different nutraceuticals. In this regards formulation of nutraceutical plays important role after using different nanoparticle like nanoemulsion, lipid, polymer micelles and polymer nano particles. The detailed formulation strategies for formulation lipophilic nutraceutical base nano formulation is shown in Fig. 8.

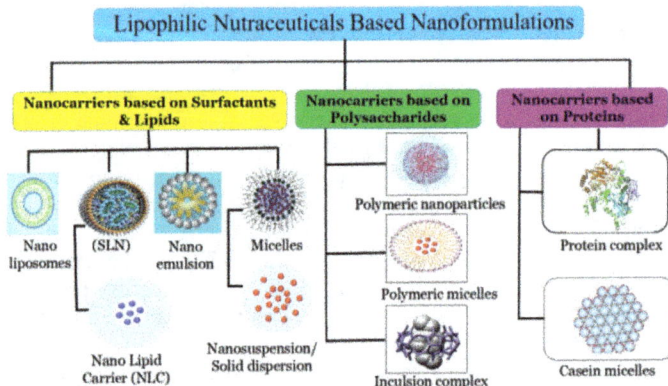

Figure 8. Nano-formulations for lipophilic nutraceuticals [24]

Applications of Emerging Nanomaterials and Nanotechnology   Materials Research Forum LLC
Materials Research Foundations 148 (2023) 276-303   https://doi.org/10.21741/9781644902554-10

The formulation mechanism has evolved significant properties like stability, solubility and dispersibility due to optimized structure and functionality. Some more significant nanostructured nutraceutical is listed in Table 3, along with properties and applications.

Table 3. Nano particles based nutraceutical

S N	Nanoparticle	Properties	Applications	References
1.	Resistant starch (lotus stem) NP	Emulsion stabilizing agent	For stabilizing the flax seed oil- water emulsion to form pickering emulsion to nanoencapsulate ferulic acid	[26]
2.	NP of *Stevia rebaudiana* extract	Natural sweetener Low calories Can control blood sugar	Alternative diet for COVID-19 diabetes patients	[27]
3.	Nano sized carbohydrate (Polysaccharide-pectin, alginate, carrageenan, agar, starch, and chitosan)	Carriers of nutraceuticals, vitamins, bioactive compounds Texture Appearance	Mouth delivery of bioactive compounds within starch nanoparticles	[28]
4.	Protein NP (casein micelles in milk)	Nutrition Appearance Delivery	Delivery in small intestine Functional ingredient in foods and beverages	
5.	Composite NP (combination of carbohydrate, protein, lipids)	Emulsifying agent	Proteins and carbohydrates have hydrophilic and lipophilic areas on their surfaces and so can act as emulsifiers that promote the formation and stabilization of lipid nanoparticles	

6.	Lipid based NP (lecithin, glyceryl monostearate, triacylglycerols)	Flavor Texture Appearance Nutrition	Carriers for lipophilic nutraceuticals and oil soluble vitamins Encapsulate flavorings, nutritious and antimicrobial substances	[29]
7.	Microencapsulated probiotics	Bioavailability of nutrients Improved digestion	dairy (yogurt and cheese), bakery (bread, cakes, biscuits), meat (fermented sausages), juices (fruits and vegetables), and others (ice cream, mayonnaise, fermented beverages)	

The basic advantages of nano manipulation of nutraceutical are the potential to improve the efficiency of plant-extracted products, reduce required quantities, eliminate side effects, and boost biological activity, as well as improve bioavailability, solubility, and stability, enhanced encapsulation, targeted and sustained delivery, and therapeutic action [29]. In this regard nutraceutical based products are in market by leading industries [29].

3.3   Packaging

Food packaging is the one sector of the industry where nanotechnology applications are beginning to live up to their promise, and nanocomposites are a rapidly growing field. Most food packaging applications have incorporated metal or oxide particles, or more commonly, nanoclays. The high surface area, mechanical strength and barrier properties of nano materials are widely used in packaging of different materials. The nano materials are used for both passive and active packaging. In general passive packaging provides outer covering with any interacting nature, however the active packaging provides interacting nature for consumption of evolved metabolites as well as releasing of gas to reduce the suppress the degradative reaction of packed item [30]. The synergistic impact of nano technology in different types of packaging applications is shown in Fig. 9.

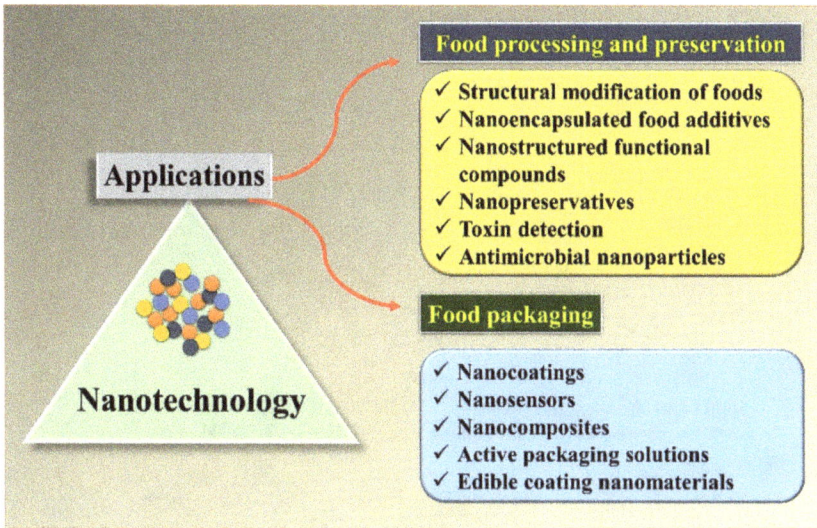

Figure 9. Illustration for use of nano technology in packaging applications[31]

Furthermore, the development of packaging is bulk process, thus, mainly hybrid materials are prepared after using both polymer and non-polymer constituents. For this purpose, both direct(ex-situ) and indirect method(in-situ) are explored along with chemical and physical treatment [30]. In direct method both the components are prepared separately and then mixed after employing solvent, heat or mechanical energy. However, in the indirect methods the monomers are polymerized in the presence of fillers after using the polymerizing agents and external energy. In the second method molecular arrangement of monomers in polymeric chain along with presence of interacting fillers yields the composite materials with novel features for advance packaging applications. For an example the polymer composite of cellulose plate and Cinnamomum zelanicum was prepared after using β-cyclodextrin and sodium caseinate as encapsulating agents. The composite exhibits controlled porosity for better storage of walnut kernel from oxidative deterioration [32]. The specific case for nano coating of fruits and vegetables are shown in Fig. 10.

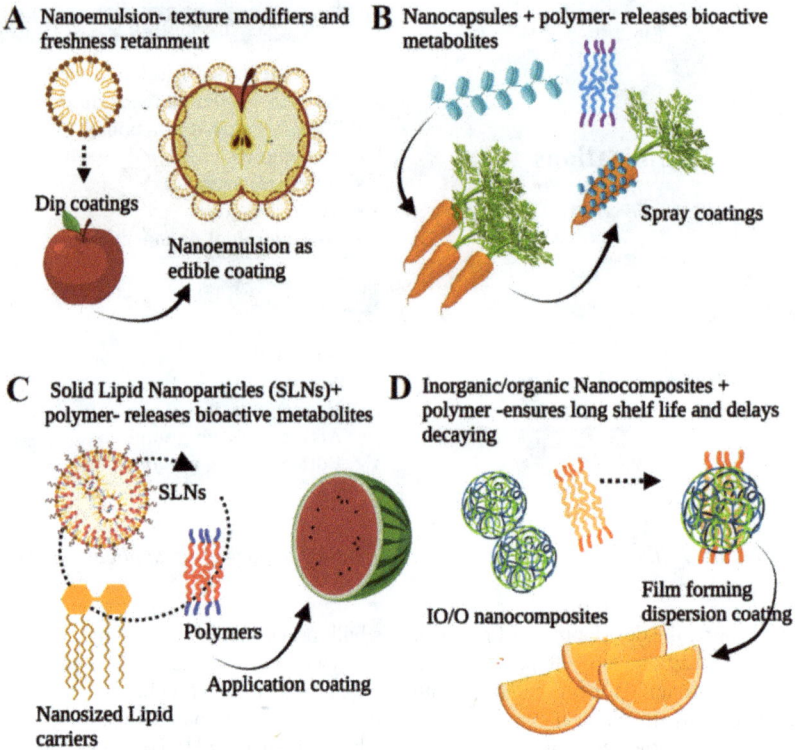

Figure 10. Representative applications of nanocoating on common fruits and vegetables [33].

Carbon dioxide emitters and oxygen scavengers are other important additives used in nano sized packaging materials and used to control the microbial growth and aerial oxidation. These types of active packaging materials are used use packing of meat, fish, milk and dairy products. The degradation of edible items develops change in pH and temperature of packets with time. Thus, the incorporation pH indicator in packaging film adds specific features to works as time and temperature indicators to monitors different properties of packed times. The simple explanation of TTI on different applications is shown in Fig. 11.

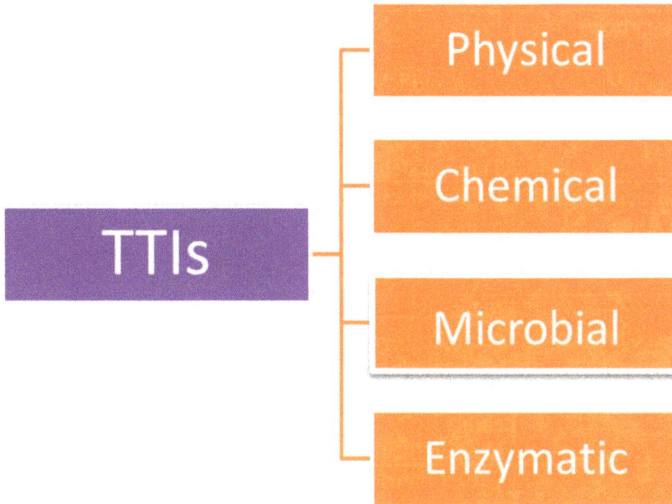

Figure 11. Importance of different types of TTI

Therefore, introducing the nanoparticles into the packaging materials has restricted microbial growth. There are various metal nanoparticles (Au, Ag, TiO2, Fe, CuO) incorporated into the packaging material for their extraordinary antimicrobial properties. These metal nanoparticles work toward the microbial toxicity in the following ways [34]:

1. uptake of metal ions through depletion of intracellular ATP

2. production of reactive oxygen species by damaging oxidative cells

3. damage to the membrane of bacteria

Some other applications of nano materials are shown in Table 4, along with properties and applications in different types of packaging.

Table 4. Nano materials in packaging applications

S N.	Nanoparticles (NPs)	Properties	Applications	References
1.	Berberine-cinnamic acid NP	Improved UV-shielding property Hydrophobicity Thermal stability	Antimicrobial nanocomposite films with extended shelf life	[35]
2.	Gallic acid grafted onto silica NP	Enhanced antioxidant property	SiO_2-GA Chitosan film-antioxidant food packaging composite film	[36]
3.	Clove essential oil loaded Chitosan-Zno hybrid NP	Enhanced tensile strength Hydrophobicity UV light blocking ability Oxygen Barrier	Clove essential oil loaded Chitosn-ZnO hybrid NP s integrated chitosan/pullulan nanocomposite film	[37]
4.	ZnO	GRAS status Antimicrobial activity	ZnO NP incorporated in materials like glass, LDPE, PP, PU, paper and chitosan for its antimicrobial activity with improved mechanical and functional properties	[38]
5.	TiO_2 (titanium dioxide)NP	Scavenger of oxygen Cost-effective Antimicrobial Durable	Films with improved surface properties TiO_2 nanotubes	[39]
6.	Ag (silver)NP	Biocidal effect High thermal resistance antimicrobial activity	Incorporated in packaging polymer to enhance shelf and halt microbial growth Edible coating containing silver nanoparticles	
7.	CuO (copper oxide) NP	Antimicrobial activity	Antimicrobial packaging material (cheese packaging)CuO NPs-chitosan nanofibers Bacterial nanofibers	

8.	Au (gold) NP	Biocidal agent Antimicrobial activity	Biofilms with biocidal action against *E.coli* and *S.auerus* Au NPs based colorimetric sensor	
9.	MgO (magnesium oxide) NP	High thermal stability Low cost Antibacterial activity	Fabricated Ag-MgO nanocomposites for antimicrobial activity	[40]
10.	Se (selenium) NP	Antioxidant Antimicrobial activity Antiviral activity Antibacterial activity	Incorporation of Se NPs on multilayer plastic substance imparted the antioxidant property Antibacterial polymeric coating	

Introduction of nanosensors in the food packaging has led to its evolution. Nano sensors are devices with distinctive measurements of a few nanometers that functions by detecting small amounts and transforming them to signals for examining. The packaging which involves incorporation of nano sensors into the packaging material is known as intelligent packaging. Such devices provide information about the internal changes that happen to the food product in terms of chemical or microbial changes [41]. One of the most intriguing uses of nanotechnology in the food industry is nanocoating on food packaging surfaces. The availability of O_2 inside packed food items provides ideal circumstances for microorganisms to flourish, compromising the quality and shelf-life of the food. For e.g. Titanium oxide nanoparticle induced with photo-indicator intelligent ink which included redox activated methylene blue dye for the detection of oxygen level. Food manufacturers can use nanotechnology to provide authenticity and track-and-trace capabilities to avoid copying, minimizing adulteration and diversion of products meant for a particular market. It works by creating complicated invisible nanobarcodes containing batch information that may be encrypted straight on packaged foods. Food safety is improved by using nanobarcode technology, which allows brand operators to track their supply chains without disclosing company information to distributors and wholesalers [42]. Different codes may be established and issued to each food item by changing the stripe ordering, allowing brand and authenticity in tracking food batches.

3.4   Recycling and reuse of used food

Recycling and even upcycling of waste food is most upcoming area in current time for food managements, value addition and maintain the food security for mankind. In is reported that 1.3 billion tons food are wasted in the form of leftover noodles, meat, bread, rice, vegetables and peels due to poor waste management and lack of literacy among different part of world. In general, the food waste is defined as used food products from various food processing industries, which cannot been recycled or consumed. Although, the food items

Applications of Emerging Nanomaterials and Nanotechnology   Materials Research Forum LLC
Materials Research Foundations 148 (2023) 276-303   https://doi.org/10.21741/9781644902554-10

are compostable items and they degrade into carbon dioxide after fungal and microbial degradation in the presence excess oxygen. However, in lack of oxygen the recycling of food like protein indicates the conversion of waste food into secondary raw materials like fertiliser, animal food and different gas and liquid organic compounds after using a biotechnical process, while the reuse indicates the reutilisation [43]. The food is a complex form of multi components compound and their recyclability also cover the conversion of polymer into monomer, it this context the nonedible cellulose extracted from peels can be converted into edible monomer like sugars, fatty acids, amino acids and phosphate after using biotechnological routes like enzymatic hydrolysis [44]. The chemical modification and chemical transition are another important strategy to wise reuse of food wastes. The conversion glucose into energy rich and functionalized molecules are depicted in Fig. 12.

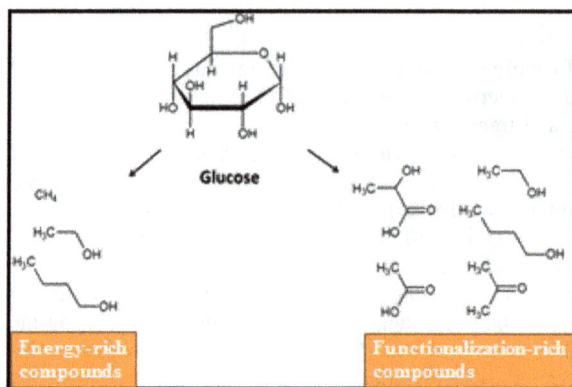

Figure 12. Chemically modified products from glucose

There are various factors involved in the degradation of food products (Fig.13). The various value added products (internal) are pH, moisture content, and enzyme activity, whereas external factors include temperature, humidity and light. However, the basic applications of nano materials are catalysts and structure modifier to facilitates the degradation. The basic routes for valorisation of food wastes are by microbial degradation and solvo thermal treatments to convert energy, methane and other organic molecules.

Applications of Emerging Nanomaterials and Nanotechnology   Materials Research Forum LLC
Materials Research Foundations 148 (2023) 276-303   https://doi.org/10.21741/9781644902554-10

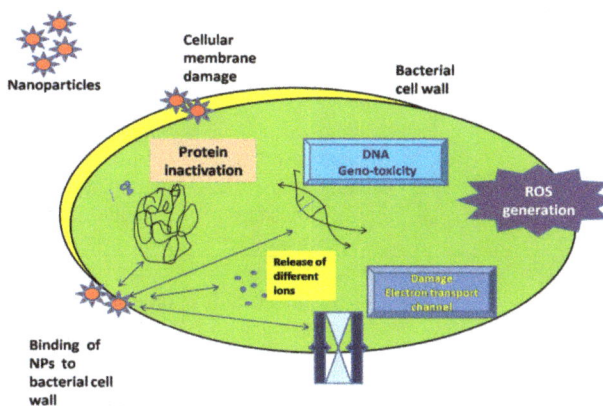

Figure 13. Nanoparticles for degradation of food items [32]

Another application of food wastes to convert into functional materials for industrial applications like super strong concrete from roots of vegetables and surfboard wax from mango seeds [45]. Some other representative examples for conversion of waste food into value added products are listed in Table 5 along with brief conditions.

Table 5. Nanotechnology based recycling of food wastes

S N.	Food waste	Properties	Applications	References
1.	Banana peel/ Sodium alginate beads	Catalytic support to metal nanoparticle	Removal of inorganic and organic water contaminants	[46]
2.	Bone waste	Extraction of calcium phosphate nanoparticles	----------------	[47]
3.	Coffee waste extract	Synthesis of Ag NP		[48]
4.	Barley agricultural waste	Synthesis of nanosilica	Industrial-technological and medical sectors	[49]
5.	Chicken feather waste	Production of keratin NP	Medical purposes	[50]
6.	Corn straw extract	Production of iron and copper NP	Promote methane production	[51]

7.	Citrus limeta peels	Production of Ag NP	Inhibits the growth of bacteria	[52]
8.	Eggshells, fish scales and shrimps	Production of calcium oxide, calcium carbonate, calcium hydroxide, chitin, chitosan, hydroxyapatite NP	biomedical, pharmaceutical, environmental remediation, fuel and food industry	[53]
9.	Industrial molasses waste	synthesis of few-layer graphene and its composites with Au and Ag NP	synthesis of other layered materials as well as other non-noble supported metallic systems	[54]
10.	*Litchi chinensis* waste (peels)	Production of Ag NP	Inhibits microbial activity	[55]

Furthermore, the importance food items stringent rules are framed by different countries and agencies to reduce waste of food to overcome the food shortage and starvation. However, in the current scenario it is required to make an effort with below schemes i.e. Figure 14 to achieve above objective.

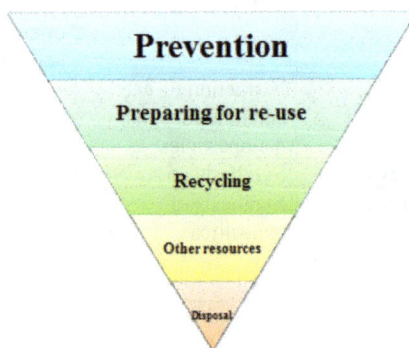

Figure 14. Scheme to control the food wastes [56].

4.   Safety issues, prospects and challenges

The precise reactivity, large surface area, dissolution and diffusion of nanomaterials than traditional materials are responsible for their quick exposure and hazards. The uses of nano materials are creating several problems. For example the migration of nano fillers like nano-clay, nano-silver, nano-titania, nano-silica, nano-zinc present in nano based polymer packaging are responsible for several toxic effects [57]. Similarly, the excess release of metal oxides from nanocomposites in water also increases toxicity in water and responsible water contaminations as well as food toxicity. Therefore, release behavior of metals from polymer composites are essential to evaluate before recommending the packaging materials for its commercial applications [58]. Further, the presence of metals raises the questions about bioavailability, digestibility and nutritional values of food items, which are still important area of investigations.

Further, the advancement in nano material-based packaging are still need to developed after incorporating nanoparticles with improved mechanical and physical properties i.e. durability, strength, flexibility, biodegradability, thermal resistivity, UV absorptivity, water vapor, and oxygen impermeability. It is reported that the presence of metal oxides in polymers improves the mechanical, barrier, and light permeability properties along with releases of metals in packed items but the migration of metal and its induced toxicity under harsh atmospheric conditions like temperature and mechanical agitations [59]. The external monitoring and indication about quality of products are still challenges for technocrats for edible and non-edible items although several indicators and data accusation systems are reported in this regards. The simple proposed scheme for advanced packaging systems is given in Figure 15.

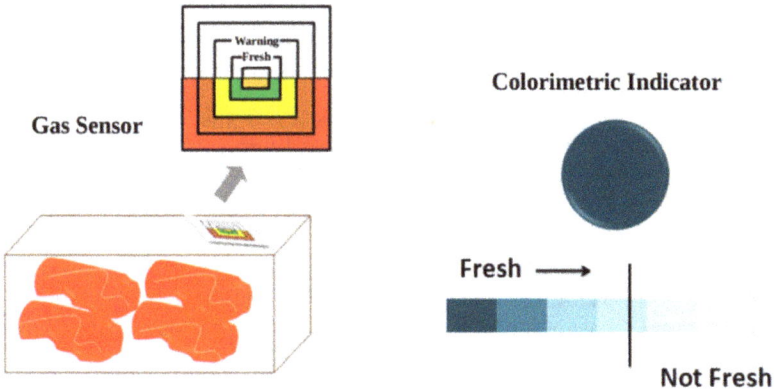

Figure 15. Illustration for advanced packaging film [59]

Conclusion

This chapter summarizes the potential of nanotechnology to improve foods, making them tastier, healthier, and more nutritious, to generate new food products, new food packaging, and storage. However, many of the applications are currently at an elementary stage, and most are aimed at high-value products, at least in the short term. Successful applications of nanotechnology to foods are limited. Nanotechnology can be used to enhance food flavor and texture, to reduce fat content, or to encapsulate nutrients, such as vitamins, to ensure they do not degrade during a product's shelf life. In addition to this, nanomaterials can be used to make packaging that keeps the product inside fresher for longer. Intelligent food packaging, incorporating nanosensors, could even provide consumers with information on the state of the food inside. Food packages are embedded with nanoparticles that alert consumers when a product is no longer safe to eat. Sensors can warn before the food goes rotten or can inform us the exact nutritional status contained in the contents. In fact, nanotechnology is going to change the fabrication of the entire packaging industry. Food nanotechnology advances offers important challenges for both government and industry. The food processing industry must ensure consumer confidence and acceptance of nanofoods. Regulatory bodies, such as FDA, should author guidance with respect to the criteria to be followed in evaluating the safety of food, food packaging, and supplement uses of nanomaterials with novel properties.

Acknowledgements

Authors are thankful to Principal, Bhaskaracharya College of Applied Sciences, University of Delhi for maintaining socio academic environment in the college. RA is also acknowledging the fellowship provided by Delhi University.

References

[1] P. Ningthoujam, B. Jena, S. Pattanayak, S. Dash, M.K. Panda, R.K. Behera, N.K. Dhal, Y.D. Singh, Nanotechnology in food science. Bio-Nano Interface: Applications in Food, Healthcare and Sustainability. Singapore: Springer Singapore (2021) 59-73. https://doi.org/10.1007/978-981-16-2516-9_4

[2] R.Dobrucka, Application of Nanotechnology in Food Packaging, J microbiol biotechnol food sci 3 (2014)353-359.

[3] O. Bashir, S.A. Bhat, A. Basharat, M. Qamar, S.A. Qamar, M. Bilal, H.M. Iqbal Nano-engineered materials for sensing food pollutants: Technological advancements and safety issues, Chemosphere 292 (2022) 133320. https://doi.org/10.1016/j.chemosphere.2021.133320

[4] S.H. Nile, V. Baskar, D. Selvaraj, A. Nile, J. Xiao, G. Kai, Nanotechnologies in food science: applications, recent trends, and future perspectives, Nano-Micro Lett (12),(2020)1-34. https://doi.org/10.1007/s40820-020-0383-9

[5] R. K. Sastry, S. Anshul, N.H. Rao, Nanotechnology in food processing sector-An assessment of emerging trends, J Food Sci Technol. 50(2013)831-841. https://doi.org/10.1007/s13197-012-0873-y

[6] T. Singh, S. Shukla, P. Kumar, V. Wahla, V.K. Bajpai, I.S. Rather, Application of nanotechnology in food science: perception and overview, Front Microbiol. (8), (2017)1501. https://doi.org/10.3389/fmicb.2017.01501

[7] A. Ali, S. Ahmed, Recent advances in edible polymer based hydrogels as a sustainable alternative to conventional polymers, J Agric Food Chem. 66(27), (2018)6940-6967. https://doi.org/10.1021/acs.jafc.8b01052

[8] S. Farris, K.M. Schaich, L. Liu, L. Piergiovanni, K.L. Yam, Development of polyion-complex hydrogels as an alternative approach for the production of bio-based polymers for food packaging applications: a review, Trends Food Sci Technol 20(8),(2009) 316-332. https://doi.org/10.1016/j.tifs.2009.04.003

[9] M.S. Umekar, G.S. Bhusari, A.K. Potbhare, A. Mondal, B.P. Kapgate, M.F. Desimone, R.G. Chaudhary, Bioinspired reduced graphene oxide based nanohybrids for photocatalysis and antibacterial applications, Curr Pharm Biotechnol 22(13),(2021)1759-1781. https://doi.org/10.2174/1389201022666201231115826

[10] V.K. Bajpai, M. Kamle, S. Shukla, D.K. Mahato, P. Chandra, S.K. Hwang, Y.K. Han, Prospects of using nanotechnology for food preservation, safety, and security, J Food Drug Anal 26(4),(2018)1201-1214. https://doi.org/10.1016/j.jfda.2018.06.011

[11] S. Murugadoss, D. Lison, L. Godderis, S. Van Den Brule, J. Mast, F. Brassinne, N. Sebaihi, P.H. Hoet, Toxicology of silica nanoparticles: an update, Arch Toxicol 91(9),(2017) 2967-3010. https://doi.org/10.1007/s00204-017-1993-y

[12] S.K. Paul, H. Dutta, S. Sarkar, L.N. Sethi, S.K. Ghosh, Nanosized zinc oxide: Super-functionalities, present scenario of application, safety issues, and future prospects in food processing and allied industries, Food Rev Int 35(6),(2019)505-535 https://doi.org/10.1080/87559129.2019.1573828

[13] A.Q. Truong-Tran, J. Carter, R.E. Ruffin, P.D. Zalewski, The role of zinc in caspase activation and apoptotic cell death, Zinc Biochemistry, Physiology, and Homeostasis: Recent Insights and Current Trends (2001)129-144 https://doi.org/10.1007/978-94-017-3728-9_7

[14] R. Bagade, A.K. Potbhare, R.G. Chaudhary, A. Mondal, M. Desimone, R. Mishra, H.D. Juneja, Microspheres/Custard-Apples Copper (II) Chelate Polymer: Characterization, Docking, Antioxidant and Antibacterial Assay, ChemistrySelect, 4 (20), (2019)6233-6244. https://doi.org/10.1002/slct.201901115

[15] Q Hu, Y Luo, Chitosan-based nanocarriers for encapsulation and delivery of curcumin: A review, Int J Biol Macromol 179,(2021)125-135 https://doi.org/10.1016/j.ijbiomac.2021.02.216

[16] S.K. Pyne, K. Paria, S.M. Mandal, P.P. Srivastav, P. Bhattacharjee, T.K. Barik, Green microalgae derived organic nanodots used as food preservative, Current

Research in Green and Sustainable Chemistry (2022)100276.
https://doi.org/10.1016/j.crgsc.2022.100276

[17] J.A. Tanna, R.G. Chaudhary, H.D. Juneja, N.V. Gandhare, A.R. Rai, Histidine capped ZnO nanoparticles: an efficient synthesis, characterization and effective antibacterial activity, BioNanoScience, (5),(2015)123-134.
https://doi.org/10.1007/s12668-015-0170-0

[18] A. Grasso, M. Ferrante, A. Moreda-Piñeiro, G. Arena, R. Magarini, G.O. Conti, C. Copat, Dietary exposure of zinc oxide nanoparticles (ZnO-NPs) from canned seafood by single particle ICP-MS: Balancing of risks and benefits for human health, Ecotoxicol Environ Saf 231(2022)113217.
https://doi.org/10.1016/j.ecoenv.2022.113217

[19] W.J. Chong, S. Shen, Y. Li, A. Trinchi, D. Pejak, I.L. Kyratzis, C. Wen, Additive manufacturing of antibacterial PLA-ZnO nanocomposites: Benefits, limitations and open challenges, J Mater Sci Technol 111(2022)120-151
https://doi.org/10.1016/j.jmst.2021.09.039

[20] X. Hou, J. Liang, X. Yang, J. Bai, M. Yang, N. Qiao, Y. Shi, Poloxamer188-based Nanoparticles Improve the Anti-oxidation and Anti-degradation of Curcumin, Food Chem (375),(2022)131674 https://doi.org/10.1016/j.foodchem.2021.131674

[21] R. Marion-Letellier, A. Amamou, G. Savoye, S. Ghosh, Inflammatory bowel diseases and food additives: to add fuel on the flames!, Nutrients 11(5), (2019)1111
https://doi.org/10.3390/nu11051111

[22] H. Zhou, D.J. McClements, Recent advances in the gastrointestinal fate of organic and inorganic nanoparticles in foods, Nanomater 12(7), (2022)1099
https://doi.org/10.3390/nano12071099

[23] S. Siemer, A. Hahlbrock, C. Vallet, DJ McClements, J. Balszuweit, J. Voskuhl, R.H. Stauber, Nanosized food additives impact beneficial and pathogenic bacteria in the human gut: a simulated gastrointestinal study, NPJ Sci Food 2(1),(2018)1-10
https://doi.org/10.1038/s41538-018-0030-8

[24] A.R. Singh, P.K. Desu, R.K. Nakkala, V. Kondi, S. Devi, M.S. Alam, P. Kesharwani, Nanotechnology-based approaches applied to nutraceuticals, Drug Deliv Transl Res (2021)1-15 https://doi.org/10.1007/s13346-021-00960-3

[25] N. Noor, A. Gani, F. Jhan, M.A. Shah, Z. ul Ashraf, Ferulic acid loaded pickering emulsions stabilized by resistant starch nanoparticles using ultrasonication: Characterization, in vitro release and nutraceutical potential, Ultrason Sonochem 84,(2022)105967 https://doi.org/10.1016/j.ultsonch.2022.105967

[26] L. Chabib, A. Suryani, S.N.P. Hakim, M.I. Rizki, F. Firmansyah, F. Romadhonsyah, IAI SPECIAL EDITION: Stevia rebaudiana as a nutraceutical for COVID-19 patients with no sugar diet during recovery and its nanoparticle application, Pharm Educ 22(2),(2022)174-179 https://doi.org/10.46542/pe.2022.222.174179

[27] H. Zhou, D.J. McClements, Recent advances in the gastrointestinal fate of organic and inorganic nanoparticles in foods, Nanomater 12(7) (2022)1099 https://doi.org/10.3390/nano12071099

[28] PM Reque, A. Brandelli, Encapsulation of probiotics and nutraceuticals: Applications in functional food industry, Trends Food Sci Technol, 114(2021)1-10 https://doi.org/10.1016/j.tifs.2021.05.022

[29] S. Manocha, S. Dhiman, A.S. Grewal, K. Guarve, Nanotechnology: An approach to overcome bioavailability challenges of nutraceuticals, J Drug Deliv Sci Techno (2022)103418 https://doi.org/10.1016/j.jddst.2022.103418

[30] J. Siddiqui, M. Taheri, A.U. Alam, M.J. Deen, Nanomaterials in smart packaging applications: a review, Small 18(1),(2022)2101171. https://doi.org/10.1002/smll.202101171

[31] M. Sahoo, S. Vishwakarma, C. Panigrahi, J. Kumar, Nanotechnology: Current applications and future scope in food, Food Frontiers 2(1), (2021)3-22. https://doi.org/10.1002/fft2.58

[32] P. Chaudhary, F. Fatima, A. Kumar, Relevance of nanomaterials in food packaging and its advanced future prospects, J Inorg Organomet Polym 30,(2020) 5180-5192. https://doi.org/10.1007/s10904-020-01674-8

[33] K. Pushparaj, W.C. Liu, A. Meyyazhagan, A. Orlacchio, M. Pappusamy, C. Vadivalagan, A.A. Robert, V.A. Arumugam, H. Kamyab, JJ Klemeš, T. Khademi, Nano-from nature to nurture: A comprehensive review on facets, trends, perspectives and sustainability of nanotechnology in the food sector, Energy 240(2022) 122732. https://doi.org/10.1016/j.energy.2021.122732

[34] A. Erfani, M.K. Pirouzifard, H. Almasi, N. Gheybi, S. Pirsa, Application of cellulose plate modified with encapsulated Cinnamomum zelanicum essential oil in active packaging of walnut kernel. Food Chem. 381(2022)132246. https://doi.org/10.1016/j.foodchem.2022.132246

[35] K. Ma, T. Zhe, F. Li, Y. Zhang, M. Yu, R. Li, L. Wang, Sustainable films containing AIE-active berberine-based nanoparticles: A promising antibacterial food packaging, Food Hydrocoll 123(2022) 107147. https://doi.org/10.1016/j.foodhyd.2021.107147

[36] W. Dong, J. Su, Y. Chen, D. Xu, L. Cheng, L. Mao, F. Yuan, Characterization and antioxidant properties of chitosan film incorporated with modified silica nanoparticles as an active food packaging, Food Chem 373(2022) 131414. https://doi.org/10.1016/j.foodchem.2021.131414

[37] T. Gasti, S. Dixit, V.D. Hiremani, R.B. Chougale, S.P. Masti, S.K. Vootla, B.S. Mudigoudra, Chitosan/pullulan based films incorporated with clove essential oil loaded chitosan-ZnO hybrid nanoparticles for active food packaging, Carbohydr Polym 277(2022)118866. https://doi.org/10.1016/j.carbpol.2021.118866

[38] P.J.P. Espitia, N.D.F.F. Soares, J.S.D.R. Coimbra, N.J. de Andrade, R.S. Cruz, E.A.A. Medeiros, Zinc oxide nanoparticles: synthesis, antimicrobial activity and food

packaging applications, Food Bioproc Tech 5(5),(2012)1447-1464. https://doi.org/10.1007/s11947-012-0797-6

[39] K.K. Dash, P. Deka, S.P. Bangar, V. Chaudhary, M. Trif, A. Rusu, Applications of inorganic nanoparticles in food packaging: A Comprehensive Review, Polymers 14(3),(2022)521. https://doi.org/10.3390/polym14030521

[40] M. Hoseinnejad, S.M. Jafari, I. Katouzian, Inorganic and metal nanoparticles and their antimicrobial activity in food packaging applications, Crit Rev Microbiol 44(2),(2018)161-181. https://doi.org/10.1080/1040841X.2017.1332001

[41] C. Xu, M. Nasrollahzadeh, M. Selva, Z. Issaabadi, R. Luque, Waste-to-wealth: Biowaste valorization into valuable bio (nano) materials, Chem Soc Rev 48(18),(2019) 4791-4822. https://doi.org/10.1039/C8CS00543E

[42] E. Uçkun Kıran, A.P. Trzcinski, Y. Liu, Platform chemical production from food wastes using a biorefinery concept, J Chem Technol Biotechnol 90(8), (2015)1364-1379. https://doi.org/10.1002/jctb.4551

[43] C.S.K. Lin, L.A. Pfaltzgraff, L. Herrero-Davila, E.B. Mubofu, S. Abderrahim, J.H. Clark, R. Luque, Food waste as a valuable resource for the production of chemicals, materials and fuels. Current situation and global perspective, Energy Environ Sci 6(2),(2013) 426-464. https://doi.org/10.1039/c2ee23440h

[44] T.M. Fagieh, E.M. Bakhsh, S.B. Khan, K. Akhtar, A.M. Asiri, Alginate/banana waste beads supported metal nanoparticles for efficient water remediation, Polymers 13(23),(2021)4054. https://doi.org/10.3390/polym13234054

[45] B. Zhang, H. Li, L. Chen, T. Fu, B. Tang, Y. Hao, J. Li, Z. Li, B. Zhang, Q. Chen, C. Nie, Recent advances in the bioconversion of waste straw biomass with steam explosion technique: A comprehensive review, Processes 10(10), (2022)1959. https://doi.org/10.3390/pr10101959

[46] N. El-Desouky, K. Shoueir, I. El-Mehasseb, M. El-Kemary, Synthesis of silver nanoparticles using bio valorization coffee waste extract: photocatalytic flow-rate performance, antibacterial activity, and electrochemical investigation, Biomass Convers Bioref (2022)1-15. https://doi.org/10.1007/s13399-021-02256-5

[47] D. Kavaz, Synthesis of silica nanoparticles from agricultural waste, Agri-Waste and Microbes for Production of Sustainable Nanomaterials. Elsevier, (2022) 121-138. https://doi.org/10.1016/B978-0-12-823575-1.00028-7

[48] M. Pakdel, Z. Moosavi-Nejad, R.K. Kermanshahi, H. Hosano, Self-assembled uniform keratin nanoparticles as building blocks for nanofibrils and nanolayers derived from industrial feather waste, J Clean Prod (2022)130331. https://doi.org/10.1016/j.jclepro.2021.130331

[49] Z. Dong, H. Guo, M. Zhang, D. Xia, X. Yin, J. Lv, Enhancing biomethane yield of coal in anaerobic digestion using iron/copper nanoparticles synthesized from corn straw extract, Fuel 319(2022)123664. https://doi.org/10.1016/j.fuel.2022.123664

[50] R.A. Omar, D. Chauhan, N. Talreja, R.V. Mangalaraja, M. Ashfaq, Vegetables waste for biosynthesis of various nanoparticles, Agri-Waste and Microbes for Production of Sustainable Nanomaterials. Elsevier, (2022) 281-298. https://doi.org/10.1016/B978-0-12-823575-1.00014-7

[51] M. Yadav, N. Pareek, V. Vivekanand, Eggshell and fish/shrimp wastes for synthesis of bio-nanoparticles, Agri-Waste and Microbes for Production of Sustainable Nanomaterials. Elsevier, (2022) 259-280. https://doi.org/10.1016/B978-0-12-823575-1.00002-0

[52] K. Shoueir, A. Mohanty, I. Janowska, Industrial molasses waste in the performant synthesis of few-layer graphene and its Au/Ag nanoparticles nanocomposites, J Clean Prod 351(2022)131540. https://doi.org/10.1016/j.jclepro.2022.131540

[53] M. Yadav, N. Pareek, V. Vivekanand, Eggshell and fish/shrimp wastes for synthesis of bio-nanoparticles, Agri-Waste and Microbes for Production of Sustainable Nanomaterials. Elsevier, (2022) 259-280. https://doi.org/10.1016/B978-0-12-823575-1.00002-0

[54] C.S.K. Lin, L.A. Pfaltzgraff, L. Herrero-Davila, E.B. Mubofu, S. Abderrahim, J.H. Clark, R. Luque, Food waste as a valuable resource for the production of chemicals, materials and fuels. Current situation and global perspective, Energy Environ Sci 6(2), (2013)426-464. https://doi.org/10.1039/c2ee23440h

[55] N. Tavker, M. Sharma, Nanosensors for intelligent food packaging, Nanosensors for Smart Agriculture, Elsevier, 2022. 737-756. https://doi.org/10.1016/B978-0-12-824554-5.00014-8

[56] A. Mondal, M.S. Umekar, G.S. Bhusari, S. Mondal, R.G. Chaudhary, M. Sami, Biogenic synthesis of metal/metal oxide nanostructured materials, Curr. Pharm. Biotechnol, 22 (2022)1782-1793. https://doi.org/10.2174/1389201022666210111122911

[57] S. Radoor, J. Karayil, J.M. Shivanna, A. Jayakumar, S.A. Varghese, R.E. Krishnankutty, J. Parameswaranpillai, S. Siengchin, Environmental and Toxicological Aspects of Nanostructures in Food Packaging, Nanotechnology-Enhanced Food Packaging (2022)361-378. https://doi.org/10.1002/9783527827718.ch15

[58] V.G.L. Souza, A.L. Fernando, Nanoparticles in food packaging: biodegradability and potential migration to food: a review, Food Packag Shelf Life 8(2016)63-70. https://doi.org/10.1016/j.fpsl.2016.04.001

[59] A. Ashfaq, N. Khursheed, S. Fatima, Z. Anjum, K. Younis, Application of nanotechnology in food packaging: Pros and Cons, J Agric Food Inf (2022)100270. https://doi.org/10.1016/j.jafr.2022.100270

Applications of Emerging Nanomaterials and Nanotechnology   Materials Research Forum LLC
Materials Research Foundations 148 (2023) 304-333   https://doi.org/10.21741/9781644902554-11

Chapter 11

Nanomaterials as Photocatalyst

Ajay K. Potbhare[1], Pavan R. Bhilkar[1], Sachin T. Yerpude[2], Rohit S. Madankar[1], Sampat R. Shingda[1], Rameshwar Adhikari[3] and Ratiram G. Chaudhary[1]*

[1]Post Graduate Department of Chemistry, Seth Kesarimal Porwal College of Arts, And Science and Commerce, Kamptee, 441001, India

[2]Post Graduate Department of Microbiology, Seth Kesarimal Porwal College of Arts, And Science and Commerce, Kamptee, 441001, India

[3]Research Centre for Applied Science and Technology (RECAST), Tribhuvan University, Kathmandu, Nepal

* chaudhary_rati@yahoo.com

Abstract

Clean and drinkable water is a big challenge in 21st century. A variety of organometallic compounds have been utilized by human being for rapid civilization and modernization. These hazardous waste discharges from the industries and directly mixed with environment especially in water reservoir and adulterate water, which is responsible for many contagious diseases. To vanquish this issue we needed an eco-friendly, safe, cost-effective nanomaterials for the degradation and removal of noxious waste. In this chapter emphasized on different nanomaterials as photocatalyst for photocatalytic performances, also critically discussed applicability of different nanomaterials for photocatalytic process comprising with types photocatalyst, light source, scavengers, trapping agents, photodegradation activity mechanism and its utility. Moreover, removal of toxic dyes, pharmaceutical drugs, agrochemical waste, heavy metal ions, and phenolic compounds have been discussed.

Keywords

Nanomaterials, Photocatalysts, Photodegradation, Pharmaceutical Drugs, Phenolic Compounds

Contents

1. Introduction

In recent advances, the scientific community has extremely focused on resolving environmental related issues, which are the major concerns of the 21st century. The biggest challenges are to get clean water owing to fast civilization and industrialization, which is the most vital resource for the existence of all beings' water pollution has specified critical environmental concern, as the industrial gradual enlargements release of wastewater streams without sufficient processing and their leaking into the natural water cycle [1]. Around the globe, an extensive use of carcinogenic organic dyes in the textiles industries increase gradually and the more amounts of polluted effluents were released into an aquatic ecosystem. It has become the chief carrier of all contaminants such as carcinogenic dye [2] effluent, heavy metals [3], and pharmaceutical waste [4] despite carrying vital natural minerals such as sodium, magnesium, calcium etc. It is mainly due to the release of unprocessed industrial wastage into the water bodies. It causes the depletion of dissolved oxygen content and results in adverse effects to aquatic creatures and mankind.

On the other hand, environmentalists are striving hard to combat the dye ill effects. The industrial discharges treatment has to be carried out to avoid the harmful effects of pollutants in water, eventually to save the planet, earth and to prevent diseases. Pesticides are one of the most essential groupings of organic pollutants present in waste water. There are numerous methodologies to treat industrial wastes such as precipitation ion-exchange, evaporation, reverse osmosis, ultrafiltration, microfiltration, solvent, extraction [5-6]. Photocatalytic degradation technique was employed for the treatment of water purification because other methods purely transfer the industrial waste from one phase to another phase *i.e.* detrimental sludge remains even after the treatment. Therefore, the photocatalytic technique is certainly more effective method to convert harmful organic impurities degradation into carbonaceous products [7]. Photocatalysis is an advanced oxidation process (AOP) that can be used for the destruction of pollutants in a simple and competent mode. It is one of the most economic and promising industrial effluent treatment processes. The p-type semiconductor nanomaterials which are being used as most effective photo-active catalysts for photocatalytic exclusion of various organic dyes, hazardous composites and venomous chemicals in water have been broadly studied. The phenolic compounds, color compounds are chemically stable so the technologies relating to UV radiation and hydrogen peroxide oxidation are not effectively deals with the degrade the colour dyes, at present photocatalytic method have paid the great attention due to its effective towards decolorization of the dyes by using semiconductor nanomaterials [8-10].

Recently, development and design of nanostructured semiconductor nanomaterials for potential applications such as environmental remediations, energy and water disinfection have been more attracting material for the degradation of various organic pollutant, pharmaceutical waste have shown effective results. Metal oxide NPs has been broadly used as an eco-friendly photocatalyst and attracted major attention because of their chemical stability, electron transfer capability, low cost, nontoxicity, and electro catalytic characteristics [11]. This p-type and n-type semiconductor metal oxide possesses a tunable band gap in the range 3.0-4.0eV, depending on the nature of the defects and density, Due to their wide band gap and high photosensitive nature, various semiconducting metal oxides such as ZnO, SnO_2, Co_3O_4, CeO_2, TiO_2, Bi_2O_3 and Fe_2O_3 are used in the photocatalytic degradation of organic pollutants [12-17].

In the photocatalytic reaction, the semiconducting photocatalyst absorbs light energy more than equal to band gap, which generate the hole, and the electrons, which further gives rise to efficient oxidizers of organic dyes. Due to the fast recombination of photo-induced CB electrons and VB holes, it is challenging to obtain efficient degradation efficiency using semiconductor photocatalyst from the aqueous solution [27]. In the case of dye pollutants, the dye acts as a photosensitizer in visible light, and on the other hand, it injects an excited electron to an electron acceptor and converted to cationic dye radical ($dye^{\bullet+}$), followed by self-degradation or decomposition. Such a photocatalytic degradation process for dyes is different from that of the conventional semiconductor photocatalysis under UV irradiation, and has been demonstrated to be an efficient approach to remove textile dyestuff from aquatic environments [28-30].

2.   Historical background of photocatalyst

Metal/metal oxides and graphene NCs have promising application as photocatalyst due to advanced arrangement of electronic structure, charge separation and transport and light absorption properties. A major breakthrough in photocatalysis research was occurred in 1972, when *Fujishima* and *Honda* discovers that TiO_2 in connection with Pt electrode cause electrochemical photocatalysis of water under UV light radiations [104]. Verity of inorganic NPs (NPs) including TiO_2, CuO, $BiOCl$, ZnO, CdO, CdS, $ZnWO_4$, SnO_2, $NiFe_2O_4$, $Bi_2O_3/BiOCl$, $MnFe_2O_4$ etc have been proposed to poses photocatalytic activity. In 1977, Nozik discovered that the combination of a noble metal in the electrochemical photolysis process, such as platinum, silver and gold, among others, could increase photoactivity, and that an external potential was not required [31-34]. Wagner and Somorjai (1980) and Sakata and Kawai (1981) delineated hydrogen production on the surface of strontium titanate ($SrTiO_3$) *via* photogeneration, and the generation of hydrogen and methane from the illumination of TiO_2 and PtO_2 in ethanol, respectively.

The photocatalysis were originally used by fujishima and Honda in 1972, the history of photocatalysis can be divided by different ways as an illustration, the relatively effective division proposed by Serpone and Emeline [36], as a function of the type of accoutrements used to carry out the photocatalytic processes is frequently cited. In particular, three different generations of photocatalysts have been proposed. The first is composed by pristine inorganic semiconductors (in the morning substantially ZnO, WO3 and TiO2 and also substantially TiO2 only), of which the photocatalytic parcels have been delved in depth, frequently with introducing experimental approaches, with the main end to give perceptivity into the medium of product and transfer of the charge carriers generated inside the material and into the nature of the reactive species operative in the delved photocatalyzed processes. The alternate generation of photocatalysts is substantially composed by doped semiconductors synthesized to push the onset of immersion toward longer wavelengths introducing intra-band gap countries, the nature and energetic positioning of which have been deeply delved , indeed though the first report of their discovery remained unnoticed for a long time,[36] while a consecutive report entered more recognition(and 16 times further citations!), being the most cited photocatalysis paper of its decade(videinfra) [37]. The doping strategies proposed were grounded on the preface of non-metal or essence centers inside the crystallographic structure of the pristine material with the end to induce color centers (localized countries located between the valence and conduction bands [38-40] and/ or oxygen vacuities with the conformation of centers where the essence cations have different oxidation countries, e.g., Ti^{3+} centres in a TiO_2 structure [42]. The third and last generation proposed is grounded on the product of mongrel heterostructures formed by two or further inorganic semiconductors [43-45]. In this case, the dynamics of the charge carriers came extremely complex, especially if the number of the accoutrements composing the mongrels is advanced than two. In some cases, these photocatalysts have shown both high effectiveness and a better capability to gather solar irradiation with respect to the pristine and doped semiconductors.

3. Different Nanomaterials used as photocatalyst

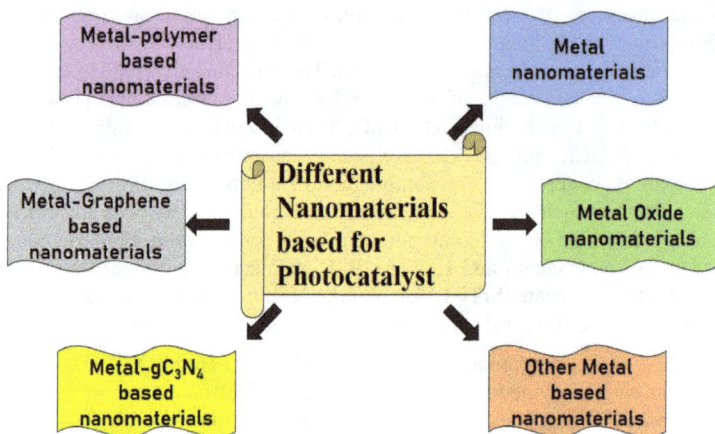

Fig. 1. Different nanomaterials used as photocatalyst

3.1 Metal-based nanomaterials

The different nanomaterials used as photocatalyst has been shown in Fig. 1. Many works are published on metal NPs as photocatalysts. For instances, Mongal *et al.,* [46] investigate the photocatalytical behaviour of silver doped TiO_2 NPs *via* convenient single step sol gel method. Doping concentration of silver significantly change the photocatalytical properties of TiO_2. In terms of efficient photocatalyst for organic compounds under UV rays and air conditions, 0.75 wt% silver doped TiO_2 shows the best performance. Further study reveals that studied doping of gold can exhibit much higher photocatalytical activity as compared to silver. Kowalska et al., [47] reported photodeposition of 2 wt% doping gold on the surface of TiO_2 with controlled temperature conditions. In comparison to silver monometallic and Au/Ag bimetallic photocatalysts, gold monometallic photocatalysts

Applications of Emerging Nanomaterials and Nanotechnology   Materials Research Forum LLC
Materials Research Foundations 148 (2023) 304-333   https://doi.org/10.21741/9781644902554-11

demonstrated significantly higher activity during alcohol dehydrogenation under UV irradiation. Montoya and Gillan [48] studied the surface modification of doping of 1 or 2 wt% cobalt, nickel and copper enhances the photocatalytical hydrogen evaluation. The photoinduced TiO_2 conduction band electrons in solution are likely to be transferred to protons in solution that are reduced to H_2 through the photodeposited 3d metal species on the TiO_2 surface. Dholam et al., [49] investigated the photocatalytical activity of chromium and iron doped TiO_2 nanocomposites (NCs) synthesized by physical and chemical method. Radio frequency magnetron sputtering and a *sol gel* method were used to synthesized nanocomposite and study the effect of different percentage of metal dopant (up to 5 wt %) for hydrogen generation under visible light irradiation. Salas et al., [50] reported a synthesis of platinum doped TiO_2 photocatalyst by using incipient wetness impregnation method. 1 wt % doping of platinum in TiO_2 decreases the band gap from 3.20 to 2.73 eV and shows favourable photocatalytical properties for water splitting.

3.2   Metal oxide-based nanomaterials

Manivel et al., [51] reported the synthesis of CuO-TiO_2 nano-catalyst, with high photocatalytical properties *via* easy impregnation method. Compared with bare TiO_2, CuO-TiO_2 nano-catalyst is much more effective in photocatalytic degradation of Acid Red 88. This nano-catalyst provides more surface-active sites for photocatalytical activity. Liu et al., [52] demonstrated the synthesis of ZnO/TiO_2 NPs *via* two different methods viz., sol gel and solid phase reaction method. Using a *sol-gel* method, the photocatalytic activity of ZnO/TiO_2 particles was nearly the same as that of pure TiO_2, whereas, the photocatalytic activity of ZnO/TiO_2 particles prepared by solid phase reaction was significantly higher. Doping of 0.5% of ZnO in mole with 500 °C calcinations shows excellent photodegradation of Rhodamine B. Using CeO_2 doped *anatase* TiO_2 with exposed (001) high energy facets, Wang et al., [53] reported selective reduction of NO by NH_3 with its catalytic properties. Initially TiO_2 was prepared hydrothermally by using its precursors and then it is exposed with (001) high energy facet. Photocatalytic activity was monitored with doping of different molar ratio of cerium into the titania (Ce/Ti molar ratio) surface. It was found that at unit molar ratio, catalyst show excellent selective catalytic reaction NO with NH_3 at high temperature. Using an easy aqueous *sol-gel* synthesis at ambient temperature, Mahy et al., [54] synthesized TiO_2 materials doped with different mol % of zirconia precursors. As a result of better charge separation between TiO_2 and ZrO_2 oxides under UV/visible light, zirconia improved the p-nitrophenol degradation in water efficiency. In addition, photocatalytic activity was evaluated on the degradation of methylene blue dye under UV-A light, the most doped sample demonstrating 4 times greater degradation compared to the pure TiO_2 sample. Gnanasekaran et al., [55] synthesized the $TiO_2@Fe_3O_4$ NCs by mixing the precipitation and sol gel method with higher surface area value 115.7 $m^2.g^{-1}$. The TiO_2 and Fe_3O_4 was prepared separately *via sol gel* and precipitation method respectively and then it mixed with isopropanol with the ratio 1:0.2:5 with constant stirring. This reaction mixture then converted to sol gel by drop wise addition of citrate solution. The $TiO_2@Fe_3O_4$ NCs were studied with a UV-abs spectrometer, revealing 2.70 eV as the band gap for the material.

3.3   Metal-polymer based nanomaterials

Using TiO$_2$ NPs as initiators to photopolymerize 2-(tert-butylamino)ethyl methacrylate and ethylene glycol dimethacrylate, novel TiO$_2$/biocidal polymer core/shell NPs were synthesized by Kong and his co-worker [56]. In the preparation, the monomers were adsorbed onto TiO$_2$ surfaces through interactions between ester groups and metal oxide surface sites. During UV light irradiation, the excited TiO$_2$ NPs produced electron-hole pairs, which simultaneously initiated the surface polymerization of the monomers on the surface. Tekin et al., [57] prepared the PVA/TiO$_2$ and PEG/iO$_2$ nanocomposites *via sol gel* method and studied their thermal, antibacterial and photocatalytic properties in detailed manner. The study compared the antibacterial and photocatalytical activity of PVA/TiO$_2$is much better than bare TiO$_2$and PEG/TiO$_2$for *E. coli* and Acid Black I dye respectively. Using ammonium persulfate as an oxidant in the presence of ultrafine grade powder of *anatase* TiO$_2$ cooled in an ice bath, Zhang et al., [58] synthesised polyaniline-*anatase* TiO$_2$ NCs in a series of different ratios of polyaniline:TiO$_2$ by 'in-situ' deposition oxidative polymerization of aniline hydrochloride and *anatase* TiO$_2$. The study reveals that photocatalytic degradation of polyaniline solid-phase products could be achieved as well as polyaniline-TiO2 NCs having potential to be used as photodegradable products. Also, the photodegradation rate is increases as the ratio of polyaniline:TiO$_2$ is decreases. Saravanan et al., [59] reported a synthesis of TiO$_2$/Chitosan NCss with different weight ratio of TiO$_2$ and chitosan by two step method. In first step, TiO$_2$ NPs prepared via sol gel method using 1:2 ratio of acetic acid and isopropanol. In next step, different weights of TiO$_2$ and chitosan were weighed and dissolve in acetic acid. The pH of solution was increased by addition on NaOH to precipitate out the TiO$_2$/Chitosan NCs. As mass of chitosan increases, the crystallinity of NCss was decreased. Fischer et al., [60] demonstrated the easy scale-up methods with highly efficient crystalline TiO$_2$ NPs on polyethersulfone microfiltration membrane to remove pollutants in a continuous way. A mixture of titanium (IV) isopropoxide and hydrochloric acid was used for the purpose. After heating and cooling to 210 °C, the membrane became thoroughly dipped into the dispersion of TiO$_2$.Photocatalytic activity of ultrasound-treated NPs was highest in degrading carbamazepine, and degradation did not decrease after nine repetitions.

4.   Synthetic approaches towards photocatalysts

4.1   Sol-gel

Among the most commonly used methods for synthesizing photocatalysts is the *sol-gel* method. Since it is possible to regulate many parameters, it is an effective way to tailor metal oxides to specific applications. These parameters include, for example, precursor composition, pH, temperature and reaction time, concentration of reagents, nature and concentration of catalysts, aging time and temperature, addition of organic additives, and water amount. *Sol-gels* are primarily used because they produce polycrystalline particles with special properties due to their homogeneous mixing of metal ions at the molecular level. Additionally, the *sol-gel* method has the advantage of incorporating different types of dopants during some stages. During the gelation stage, active dopants are introduced into the sol, allowing direct interaction between them and the support, enhancing the photocatalytic properties of the material [61]. As defined by the *sol-gel* method, it involves the inorganic polymerization reaction induced by water to convert a precursor solution into an inorganic solid; an aqueous or alcoholic mixture of metal-organic salts (alkoxides) or inorganic salts (chloride, nitrate, sulfate, acetate, etc.) is used as a precursor. After preparing a homogeneous solution, the next steps in the *sol-gel* process are: (a) converting the homogeneous solution into a sol by using a suitable reagent (generally water with or without acid/base), (b) aging, (c) shaping, and (d) thermal treatment/sintering [62]. As a result of hydrolysis and condensation, M-O-M bonds are formed inorganic polymers, which are further condensed to form gels. Aerogels can be synthesized by drying the gels under extremely harsh circumstances. A xerogel is fabricated when the gel is dried at room temperature. The desired material is then produced by thermally treating the gel, and it can then be formed into a variety of shapes such monoliths, films, fibres, and monosized powders. Using titanium (IV) n-butoxide as a precursor, Candrappa K. G. et al., and Chen Y. C. et al., synthesised TiO_2 using the *sol-gel* method [63, 64]. Depending on the synthesis conditions and calcination temperature, *anatase* or rutile phase can be produced. TiO_2 was synthesised by the *sol-gel* method from two different alkoxide precursors, titanium (IV) n-butoxide and titanium (IV) isopropoxide, and Chen Z et al., investigated the impact of the nature of the oxide precursors throughout this process [65]. Also, the *sol-gel* approach was used by Clayton Farrugian et al., to fabricate W, Ag, and W/Ag co-doped TiO_2 nanopowders. [66]. Choi J et al., used the *sol-gel* method to blend gold particles into $Ti_{1x}Zn_xO_2$ films [67].

4.2   Hydrothermal

Any heterogeneous reaction occurring under high pressure and temperature circumstances with aqueous solvents or mineralizers is typically referred to as hydrothermal. With the reaction taking place in aqueous solutions, this synthetic process is often carried out in steel pressure containers called autoclaves with or without Teflon liners under regulated temperature and/or pressure. It is possible to raise the temperature beyond the point at which water starts to boil and achieve vapour saturation. The internal pressure generated by the autoclave is mostly determined by its temperature and the volume of solution supplied [68]. The experimental temperature and pressure settings and the corrosion resistance in that pressure-temperature range in a certain solvent or hydrothermal fluid are the most crucial factors for choosing an appropriate autoclave. The corrosion resistance is a key consideration in the selection of the autoclave material if the reaction is occurring directly in the vessel. High-strength alloys, such as stainless steel, iron, nickel, cobalt-based superalloys, titanium and its alloys, are the most effective materials for resisting corrosion [69]. By hydrothermally treating peptized precipitates of a titanium precursor with water, Han C. et al., synthesized TiO_2 NPs [70]. Through hydrothermal processing, Horikoshi S. et al., synthesised phosphorus-doped TiO_2 (P-TiO_2) having a mesoporous structure [71]. Well-crystallized mesoporous P-doped titania NPs were developed by Hu L. et al., using both hydrothermal and *sol-gel* techniques [72]. Alternately, using a combination of the *sol-*

gel and hydrothermal techniques, ternary Bi_2WO_6 photocatalyst has been formed as microspheres. In contrast, Bi_2WO_6 was also produced directly by a hydrothermal reaction without using the *sol-gel* method. *Sol-gel*-hydrothermal (SH) Bi_2WO_6 was composed of hollow monodispersed hierarchical microspheres, whereas hydrothermal (H) Bi_2WO_6 has an uneven platelike structure. The results of the degradation of methylene blue (MB) in the presence of the various Bi_2WO_6 catalysts revealed that SH-Bi_2WO_6 demonstrates higher photocatalytic activity over H-Bi_2WO_6. It was determined that SH-improved Bi_2WO_6's photocatalytic activity was due to the unique hierarchical structure that was developed by the fusion of *sol-gel* and hydrothermal processes [73].

4.3 Solvothermal

The hydrothermal method is a derivative of the solvothermal method, which uses a nonaqueous solvent. Due to the wide range of organic solvents with high boiling points that can be chosen, the reaction temperature can be increased to much greater levels than in the hydrothermal approach. In general, the solvothermal technique is superior to the hydrothermal method for controlling the photocatalyst particle size, shape, distribution, and crystallinity [74]. With or without the use of surfactants, the solvothermal approach has been used to create TiO_2 NPs and nanorods. It has been proven to be a practical method for the synthesis of a range of NPs with narrow size distribution and dispersity. Otherwise, the solvent has a significant impact on the crystal shape. The solubility, reactivity, and diffusion behaviour of the reactants can be affected by solvents with various physical and chemical properties; in particular, the solvent's polarity and coordinating ability can affect the morphology and crystallisation behaviour of the end products. A high concentration of ethanol not only has the ability to modify the solvent's polarity but also has a significant impact on the reactant particles' potential values and raises the viscosity of the solution. For instance, instead of nanowires, short and wide flakelike structures of TiO_2 were produced in the absence of ethanol. TiO_2 nanorods were synthesized by Kappe using chloroform [75]. It is typically easier to regulate the size, crystallinity, and agglomeration behaviour of the NPs using nonaqueous solvothermal methods. Since there are more active sites accessible for photocatalytic processes, mesoporous TiO_2 microspheres with rough surfaces often exhibit high specific surface area, which is a crucial characteristic for the photocatalytic activity. Solvothermal techniques were used to create a number of distinct types of mesoporous TiO_2 samples in various sizes [76, 77, 78]. Using tetrabutyl titanate as a precursor in a polyethylenimine solution diluted with 100% ethanol, hierarchical mesoporous TiO_2 microspheres with high crystallinity and high BET specific surface area were synthesized in this context. Methyl orange (MO) and phenol aqueous solutions were degraded using the produced TiO_2 microspheres, and the activity of this process was assessed. The solvent's physicochemical behaviour qualities are altered in solvothermal circumstances, which involve high pressure. A nanocatalyst made of bismuth oxyiodide (BiOI) by Li X et al., has high photocatalytic activity when exposed to visible light [79]. According to a report by Mera et al., they developed BiOI microspheres using a solvothermal process (2014) [80].

4.4   Chemical vapor deposition (CVD)

Chemical vapor deposition (CVD) is another method that has been exploited for the fabrication of supported titanium dioxide materials. In CVD, thin films prepared on glass, metallic substrates, or semiconductive substrates, composites fabricated from activated carbon/titanium dioxide, amine polymers synthesized from titanium dioxide, and titanium dioxide nanorods fabricated from titanium precursor using CVD method. A chemical vapor deposition technique combines physical and chemical methods of producing materials. Due to their simplicity of control over synthesis, narrow particle size distribution, and low cost, chemical methods are the most popular method of synthesis of photocatalysts. While physical methods have been less explored, they are particularly relevant to large scale production of materials because highly specialized equipment and operators are required. High purity materials can be produced by CVD; besides, it does not require post-heat treatment to improve crystallinity. A few of these techniques include atmospheric pressure chemical vapor deposition (APCVD), plasma-enhanced chemical vapor deposition (PECVD), metal-organic chemical vapor deposition (MOCVD), which uses metal-organic precursors, and hybrid physical chemical vapor deposition (HPCVD). In cases requiring conformal deposition of the material, CVD is considered one of the most precise techniques. Molecular precursors are needed for CVD film growth. In order to transport precursor molecules, a chain of inert or reactive gas used. Chemical reactions in the gas phase near the surface convert them into a thin solid film of the desired material. The deposition reaction in CVD technologies is driven by hot carrier gases. TiO_2 thin films were synthesized by Sarantopoulos et al., utilising borosilicate glass and silicon wafers as substrates and titanium (IV) isopropoxide as the molecular precursor in an isothermal, isobaric, horizontal low-pressure CVD reactor. Visible light assisted photocatalytic activity of various dyes using Boron-doped TiO2 nanotubes reported by Q. Zhang et. al. The synthesis of TiO_2 nanotubes was achieved by using electrochemical anodization technique [82]. Zhu Y. et al., synthesized B-doped samples displayed stronger absorption in both UV and visible ranges. Additionally, the CVD techniques may quickly and at low deposition temperatures create coating materials that are single-layer, multilayer, composite, nanostructured, and functionally graded with precise dimension control [83]. Foster et al., (2010) provided information on the atmospheric pressure thermal CVD development of TiO_2/Ag and TiO_2/CuO thin films. Even against MRSA (methicillin-resistant *staphylococcus aureus*) and a few other infections, the movies had good microbicidal action [84]. The degradation of methylene blue was tested using a modified CVD technique to create V_2O_5/TiO_2 NCs particles and films, according to Lianjie Zhou et al., [85,86]. By using CVD, ZnO and other chalcogenide films have been created.

4.5   Green Techniques

Green synthesis technique is an eco-friendly approach for the synthesis of photocatalyst or NMs. This exploited natural or biological resources for the synthesis of NMs (Fig. 2) This technique has nontoxic, efficient and environmentally benign. In a recent time, scientific community have been described to accomplish the success of synthesis of NMs from plant extracts like leaves, roots, stem, fruits, peels, and seeds [87-89]. The notable success in this field has opened new route to advance "greener" approaches for the manufacturing of NMs with impeccable structural properties using milder and cheaper initial constituents [90].

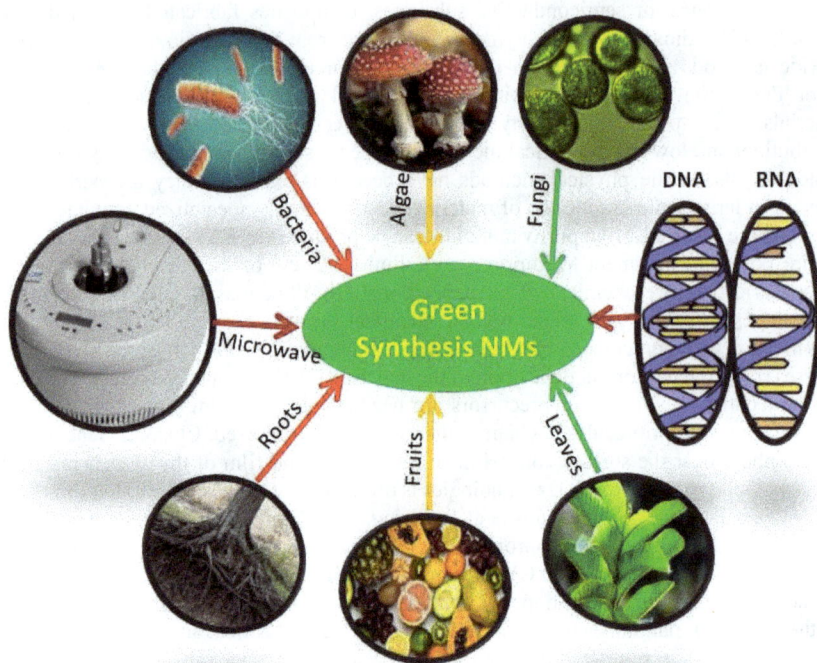

Figure 2. Green synthesis of photocatalysts

Indeed, the biogenic or green technique is commendable approach in the area of nanoscience and nanotechnology due to the concern of cost-effectiveness, fast, use of non-hazardous biochemicals, and their implausible properties. However, the conventional methods are being widely used for fabrication of nanomaterials like metal, metal oxide, metal oxides-polymer NCs, quantum dots, and carbon-based NMs [91-93], although, these methods encompasses with several pitfalls like expensive reagents, harmful chemicals, ominous effects on the medical equipment, and environment hazardousness. Considering the precedent apprehensions, exploration interest is being paid to develop a green, sustainable, non-toxic, inexpensive, and biocompatible technique. One of the best green synthesis techniques of nanomaterials; is a biogenic synthesis, it is a substitute technique to traditional or orthodox methods [94-97]. The biogenic synthesis of NMs involves bacteria, algae, fungi, enzymes, animal mass cultures, yeasts, and plant extracts. Among

Applications of Emerging Nanomaterials and Nanotechnology   Materials Research Forum LLC
Materials Research Foundations 148 (2023) 304-333   https://doi.org/10.21741/9781644902554-11

these, plant/microbial-assisted synthesis is convenient, rapid, cheaper, green, easy to handle, and non-hazardous approach. The biochemicals present in extract of plants can play disparate roles during fabrication NMs like reducing, surfactant, capping, and stabilizing agents [98-99].

5.   Process of photocatalytic activity

5.1   Types of photocatalyst used in photodegradation activity

The photocatalytic responses can be distributed into two types on the base of appearance of the physical state of reactants. Homogeneous photocatalysis, when both the semiconductor and reactant are in the same phase, i.e. gas, solid, or liquid, similar photocatalytic responses are nominated as homogeneous photocatalysis [100]. Different photocatalysts likewise ZnO, ZnS, CdS, SrO_2, WO_3 and Fe- TiO_2 have been used for treatment of air adulterants. The most popular among them is TiO_2 due to its low cost, high stability, environmental benevolence. It's a semiconductor which is used for print-convinced redox responses for declination of VOCs [101].

5.2   Source of light

Ultraviolet LED (UVLEDs) are mainly employed for the photocatalytic degradation of organic pollutants present in air and water. Recent findings have shown that visible LEDs, like blue, red, green, and white, can also be used for photocatalytic applications [102]. Photodegradation is degradation of a photodegradable molecule caused by the absorption of photons, particularly those wavelengths found in sunlight, such as infrared radiation, visible light, and ultraviolet light. However, other forms of electromagnetic radiation can cause photodegradation [103]. Methylene blue is brought into contact with the photocatalytic active surface of a test sample that is irradiated through the supernatant solution (320 nm << 400 nm). The solution is decolourated in the process. Throughout the measurement, the colour concentration of the solution is measured by UV-vis spectroscopy [104]. Photocatalytic semiconductors can be prepared in the form of powders, fibers, and films by different synthetic methods including *sol-gel* process, hydrothermal and solvothermal techniques, direct oxidation reactions, sonochemical method, microwave method, chemical vapor deposition method, and electrodeposition method [105].

5.3   Mechanism of photocatalytic degradation

The dye degradation by GMO photocatalyst as depicted in exemplifies oxidation process, adopted by the reactive oxygen species, for example hydroxyl radicals, superoxide, anions, and singlet oxygen species that formed in the reactor. Photocatalytic oxidation mechanism using semiconducting material can by summarized as:

5.3.1 Photo excitation

In photocatalytic reaction, electron promoted from the filled valance band (VB) of photocatalyst to the conduction band (CB) because of light irradiation. The absorbed photon has energy (hυ) are either greater than the band gap of the semiconductor photocatalyst. The excitation process leaves behind a hole in the h^+(VB) thus electron and hole pair(e^-/h^+) is generated as shown below in equation.

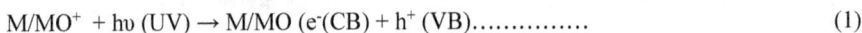

$$M/MO^+ + h\upsilon \ (UV) \rightarrow M/MO \ (e^-(CB) + h^+ \ (VB)\ldots\ldots\ldots\ldots \tag{1}$$

5.3.2 Ionization of water

The photo-generated holes at the valance band then react with water to produce OH^- radical.

$$H_2O \ (ads) + h^+(VB) \rightarrow M/MO + OH^- \ (ads) + H^+\ldots\ldots\ldots\ldots \tag{2}$$

The produced hydroxyl/superoxide are radical anion further react with remaining water molecules to produce the active hydroxyl ion exclusively which has been acted as a strong oxidized agent. Owing to their strong oxidation property, the toxic dye molecules have been break down and resulted corresponding CO_2 and H_2O.

5.3.3 Oxygen ion sorption

The photogenerated hole (h_{vB}^+) reacts with surface bound water or OH- to produce the hydroxyl radical, electron in the conduction (e_{CB}^-) is taken up by the oxygen in order to generate anionic superoxide radical (O_2^-). This superoxide ion may not only take part in the further oxidation process but also prevent the electron hole recombination, thus, sustaining electron neutrality within the graphene based NCs.

$$O_2 \ (ads) + \quad M/MO \ (e^-(CB) \quad \rightarrow \quad M/MO + O_2^-(ads)\ldots\ldots\ldots\ldots \tag{3}$$

5.3.4 Protonation of superoxide

The superoxide (O^{2-}) produced gets protonated forming hydroxyl radical (HO^{2-}) and then subsequently H_2O_2 further dissociates into highly reactive hydroxyl ions (OH^-).

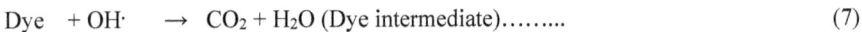

$$O_2^-(ads) + H^+ \quad \rightarrow \quad HOO^-(ads) \ldots\ldots\ldots\ldots\ldots\ldots\ldots\ldots \tag{4}$$

$$2HOO^-(ads) \quad \rightarrow \quad H_2O_2 \ (ads) + O_2\ldots\ldots\ldots\ldots\ldots\ldots\ldots \tag{5}$$

$$H_2O_2 \ (ads) \quad \rightarrow \quad 2OH^-(ads) \ldots\ldots\ldots\ldots\ldots\ldots\ldots \tag{6}$$

$$Dye \ + OH^\cdot \quad \rightarrow \quad CO_2 + H_2O \ (Dye \ intermediate)\ldots\ldots\ldots \tag{7}$$

$$\text{Dye} + h^+ \text{(VB)} \quad \rightarrow \quad \text{Oxidation product}\dots\dots\dots\dots\dots \qquad (8)$$

$$\text{Dye} + e^-\text{(CB)} \quad \rightarrow \quad \text{Reduction product}\dots\dots\dots\dots\dots \qquad (9)$$

Both oxidation and reduction process take place on the photo excited semiconductor photocatalyst.

5.3.5 Scavengers and trapping agent in photocatalysis

The part of hole scavengers in photocatalysis is frequently regarded to be removing the holes in photocatalysts by giving electrons to them. Mureithi et al., show that the effect of hole scavengers goes beyond junking of holes to affecting the response time, pathways, and products in photocatalytic responses. Utmost photocatalytic nanosystems reported so far employ a hole scavenger a sacrificial chemical which is oxidized in the photocatalytic process and therefore removes the hole from the nanostructure [106]. Fastening on the influence of different species, similar as scavengers (t- butanol, formic acid, methanol, p-benzoquinone, oxalate, superoxide dismutase, and azide), snooping species (sulfite, dichromate, bromate, carbonate, chloride, and iodide) and inorganic ions (nitrate, sulfate, and phosphate), this work delved the product of hydroxyl revolutionaries and singlet oxygen during TiO_2/UVA responses. Electron paramagnetic resonance spectroscopy (EPR) was applied to probe revolutionaries formed in the presence of each snooping/ scavenger species. Some scavengers and snooping species were studied during phenol declination, chosen as a model substrate. All species, except bromate, hindered the declination. Para-benzoquinone showed an increased hydroxyl radical product, attributed to the print- reduction of quinones. revolutionaries other than hydroxyl revolutionaries, similar as carbon dioxide, hydroxymethyl, azide, and semiquinone, were linked in the presence of oxalate, methanol, azide, andpara-benzoquinone, independently. Some of these revolutionaries can conceivably interact with organic substrates due to their reduction eventuality; as a result, a critical interpretation must be done when these species are added to a miscellaneous photocatalysis process.

6.   Photodegradation & photoreduction of toxic compounds

6.1   Photocatalysis

In recent years, a great deal of effort has been dedicated to solve the widespread pollution caused by effluents from urban and agricultural industries, which includes bio recalcitrant and organic pollutants [107]. Development in industrialization and continuous progression of newer technologies coupled with poor environmental policies have led to the generation of toxic effluents in huge amount which are discarded into natural water bodies. The main component of this chemical discharge from textile, chemical, petroleum and many other industries are azo dyes. These toxic dyes not only disturb the aquatic ecosystem but also has a devastating effect on human life and terrestrial biota. These dyes are mutagenic, carcinogenic, and highly toxic to CNS and can cause respiratory diseases, cancer, genetic disorders, chromosomal abnormalities and many acute disorders. Moreover, these dyes remain in environment for longer periods of time without degrading [108]. In view of this situation, it has become imperative to reduce and remove these chemical pollutants (synthetic dyes) from environment. In this regard, semiconductor assisted photocatalysis has received global attention, since these dyes are resistant to biological and physical treatment methods [109].

Heterogeneous semiconductor photocatalysts can absorb solar radiation to generate electron-hole pairs which can accelerate the remediation of these dyes. A simple mechanism of heterogeneous photocatalysis includes absorption of light by semiconductor, leading to the excitation of electrons from valence band to conduction band leaving behind holes in VB. Figs. 3 & 4 depicts the process of dye degradation on a semiconductor photocatalyst involving direct and indirect pathways. In the indirect pathway, dye is degraded by reactive oxygen species formed on the catalyst surface. Photogenerated holes react with adsorbed water or hydroxyl ion to generate hydroxyl radicals whereas electrons in CB reacts with oxygen to form superoxide radicals. These reactive oxygen species quickly, and non-selectively reduces the pollutants, including dyes, antibiotics, drugs, and phenolic compounds. In the direct pathway under light irradiation, shaded organic dyes such as RhB, EBT, IC, TB, MO, CR and MB can also be excited in the process called photosensitization, where dye is excited from the ground state to the excited state. Charge transfer occurs from the highest occupied molecular orbital (HOMO) of excited state dye to the conduction band of the photocatalyst providing more electrons to generate superoxide radicals, as a consequence enhancing the photocatalytic degradation activity [110]. Still, the quick recombination rate of photogenerated electron – hole dyads within photocatalytic accoutrements results in its low proficiency, therefore limiting its practical operations. Thus, repression of recombination of charge carriers is crucial for the improvement of photocatalytic exertion of semiconductor PCs. Besides the conventional doping and adding co-sorbents, carbon – semiconductor mongrel accoutrements have come a new class of photocatalysts, which lately has attracted a lot of attention [111]. Among carbonaceous materials, graphene has received attention due to its unique properties, such as high charge-carrier mobility, high thermal and electronic conductivity, and high specific surface area. These unique properties of graphene make it an ideal support material for semiconductor photocatalysis. Furthermore, interaction of graphene with semiconductor materials give rise to unique properties such as an extended light absorption and decreased charge carrier recombination rates along with high stability [106]. Therefore graphene-semiconductor hybrid materials have been widely used for the degradation of organic pollutants, antibiotics, organic compound, photocatalytic hydrogen generation and photocatalytic disinfection. In this segment the main applications of graphene-based semiconductor photocatalysts like PANI, NiO, TiO_2, Fe_3O_4, and ZnO etc., are briefly summarized.

Applications of Emerging Nanomaterials and Nanotechnology   Materials Research Forum LLC
Materials Research Foundations 148 (2023) 304-333   https://doi.org/10.21741/9781644902554-11

6.2   Photocatalytic degradation of dyes

Nowadays, we are facing huge inadequacy of pure and potable water and its accessibility is becoming more and more luxurious day by day. This is due to the disposal of several toxic wastes from many industries, pharmaceutical laboratories and agricultural field. Among these effluents the inorganic and organic dyes are issue of major concerned. Scientists are focusing on biodegraded dyes and dye derived components. Biopolymers are the most preferable choice for the degradation of dyes as their enhanced catalytic activities are reported by so many researchers so far. Millions of people in the developing countries decease due to unavailability of clean and pure drinking water and safe hygienic environment. We need an eco-friendly, cheap and easy process to biodegrade these toxic dyes for purification of water as day by day the sources for portable water is vanishing and turning into the ponds of eutrophicated, toxic, dirty and polluted water bodies. The pollutants which are majorly found in water bodies consist of pesticides, herbicides, dyes, surfactants etc., [112]. To overcome these problem, recent past year's graphene-based photocatalyst have gained significant importance for the photocatalytic application. Due to their high photosensitivity, high surface area and excellent optical, mechanical and electrical properties and stability, eco-friendly tactic, nontoxic, low cost and enhanced catalytic properties, they are being widely used in water cleansing, photodegradation of toxic dyes and chemical effluents (Figs. 3&4) [113]. Chaudhary *et al.,* synthesized ZnO-rGO NCs for photodegradation of methylene dye using visible irradiation, study revels methylene blue dye was degraded about 85% in 70 min by using ZnO-rGO NCs [114]. Sonkusare *et al.,* microwave-mediated synthesis, photocatalytic degradation activity of α-Bi_2O_3 microflowers/novel γ-Bi_2O_3 microspindles, theα-Bi_2O_3 MFs show enhanced photocatalytic activity than γ-Bi_2O_3 MSs under UV light irradiations [115]. Rahimi *et al.,* projected rGO/NiO nanowires for photocatalytic disintegration of methyl orange (MO). For Photocatalytic decomposition of MO requires both hydroxide and superoxide radicals [116]. Mitra *et al.,* have synthesized a new PCs based on Polyaniline/reduced graphene oxide (PANI/rGO) composites by oxidative polymerization. They have reported that 5 wt % rGO(PG_2) can be engaged with outstanding photocatalytic efficiency percentage 99.68, 99.35, and 98.73 for MG, RhB, and CR within 15, 30, and 40 min, respectively. The experimental results showed that rGO can reduce accretion of PANI and increased surface area which enriches the photocatalytic activity [117].

In another study, Elshypani *et al.,* [118] described solid state mediated magnetite zinc oxide (MZ) (Fe_3O_4/ZnO) with different ratios of reduced graphene oxide (rGO). Results revealed that the photodegradation activity of Magnetite Zing Graphene MZG was more than 98.5% against Methylene Blue. This was due to synergic effect of magnetite and zinc oxide in attendance of rGO.

Hamed *et al.,* [119] have reported Gr/Pd/TiO_2-NPs and Gr/Pd/TiO_2-NWs were prepared *via* combination of hydrothermal & photodeposition method. They have compared degradation capacity of TiO_2-NWs, Gr/Pd/TiO_2-NPs and Gr/Pd/TiO_2-NWs towards Rhodamine B dye in UV light irradiation. Results showed that Gr/Pd/TiO_2-NWs have high excellent photocatalytic activity due to high surface area. Zhang *et al.,* [120] have studies the photodegradation of MB in water by $CuO.ZnO.Fe_2O_3/rGO$ and $CuO.ZnO.Fe_2O_3/CNT$. The degradation efficacy of composite with CNT is better than rGO.

Chandra *et al.,* demonstrated that the photodegradation profile of eosin, methylene blue, and Rhodamine B is almost 80% in the presence of graphene-Mn_2O_3 NCs [121]. *Fu* and co-workers have been prepared magnetically separable $ZnFe_2O_4$/graphene PCs by a facile hydrothermal method. The PCs displayed 88% of MB degradation by adding H_2O_2 as a scavenging agent in 5 min and reached up to 99% at 90 min in irradiation of visible light. It assists a dual function as photoelectron-chemical degrader and generator of hydroxyl radicals *via* photo electrochemical decomposition of H_2O_2 [122].

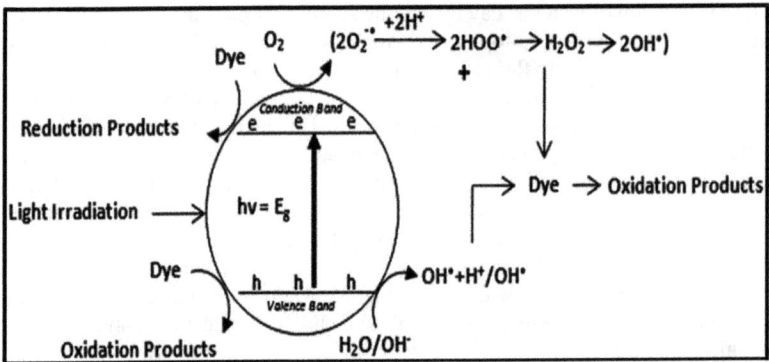

Fig. 3. Indirect dye degradation

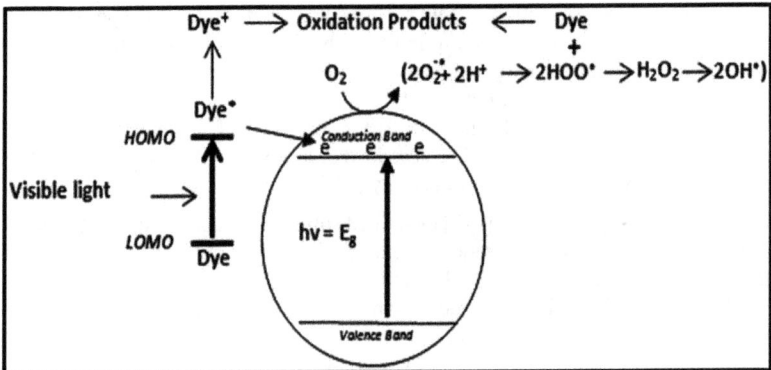

Fig. 4. Direct dye degradation

6.3   Degradation of antibiotics and pharmaceutical drugs

Besides adsorption, antibiotics can be efficiently decayed or perished into non-toxic small molecular species under the sun, visible light and ultraviolet (UV) light, due to the presence of active groups (e.g. superoxide ions (O^{2-}) and hydroxyl revolutionaries (.OH) produced by photocatalysts. So photocatalytic declination is one of the most operative, green, and generally used ways for antibiotics pollutants junking in the terrain. Graphene, as a promising PCs, has been extensively explored for photo catalytic declination of antibiotic poisons in water, due to their large specific face area for invariant distribution, narrow band- gap powers and exceptional electrical conductivity in storehouse and fast transport of electrons, and low manufacturing cost for large scale products. Therefore, graphene is frequently combined with other photocatalysts to form new photocatalysts in order to overcome these downsides and ameliorate the catalytic performance of antibiotics. In recent times, colourful attempts have been devoted to designing and fabricating graphene-grounded semiconductor photocatalyst in order to increases the declination capacity of antibiotic adulterants. In this regard, Song et al., set MnO_2/ grapheme NCs by an In-situ hydrothermal system, and it successfully removed up to 99.4 of the tetracycline residues in pharmaceutical wastewater [123]. In another study, Shanavas et al., examined the declination capacity of ibuprofen and tetracycline motes efficiently under visible light irradiation within 90 min using ternary Cu/ $Bi_2Ti_2O_7$/ rGO NCs. The attained results suggested that the Cu NPs and the rGO wastes play a major part in the photocatalytic capability of Cu/ $Bi_2Ti_2O_7$/ rGO PCs by acting as charge carrier fowlers and the repression of e −- h brace recombination [124].

6.4   Photocatalytic degradation of phenolic compound

Principally organic colourings have been substantially named as a model emulsion to estimate the photocatalytic exertion of graphene- grounded NCs [125]. Among the dangerous organic pollutants set up in artificial backwaters, phenol and phenolic composites are veritably important [126]. The discharge quantum of phenol in water must be maintained at 0.1 −1 mg/ L(ppm) grounded on the environmental protection rules of the Pollution Control Board (1992) thus, semiconductor PCs displayed bettered photocatalytic declination of phenol under dissembled solar/ UV/ Visible irradiation after objectification of graphene. Graphene- grounded TiO_2/ ZnO NCs were prepared via hydrothermal system by Malekshoar et al., [127] the photocatalytic profile of the coupled NCs with an optimized rate of the mechanisms was explored. The results reveal that the coupled mixes (Graphene-ZnO/TiO_2) with 0.95 to 0.05 rate) outpaced, when compared to a single compound by a factor of 2. Also, the parametric study was carried out to optimize the response conditions. Experimental results showed that 1 h of 100 mW/ cm2 solar irradiation was needed to degrade 40 ppm phenol at neutral pH, when1.25 g/ L of coupled ZnO-G/ TiO_2-G mixes Li. Fen Chiang et al., synthesized Cu–TiO_2 nanorods for bisphenol degradation under UV-visible light irradiation [128].

Conclusion

In a conclusion, the nanomaterials are the alternative for degradations of various organic pollutants from the contaminated waters. There are several types of nanomaterials can be used us active photocatalysts. The better catalytic activity of the nanomaterials is depends on its optical properties. Every nanophotocatalyst must have ideal bandgap. A simple mechanism of heterogeneous photocatalysis includes absorption of light by photocatalyst, leading to the excitation of electrons from valence band to conduction band leaving behind holes in valence band. A heterogeneous photocatalysts can absorb light radiation to generate electron-hole pairs which can accelerate the remediation of these dyes or macromolecules. The photocatalytic degradation of toxic compounds can be processed under different light irradiations, for instance antibiotics dyes, drugs, and phenolic compounds can be efficiently decayed or perished into non-toxic small molecular species under the sun, visible light and ultraviolet.

References

[1] Li, Zhuo & Yao, Yagang & Lin, Ziyin & Moon, Kyoung-sik & Lin, Wei & Wong, C.P.. (2010). Ultrafast, dry microwave synthesis of graphene sheets. Journal of Materials Chemistry, 20, (2010) 4781-4783. https://doi.org/10.1039/c0jm00168f

[2] G. Demazeau, Solvothermal processes: a route to the stabilization of new materials J. Mater. Chem. 9 (1999) 15-18. https://doi.org/10.1039/a805536j

[3] H. Wang, J. T. Robinson, X. Li, H. Dia, Simultaneous Nitrogen Doping and Reduction of Graphene Oxide J. Am. Chem. Soc. 131 (2009) 9910-9911. https://doi.org/10.1021/ja904251p

[4] Y. Zhou, Q. Bao, L. A. L. Tang, Y. Zhong, K. P. Loh, Hydrothermal Dehydration for the "Green" Reduction of Exfoliated Graphene Oxide to Graphene and Demonstration of unable Optical Limiting Properties Chem. Mater. 21 (2009) 2950- 2956. https://doi.org/10.1021/cm9006603

[5] R. Wang, Y. Wang, C. Xu, J. Sun, L. Gao, Facile one-step hydrazine-assisted solvothermal synthesis of nitrogen-doped reduced graphene oxide: reduction effect and mechanisms., RSC Adv. 3 (2013) 1194-1200. https://doi.org/10.1039/C2RA21825A

[6] S. Dubin, S. Gilje, K. Wang, V. C. Tung, K. Cha, A. S. Hall, A one-step, solvothermal reduction method for producing reduced graphene oxide dispersions in organic solvents., ACS Nano, 4, (2010) 3845-3852. https://doi.org/10.1021/nn100511a

[7] D. Zhou, Q. -Y. Cheng, B. -H. Han, Solvothermal synthesis of homogeneous graphene dispersion with high concentration. .Carbon, 49 (2011) 3920-3927. https://doi.org/10.1016/j.carbon.2011.05.030

[8] J. Xiao, W. Lv, Z. Xie, Environment-friendly reduced graphene oxide as a broad-spectrum adsorbent for anionic and cationic dyes via π-π interaction. J. Mater. Chem. A 4 (2016) 12126- 12135 https://doi.org/10.1039/C6TA04119A

[9] O. A. Rahman, V. Chellasamy, N. Ponpandian, S. Amirthapandian, B. Panigrahi, P. Thangadurai, RSC Adv. 4 (2014) 56910-56917. https://doi.org/10.1039/C4RA06203E

[10] M. Fernandez-Merino, L. Guardia, J. Paredes, S. Villar-Rodil, P. Solis-Fernandez, A. Martinez-Alonso, Vitamin C Is an Ideal Substitute for Hydrazine in the Reduction of Graphene Oxide Suspensions J. Phys. Chem. C 114 (2010) 6426- 32. https://doi.org/10.1021/jp100603h

[11] H. L. Guo, X. F. Wang, Q. Y. Qian, A Green Approach to the Synthesis of Graphene Nanosheets., ACS Nano 3 (2009) 2653-2659. https://doi.org/10.1021/nn900227d

[12] C. Bosch-Navarro, E. Coronado, C. Martí-Gastaldo, J. Sánchez-Royo, M.G. Gómez, Influence of the pH on the synthesis of reduced graphene oxide under hydrothermal conditions., Nanoscale 4 (2012) 3977-82. https://doi.org/10.1039/c2nr30605k

[13] Z. Bo, X. Shuai, S. Mao, H. Yang, J. Qian, J. Chen, Green preparation of reduced graphene oxide for sensing and energy storage applications. Sci. rep. (2014) 4-15. https://doi.org/10.1038/srep04684

[14] S. Liu, J. Tian, L. Wang, X. Sun, A method for the production of reduced graphene oxide using benzylamine as a reducing and stabilizing agent and its subsequent decoration with Ag NPs for enzymeless hydrogen peroxide detection, Carbon 49 (2011) 3158-64. https://doi.org/10.1016/j.carbon.2011.03.036

[15] X. Fan, W. Peng, Y. Li, X. Li, S. Wang, G. Zhang, Deoxygenation of Exfoliated Graphite Oxide under Alkaline Conditions: A Green Route to Graphene Preparation. Adv. Mater. 20 (2008) 4490-3 https://doi.org/10.1002/adma.200801306

[16] C. Chua, A. Ambrosi, M. Pumera, Graphene oxide reduction by standard industrial reducing agent: thiourea dioxide., J. Mater. Chem. 22 (2012) 11054-61 https://doi.org/10.1039/c2jm16054d

[17] D.N. Tran, S. Kabiri, D. Losic, A green approach for the reduction of graphene oxide nanosheets using non-aromatic amino acids., Carbon 76 (2014) 193-202. https://doi.org/10.1016/j.carbon.2014.04.067

[18] A. Esfandiar, O. Akhavan, A. Irajizad, Melatonin as a powerful bio-antioxidant for reduction of graphene oxide., J. Mater. Chem. 21 (2011) 10907-14. https://doi.org/10.1039/c1jm10151j

[19] D. Wan, C. Yang, T. Lin, Y. Tang, M. Zhou, Y. Zhong,. Low-temperature aluminum reduction of graphene oxide, electrical properties, surface wettability, and energy storage applications. ACS Nano 6 (2012) 9068-78. https://doi.org/10.1021/nn303228r

[20] X. Mei, J. Ouyang, Ultrasonication-assisted ultrafast reduction of graphene oxide by zinc powder at room temperature. Carbon 49 (2011) 5389-97 https://doi.org/10.1016/j.carbon.2011.08.019

[21] M. Aunkor, I. M. Mahbubul, R. Saidur, H. Metselaar The green reduction of graphene oxide., RSC Adv. 6 (2016) 27807-27828 https://doi.org/10.1039/C6RA03189G

[22] M. Khan, A. Al-Marri, M. Khan, N. Mohri, S. Adil, A. Al-Warthan, M. Siddiqui, H. Alkhathlan, R. Berger, W. Tremel, Ismail Cer. Internat. l 45 (2019) 23857-23868

[23] N. Elavarasan, S. Prakasha, K. Kokilaa, C. Thirunavukkarasuc, V. Sujatha, New J. Chem. 44 (2020) 2166-2179

[24] J. Li, G. Xiao, C. Chen, R. Li, D. Superior dispersions of reduced graphene oxide synthesized by using gallic acid as a reductant and stabilizer., J. Mater. Chem. A. 1 (2013) 1481-87. https://doi.org/10.1039/C2TA00638C

[25] T. Kuila, S. Bose, P. Khanra, A.K. Mishra, N.H. Kim, J.H. Lee, A green approach for the reduction of graphene oxide by wild carrot root. Carbon, 50 (2012) 914-921. https://doi.org/10.1016/j.carbon.2011.09.053

[26] G. Wang, F. Qian, C. Saltikov, Y. Jiao, Y. Li, Nano Res. 4 (2011) 563-70. https://doi.org/10.1007/s12274-011-0112-2

[27] A. K. Potbhare, M. S. Umekar, P. B. Chouke, M. B. Bagade, S. K. Tarik Aziz, A. A. Abdala, R. G. Chaudhary, Bioinspired graphene-based silver NPs: Fabrication, characterization and antibacterial activity. Materials Today: Proceedings 29 (2020) 720-725. https://doi.org/10.1016/j.matpr.2020.04.212

[28] O. Akhavan, E. Ghaderi, E. Abouei, S. Hatamie, E. Ghasemi, Accelerated differentiation of neural stem cells into neurons on ginseng-reduced graphene oxide sheets. Carbon 66 (2014) 395-406. https://doi.org/10.1016/j.carbon.2013.09.015

[29] D. Suresh, H. Nagabhushana, S. Sharma, Clove extract mediated facile green reduction of graphene oxide, its dye elimination and antioxidant properties.Mater. Lett. 142 (2015) 4-6. https://doi.org/10.1016/j.matlet.2014.11.073

[30] B. Feng, J. Xie, C. Dong, S. Zhang, G. Cao, X. Zhao, From graphite oxide to nitrogen and sulfur co-doped few-layered graphene by a green reduction route via Chinese medicinal herbs.RSC Adv. 4 (2014) 17902-17907. https://doi.org/10.1039/c4ra01985g

[31] O. G. Akhavan, Escherichia coli bacteria reduce graphene oxide to bactericidal graphene in a self-limiting manner. Carbon 50 (2012) 1853-60 https://doi.org/10.1016/j.carbon.2011.12.035

[32] P. Khanra, T. Kuila, N.H. Kim, S.H. Bae, D.-s. Yu, J.H. Simultaneous bio-functionalization and reduction of graphene oxide by baker's yeast. Chem. Eng. J. 183 (2012) 526-33. https://doi.org/10.1016/j.cej.2011.12.075

[33] Z. Bo, X. Shuai, S. Mao, H. Yang, J. Qian, J. Chen, J. Yan, K. Cen, Green preparation of reduced graphene oxide for sensing and energy storage applications. Sci. Rep. 4 (2014) 4684. https://doi.org/10.1038/srep04684

[34] H.-J. Chu, C.-Y. Lee, N.-H. Tai., Green reduction of graphene oxide by Hibiscus sabdariffa L. to fabricate flexible graphene electrode. Carbon, 80 (2014) 725-733. https://doi.org/10.1016/j.carbon.2014.09.019

[35] N.Serpone, A.V. Emeline, Semiconductor Photocatalysis-Past, Present, and Future Outlook. J. Phys. Chem. Lett. 3 (2012),673-677. https://doi.org/10.1021/jz300071j

[36] S. Sato, Photocatalytic activity of NOx-doped TiO2 in the visible light region. Chem. Phys. Lett. 123 (1986), 126-128. https://doi.org/10.1016/0009-2614(86)87026-9

[37] R. Asahi, T. Morikawa, T. Ohwaki, K. Aoki, Y.Taga, Visible-Light Photocatalysis in Nitrogen-Doped Titanium Oxides. Science,293 (2001), 269-271. https://doi.org/10.1126/science.1061051

[38] J. Biedrzycki, S. Livraghi, E. Giamello, S. Agnoli, G. Granozzi, Fluorine- and niobium-doped TiO2: Chemical and spectroscopicproperties of polycrystalline n-type-doped *anatase*. J. Phys.Chem. C,118 (2014), 8462-8473. https://doi.org/10.1021/jp501203h

[39] A.M. Czoska, S. Livraghi, M. Chiesa, E. Giamello, S. Agnoli, G. Granozzi, E. Finazzi, C. Di Valentiny, G. Pacchioni, Thenature of defects in fluorine-doped TiO2. J. Phys. Chem. C, 112(2008), 8951-8956. https://doi.org/10.1021/jp8004184

[40] C. Di Valentin, E. Finazzi, G. Pacchioni, A. Selloni, S. Livraghi, M.C. Paganini, E. Giamello, N-doped TiO2: Theory and experiment. Chem. Phys., 339(2007), 44-56. https://doi.org/10.1016/j.chemphys.2007.07.020

[41] U. Diebold, The surface science of titanium dioxide. Surf. Sci. Rep.,48 (2003), 53-229. https://doi.org/10.1016/S0167-5729(02)00100-0

[42] N. Liu, C. Schneider, D. Freitag, M. Hartmann, U. Venkatesan, J. Müller, E. Spiecker, P. Schmuki, Black TiO2 Nanotubes:Cocatalyst-Free Open-Circuit Hydrogen Generation. Nano Lett., 14(2014), 3309-3313. https://doi.org/10.1021/nl500710j

[43] P. Garcia-Muñoz, F. Fresno, J. Ivanez, D. Robert, N. Keller, Activity enhancement pathways in LaFeO3@TiO2 heterojunction photocatalysts for visible and solar light driven degradation of myclobutanil pesticide in water. J. Hazard. Mater., 400(2020), 123099. https://doi.org/10.1016/j.jhazmat.2020.123099

[44] C. Ling, C. Yue, R. Yuan, J. Qiu, F.Q. Liu, J.J. Zhu, Enhanced removal of sulfamethoxazole by a novel composite of TiO2 nanocrystals in situ wrapped-Bi2O4 microrods under simulated solar irradiation. Chem. Eng. J., 384(2020), 123278. https://doi.org/10.1016/j.cej.2019.123278

[45] M. Tobajas, C. Belver, J.J. Rodriguez, Degradation of emerging pollutants in water under solar irradiation using novel TiO2-ZnO/clay nanoarchitectures. Chem. Eng. J., 309(2017), 596-606. https://doi.org/10.1016/j.cej.2016.10.002

[46] S. I. Mogal, V.G. Gandhi, M. Mishra, S. Tripathi, T. Shripathi, P.A. Joshi, D.O. Shah, Single-step synthesis of silver-doped titanium dioxide: influence of silver on structural, textural, and photocatalytic properties. Industrial & Engineering Chemistry Research, 53(14) (2014), 5749-5758. https://doi.org/10.1021/ie404230q

[47] E. Kowalska, M. Janczarek, L. Rosa, S. Juodkazis, B. Ohtani, Mono-and bi-metallic plasmonic photocatalysts for degradation of organic compounds under UV and visible light irradiation. Catalysis Today, 230(2014),131-137. https://doi.org/10.1016/j.cattod.2013.11.021

[48] A.T. Montoya, E. G. Gillan, Enhanced photocatalytic hydrogen evolution from transition-metal surface-modified TiO2. ACS omega, 3(3) (2018). 2947-2955. https://doi.org/10.1021/acsomega.7b02021

[49] R. Dholam, N. Patel, M. Adami, A. Miotello, Hydrogen production by photocatalytic water-splitting using Cr-or Fe-doped TiO2 composite thin films photocatalyst. International Journal of Hydrogen Energy, 34(13) (2009). 5337-5346. https://doi.org/10.1016/j.ijhydene.2009.05.011

[50] S. E. Salas, B. S. Rosales, H. de Lasa,. Quantum yield with platinum modified TiO2 photocatalyst for hydrogen production. Applied Catalysis B: Environmental, 140 (2013) 523-536. https://doi.org/10.1016/j.apcatb.2013.04.016

[51] A. Manivel, S. Naveenraj, P.S. Sathish Kumar, S. Anandan, CuO-TiO2 nanocatalyst for photodegradation of acid red 88 in aqueous solution. Science of Advanced Materials, 2(1) (2010), 51-57. https://doi.org/10.1166/sam.2010.1071

[52] G. Liu, X. Zhang, Y. Xu, X. Niu, L. Zheng, X. Ding,. The preparation of Zn2+-doped TiO2 NPs by *sol-gel* and solid phase reaction methods respectively and their photocatalytic activities. Chemosphere, 59(9) (2005), 1367-1371. https://doi.org/10.1016/j.chemosphere.2004.11.072

[53] H. Wang, S. Cao, Z. Fang, F. Yu, Y. Liu, X. Weng, Z. Wu,. CeO2 doped *anatase* TiO2 with exposed (001) high energy facets and its performance in selective catalytic reduction of NO by NH3. Applied Surface Science, 330 (2015), 245-252. https://doi.org/10.1016/j.apsusc.2014.12.163

[54] J. G. Mahy, S. D. Lambert, R. G. Tilkin, C. Wolfs, D. Poelman, F. Devred, S. Douven,. Ambient temperature ZrO2-doped TiO2 crystalline photocatalysts: Highly efficient powders and films for water depollution. Materials Today Energy, 13 (2019), 312-322. https://doi.org/10.1016/j.mtener.2019.06.010

[55] L. Gnanasekaran, R. Hemamalini, S. Rajendran, J. Qin, M. L. Yola, N. Atar, F. Gracia,. Nanosized Fe3O4 incorporated on a TiO2 surface for the enhanced

photocatalytic degradation of organic pollutants. Journal of Molecular Liquids, 287 (2019), 110967. https://doi.org/10.1016/j.molliq.2019.110967

[56] H. Kong, J. Song, J. Jang,. Photocatalytic antibacterial capabilities of TiO2– biocidal polymer NCss synthesized by a surface-initiated photopolymerization. Environmental science & technology, 44(14) (2010), 5672-5676. https://doi.org/10.1021/es1010779

[57] D. Tekin, D. Birhan, H. Kiziltas,. Thermal, photocatalytic, and antibacterial properties of calcinated nano-TiO2/polymer composites. Materials Chemistry and Physics, 251 (2020), 123067. https://doi.org/10.1016/j.matchemphys.2020.123067

[58] L. Zhang, P. Liu, Z. Su,. Preparation of PANI-TiO2 NCss and their solid-phase photocatalytic degradation. Polymer degradation and stability, 91(9) (2006), 2213-2219. https://doi.org/10.1016/j.polymdegradstab.2006.01.002

[59] R. Saravanan, J. Aviles, F. Gracia, E. Mosquera, V. K. Gupta,. Crystallinity and lowering band gap induced visible light photocatalytic activity of TiO2/CS (Chitosan) NCss. International journal of biological macromolecules, 109 (2018), 1239-1245. https://doi.org/10.1016/j.ijbiomac.2017.11.125

[60] K. Fischer, P. Schulz, I. Atanasov, A. Abdul Latif, I. Thomas, M. Kühnert, A. Prager, J. Griebel, A. Schulze,. Synthesis of high crystalline TiO2 NPs on a polymer membrane to degrade pollutants from water. Catalysts, 8(9) (2018), 376. https://doi.org/10.3390/catal8090376

[61] K. Byrappa, M. Yoshimura, Handbook of hydrothermal technology. A Technology for Crystal Growth and Materials Processing. Noyes, New York 96 (2001). https://doi.org/10.1016/B978-081551445-9.50003-9

[62] I. Medina-Ramı'rez, J. O. Carneiro,V. Teixeira, A. L. Portinha, L. Dupa'k, A. Magalhaes, P. Coutinho, Study of the deposition parameters and Fe-dopant effect in the photocatalytic activity of TiO2 films prepared by dc reactive magnetron sputtering. Vacuum, 78 (2005),37-46. https://doi.org/10.1016/j.vacuum.2004.12.012

[63] J. O. Carneiro, V. Teixeira, A. L. Portinha, A. Magalhaes, P. Coutinho, C. J. Tavares, R. Newton, Iron-doped photocatalytic TiO2 sputtered coatings on plastics for self-cleaning applications. Mater Sci Eng B, 138 (2007),144-150. https://doi.org/10.1016/j.mseb.2005.08.130

[64] K. G. Chandrappa, T. V. Venkatesha, Electrochemical synthesis and photocatalytic property of zinc oxide NPs. Nano-Micro Lett, 4 (2012),14-24. https://doi.org/10.1007/BF03353686

[65] Y. C. Chen, S. L. Lo, Effects of operational conditions of microwave-assisted synthesis on morphology and photocatalytic capability of zinc oxide. Chem Eng J, 170 (2011),411-418. https://doi.org/10.1016/j.cej.2010.11.057

[66] Z. Chen, W. Li, W. Zeng, M. Li, J. Xiang, Z. Zhou, J. Huang, Microwave hydrothermal synthesis of nanocrystalline rutile. Mater Lett, 62 (2008),4343-4344. https://doi.org/10.1016/j.matlet.2008.07.024

[67] B. Chen, J. Hou, K. Lu, Formation mechanism of TiO2 nanotubes and their applications in photoelectrochemical water splitting and super-capacitors. Langmuir, 29 (2013), 5911-5919. https://doi.org/10.1021/la400586r

[68] J. Choi, S. H. Cho, T. H. Kim, S. W. Lee, Comparison of sonochemistry method and sol-gel method for the fabrication of TiO2 powder. Mater Sci Forum, 695 (2011), 109-112. https://doi.org/10.4028/www.scientific.net/MSF.695.109

[69] R. N. Geyde, Organic synthesis using microwaves in homogeneous media. In: Loupy A (ed) Microwaves in organic synthesis. Wiley-VCH, Weinheim, (2002), 115-146 https://doi.org/10.1002/3527601775.ch4

[70] C. Gionco, A. Battiato, E. Vittone, M. C. Paganini, E. Giamello, Structural and spectroscopic properties of high temperature prepared ZrO2-TiO2 mixed oxides. J Solid State Chem, 201 (2013) 222-228. https://doi.org/10.1016/j.jssc.2013.02.040

[71] C. Han, R. Luque, D. D. Dionysiou, Facile preparation of controllable size monodisperse anatase titania NPs. Chem Commun, 48 (2012) 1860-1862. https://doi.org/10.1039/C1CC16050H

[72] S. Horikoshi, S. Sakamoto, N. Serpone, Formation and efficacy of TiO2/AC composites prepared under microwave irradiation in the photoinduced transformation of the 2-propanol VOC pollutant in air. Appl Catal B Environ, 140-141 (2013), 646-665. https://doi.org/10.1016/j.apcatb.2013.04.060

[73] L. Hu, K. Huo, R. Chen, B. Gao, J. Fu, P. K. Chu, Recyclable and high-sensitivity electrochemical biosensing platform composed of carbon-doped TiO2 nanotube arrays. Anal Chem, 83 (2011), 8138-8144. https://doi.org/10.1021/ac201639m

[74] N. Jin, Y. Yang, X. Luo, Z. Xia, Development of CVD Ti-containing films. Prog Mater Sci, 58 (2013), 1490-1533. https://doi.org/10.1016/j.pmatsci.2013.07.001

[75] Y. S. Jung, K. H. Kim, T. Y. Jang, Y. Tak, S. H. Baeck, Enhancement of photocatalytic properties of Cr2O3-TiO2 mixed oxides prepared by sol-gel method. Curr Appl Phys, 11 (2011), 358-361. https://doi.org/10.1016/j.cap.2010.08.001

[76] C. O. Kappe, Unraveling the mysteries of microwave chemistry using silicon carbide reactor technology. Acc Chem Res, 46(7) (2013), 1579-1587. https://doi.org/10.1021/ar300318c

[77] H. J. Kitchen, S. R. Vallance, J. L. Kennedy, N. Tapia-Ruiz, L. Carassiti, A. Harrison, A.G. Whittaker, T. D. Drysdale, S. W. Kingman, D. H. Gregory, Modern microwave methods in solid-state inorganic materials chemistry: from fundamentals to manufacturing. Chem Rev, 114 (2014),1170- 1206. https://doi.org/10.1021/cr4002353

[78] S. Komarneni, R. K. Rajha, H. Katsuki, Microwave-hydrothermal processing of titanium dioxide. Mater Chem Phys, 61(1999), 50-54. https://doi.org/10.1016/S0254-0584(99)00113-3

[79] S. Komarneni, S. Esquivel, Y. D. Noh, S. Sitthisang, J. Tantirungrotechai, H. Li, S. Yin, T. Sato, H. Katsuki, Novel synthesis of nanophase *anatase* under conventional-and microwave-hydrothermal conditions: DeNOx properties. Ceram Int, 40 (2014), 2097-2102. https://doi.org/10.1016/j.ceramint.2013.07.123

[80] X. Li, L. Wang, X. Lu, Preparation of silver-modified TiO2 via microwave-assisted method and its photocatalytic activity for toluene degradation. J Hazard Mater, 177 (2010), 639-647. https://doi.org/10.1016/j.jhazmat.2009.12.080

[81] L. Li, X. Qin, G. Wang, L. Qi, G. Du, Z. Hu, Synthesis of *anatase* TiO2 nanowires by modifying TiO2 NPs using the microwave heating method. Appl Surf Sci, 257 (2011), 8006-8012. https://doi.org/10.1016/j.apsusc.2011.04.073

[82] Q. Zhang, K. Zhang, D. Xu, G. Yang, H. Huang, F. Nie, C. Liu, S. Yang, CuO nanostructures: synthesis, characterization, growth mechanisms, fundamental properties, and applications. Prog Mater Sci, 60 (2014), 208-337. https://doi.org/10.1016/j.pmatsci.2013.09.003

[83] Umekar, M. S., Bhusari, G. S., Potbhare, A. K., Mondal, A., Kapgate, B. P., Desimone, M. F., & Chaudhary, R. G. (2021). Bioinspired reduced graphene oxide based nanohybrids for photocatalysis and antibacterial applications. Current Pharmaceutical Biotechnology, 22(13), 1759-1781. https://doi.org/10.2174/1389201022666201231115826

[84] W. Zhu, G. Wang, X. Hong, X. Shen, D. Li, X. Xie, Metal nanoparticle chains embedded in TiO2 nanotubes prepared by one-step electrodeposition. Electrochim Acta, 55 (2009), 480-484. https://doi.org/10.1016/j.electacta.2009.08.059

[85] L. Zhu, K. Liu, H. Li, Y. Sun, M. Qiu, Solvothermal synthesis of mesoporous TiO2 microspheres and their excellent photocatalytic performance under simulated sunlight irradiation. Solid State Sci, 20 (2013), 8-14. https://doi.org/10.1016/j.solidstatesciences.2013.02.026

[86] S. Thangavel, G. Venugopal, K. Jae. Enhanced photocatalytic efficacy of organic dyes using β-tin tungstate-reduced graphene oxide NCss. Mater. Chem. Phy. 145 (2014), 108-115. https://doi.org/10.1016/j.matchemphys.2014.01.046

[87] PB Chouke, T Shrirame, AK Potbhare, A Mondal, AR Chaudhary, S. Mondal, R. Sharma, RG Chaudhary, Bioinspired metal/metal oxide NPs: A road map to potential applications, Materials Today Advances, 16 (2022) 100314. https://doi.org/10.1016/j.mtadv.2022.100314

[88] P.B. Chouke, A.K. Potbhare, N.P. Meshram, M.M. Rai, K.M. Dadure, K Chaudhary, A.R. Rai, M Desimone, R.G. Chaudhary, D.T. Masram, ACS Omega, 7 (2022) 6869−6884. https://doi.org/10.1021/acsomega.1c06544

[89] G. Kahrilas, L. Wally, S. Fredrick, M. Hiskey, A. Prieto, J. Owens, ACS Sustain. Chem. Eng., 2 (2014) 367-376. https://doi.org/10.1021/sc4003664

[90] R.G. Chaudhary, P.B. Chouke, R. Bagade, A.K. Potbhare, K.M. Dadure, Molecular docking and antioxidant activity of Cleome simplicifolia assisted synthesis of cerium oxide NPs, Mater. Today: Proc. 29 (2020) 1085-1090. https://doi.org/10.1016/j.matpr.2020.05.062

[91] R.G. Chaudhary, V. Sonkusare, G. Bhusari, A. Mondal, D. Shaik, H.D. Juneja. microwave-mediated synthesis of spinel cual2o4 NCss for enhanced electrochemical and catalytic performance. Res. Chem. Intermed., 2017, 44, 239-2060. https://doi.org/10.1007/s11164-017-3213-z

[92] R.G. Chaudhary, JA. Tanna, N.V. Gandhare, A.R. Rai, S.Yerpude, H.D. Juneja. Copper NPs catalyzed an efficient one-pot multicomponents synthesis of chromenes derivatives and its antibacterial activity. J. Expt. Nanosci., 2016, 11, 884-890 https://doi.org/10.1080/17458080.2016.1177216

[93] R.G. Chaudhary, J.A. Tanna, N.V. Gandhare, A.R. Rai, H.D. Juneja. Histidine Capped ZnO NPs: An efficient synthesis, characterization and effective antibacterial activity. BioNanoScience, 2015, 5, 123-134 https://doi.org/10.1007/s12668-015-0170-0

[94] M. S. Umekar, G. S. Bhusari, A. K. Potbhare, R.G. Chaudhary, Decorated Reduced Graphene Oxide Nanohybrid by Clerodendrum Infortunatum, Emer. Mater. Res. 10 (2021) 75-84. https://doi.org/10.1680/jemmr.19.00175

[95] K. M. Dadure, A.K. Potbhare, D. Mahapatra, A. Haldar, R.G. Chaudhary, Utilization of Mother Nature's Gift for the Biofabrication of Copper/ Copper Oxide NPs for Therapeutic Applications, Jord. J. Phy. 15 (2022) 67-79. https://doi.org/10.47011/15.1.12

[96] P. B. Chouke, A. K. Potbhare, G. S. Bhusari, S. Somkuwar, R.K. Mishra, RG Chaudhary, Green fabrication of Zinc oxide nanospheres by Aspidopterys Cordata for effective antioxidant and antibacterial activity, Adv. Mater. Lett. 10 (2019) 355-360. https://doi.org/10.5185/amlett.2019.2235

[97] A.K. Potbhare, R.G. Chaudhary, P.B. Chouke, S.T. Yerpude, V. Sonkusare, A. Mondal, A.R Rai, H.D. Juneja, Phytosynthesis of nearly monodisperse CuO nanospheres using Phyllanthus Reticulatus/Conyza Bonariensis and its antioxidant/antibacterial assays, Mater. Sci. Eng. C 99 (2019) 783-793. https://doi.org/10.1016/j.msec.2019.02.010

[98] M. S. Umekar, R.G. Chaudhary, G. S. Bhusari, A Mondal, A.K. Potbhare, M Sami, Bioinspired graphene-based silver NPs: Fabrication, characterization and antibacterial activity, Mater. Today: Proc. 29 (2020) 709-714. https://doi.org/10.1016/j.matpr.2020.04.169

[99] A.K. Potbhare, R.G. Chaudhary, M. S. Umekar, P.B. Chouke, M.B. Bagade, SK. Aziz, A Abdala, Bioinspired graphene-based silver NPs: Fabrication, characterization and antibacterial activity, Mater. Today: Proc. 29 (2020) 720-725. https://doi.org/10.1016/j.matpr.2020.04.212

[100] Umekar, M. S., Bhusari, G. S., Bhoyar, T., Devthade, V., Kapgate, B. P., Potbhare, A. P., Chaudhary R. G., & Abdala, A. A. (2023). Graphitic Carbon Nitride-based Photocatalysts for Environmental Remediation of Organic Pollutants. Current Nanoscience, 19(2), 148-169. https://doi.org/10.2174/1573413718666220127123935

[101] A. Isari, A. Payan, M. Fattahi, S. Jorfi, B. Kakavandi. Photocatalytic degradation of rhodamine B and real textile wastewater using Fe-doped TiO2 anchored on reduced graphene oxide (Fe-TiO2/rGO): Characterization and feasibility, mechanism and pathway studies. Appl. Surf. Sci. 462 (2018), 549-564. https://doi.org/10.1016/j.apsusc.2018.08.133

[102] U. Gulati, R. Chinna, D. Rawat, Reduced Graphene Oxide Supported Copper Oxide NCss from a Renewable Copper Mineral Precursor: A Green Approach for Decarboxylative C(sp3)-H Activation of Proline Amino Acid To Afford Value-Added Synthons. ACS Sustain. Chem. Eng. 6 (2018), 10039-10051. https://doi.org/10.1021/acssuschemeng.8b01376

[103] C. T. Chou, F. H. Wang, J. Vac. Sci. Technol. 36 (2018) 122-131.

[104] Chouke, P. B., Dadure, K. M., Potbhare, A. K., Bhusari, G. S., Mondal, A., Chaudhary, K., ... & Masram, D. T. (2022). Biosynthesized δ-Bi2O3 NPs from Crinum viviparum Flower Extract for Photocatalytic Dye Degradation and Molecular Docking. ACS Omega 7 (2022) 20983-20993. https://doi.org/10.1021/acsomega.2c01745

[105] Sonkusare, V. N., Chaudhary, R. G., Bhusari, G. S., Mondal, A., Potbhare, A. K., Mishra, R. K., ... & Abdala, A. A. (2020). Mesoporous octahedron-shaped tricobalt tetraoxide NPs for photocatalytic degradation of toxic dyes. ACS Omega, 5(14), 7823-7835. https://doi.org/10.1021/acsomega.9b03998

[106] M. Khan, M. Khan, M. Cho, Recent progress of metal-graphene nanostructures in photocatalysis. Nanoscale 10 (2018), 9427-9440. https://doi.org/10.1039/C8NR03500H

[107] X. Li, J. Yu. M. Jaroniec, Hierarchical photocatalysts. Chem. Soc. Rev. 45 (2016), 2603-2636. https://doi.org/10.1039/C5CS00838G

[108] X. Lü, J. Shen, J. Wang, Z. Cui, J. Xie, Highly efficient visible-light photocatalysts: reduced graphene oxide and C 3 N 4 nanosheets loaded with Ag NPs. RSC Adv. 5 (2015), 15993-15999. https://doi.org/10.1039/C4RA12395F

[109] A. Ajmal, I. Majeed, R. N. Malik, H. Idriss, M. A. Nadeem, Principles and mechanisms of photocatalytic dye degradation on TiO2 based photocatalysts: a

comparative overview. RSC Adv. 4 (2014), 37003-37026.
https://doi.org/10.1039/C4RA06658H

[110] N. Saqib, R. Adnan, I. Shah, A mini-review on rare earth metal-doped TiO2 for photocatalytic remediation of wastewater. Environ. Sci. Poll. Res. 23 (2016), 15941-15951.10.1007/s11356-016-6984-7 https://doi.org/10.1007/s11356-016-6984-7

[111] R. Molinari, C. Lavorato. P. Argurio, Photocatalytic reduction of acetophenone in membrane reactors under UV and visible light using TiO2 and Pd/TiO2 catalysts. Chem. Eng. J. 274 (2015), 307-316. https://doi.org/10.1016/j.cej.2015.03.120

[112] H. M. El-Bery, Y. Matsushita. A. Abdel-moneim, Fabrication of efficient TiO2-RGO heterojunction composites for hydrogen generation via water-splitting: Comparison between RGO, Au and Pt reduction sites. Appl. Surf. Sci. 423 (2017), 185-196. https://doi.org/10.1016/j.apsusc.2017.06.130

[113] Ocampo-Pérez, R., Sánchez-Polo, M., Rivera-Utrilla, J., & Leyva-Ramos, R. (2011). Enhancement of the catalytic activity of TiO2 by using activated carbon in the photocatalytic degradation of cytarabine. Applied Catalysis B: Environmental, 104(1-2), 177-184. https://doi.org/10.1016/j.apcatb.2011.02.015

[114] Sonkusare, V. N., Chaudhary, R. G., Bhusari, G. S., Rai, A. R., & Juneja, H. D. (2018). Microwave-mediated synthesis, photocatalytic degradation and antibacterial activity of α-Bi2O3 microflowers/novel γ-Bi2O3 microspindles. Nano-Structures & Nano-Objects, 13, 121-131. https://doi.org/10.1016/j.nanoso.2018.01.002

[115] Chaudhary, R. G., Potbhare, A. K., Aziz, S. T., Umekar, M. S., Bhuyar, S. S., & Mondal, A. (2021). Phytochemically fabricated reduced graphene Oxide-ZnO NCs by Sesbania bispinosa for photocatalytic performances. Materials Today: Proceedings, 36, 756-762. https://doi.org/10.1016/j.matpr.2020.05.821

[116] K. Rahimi, H. Zafarkish, A. Yazdani, Mater, Des. 144 (2018) 214-221. https://doi.org/10.1016/j.matdes.2018.02.030

[117] Chaudhary, R. G., Sonkusare, V., Bhusari, G., Mondal, A., Potbhare, A., Juneja, H., D, Abdala, A., & Sharma, R. (2023). Preparation of mesoporous ThO2 NPs: Influence of calcination on morphology and visible-light-driven photocatalytic degradation of indigo carmine and methylene blue. Environmental Research, 115363. https://doi.org/10.1016/j.envres.2023.115363

[118] R. Elshypany, H. Selim, K. Zakaria, A. H. Moustafa, S. A. Sadeek, S. I. Sharaa, P. Raynaud, A. A. Nada, Molecules. 14 (2021) 2269. https://doi.org/10.3390/molecules26082269

[119] S. Hamed, K. Hossain, S. Masoud, S. Niasari, D. Mortazavi, J. Coll. Interface Sci. 498 (2017) 423-432. https://doi.org/10.1016/j.jcis.2017.03.078

[120] H. Zhang, X. Lv, Y. Li, Y. Wang, J. Li. ACS Nano. 4 (2010) 380-386. https://doi.org/10.1021/nn901221k

[121] S. Chandra, P. Das, S. Bag P. Pramanik, P. Pramanik. Mater. Sci. Eng.B 177 (2012) 855-861 https://doi.org/10.1016/j.mseb.2012.04.006

[122] Yongsheng Fu and Xin Wang, Magnetically Separable ZnFe2O4-Graphene Catalyst and its High Photocatalytic Performance under Visible Light Irradiation, Ind. Eng. Chem. Res. 2011, 50, 12, 7210-7218. https://doi.org/10.1021/ie200162a

[123] M. Shang, W. Wang, L. Zhou, S. Sun, W. Yin. J. Hazard. Mater. 172 (2009) 338-344. https://doi.org/10.1016/j.jhazmat.2009.07.017

[124] S. Shanavas, A. Priyadharsan, E.I. Gkanas, R. Acevedo, P.M. Anbarasan (2019) High efficient catalytic degradation of tetracycline and ibuprofen using visible light driven novel Cu/Bi2Ti2O7/rGO NCs: Kinetics, intermediates and mechanism, Journal of Industrial and Engineering Chemistry, 72, 512-528 https://doi.org/10.1016/j.jiec.2019.01.008

[125] B. Li, H. Cao. ZnO@graphene composite with enhanced performance for the removal of dye from water.,J. Mater. Chem. 21 (2011), 3346. https://doi.org/10.1039/C0JM03253K

[126] S. Lathasree, A.N. Rao, B. SivaSankar, V. Sadasivam, K. Rengaraj. Heterogeneous photocatalytic mineralisation of phenols in aqueous solutions. J. Mol. Catal. A: Chem. 223 (2004), 101. https://doi.org/10.1016/j.molcata.2003.08.032

[127] G. Malekshoar, K. Pal, Q. He, A. Yu, A. K. Ray, Enhanced Solar Photocatalytic Degradation of Phenol with Coupled Graphene-Based Titanium Dioxide and Zinc Oxide., Ind. Eng. Chem. Res. 53 (2014), 18824-18832. https://doi.org/10.1021/ie501673v

[128] Chang, L. F., Doong, R. A. (2014). Cu-TiO2 nanorods with enhanced ultraviolet- and visible-light photoactivity for bisphenol A degradation. Journal of hazardous materials, 277, 84-92.. https://doi.org/10.1016/j.jhazmat.2014.01.047

Keyword Index

About the Editors

Prof. N.B. Singh, former Head Chemistry Department and Former Dean Faculty of Science, DDU Gorakhpur University, Gorakhpur is presently Emeritus professor at Department of Chemistry and Biochemistry and Research Development Cell, Sharda University, Greater Noida, UP. Dr. Singh received the most prestigious Alexander von Humboldt fellowship (Germany) in 1977 and has worked at different universities in Germany. His main areas of research are Solid state chemistry, materials science, cement chemistry, thermodynamics, and water purification. He has published more than 300 research articles, 8 books, and more than 50 book chapters. Prof. Singh is president of Indian Association for Solid State Chemists and Allied Scientists Jammu and received many awards. His name appeared twice in the list of 2% scientists of world declared by Stanford University, USA.

Dr. Md. Abu Bin Hasan Susan, a Professor of Chemistry at Dhaka University, Bangladesh, received his Ph.D. degree from Yokohama National University, Japan in 2000 as a MEXT scholar and was awarded VBL, JSPS, CREST, and Bridge postdoctoral fellowships. He has 176 articles including 18 book chapters. He collaborates with 14 renowned laboratories and visited eleven countries. He was nominated for awards from FACS, in 2005 and ISESCO, in 2010. He received the Dean's Award of the Faculty of Science, Dhaka University (2011), the UGC Awards of Bangladesh in 2011 and 2013, the United Group Paper Awards in 2016 and 2017, and the Bangladesh Academy of Sciences

Gold Medal in 2012. He received the Silver Medal of the Society of Promotion of Education and Science, India in 2019. He is an *Associate Editor* of Spectrum of Emerging Sciences, an *Editorial Board Member* of the Journal of Bangladesh Academy of Sciences and Universal Journal of Electrochemistry, and the *Journal of Scientific and Technical Research*. He served as the National Representative of Division I of IUPAC. He has been a Fellow (2018) of the Bangladesh Academy of Sciences and IUPAC (2016). His citations are 10,901, *h*-index- 30 and *i*-10 index- 79.

Dr. Ratiram Gomaji Chaudhary is presently working as Associate Professor and Head, Department of Chemistry, Seth Kesarimal Porwal College of Arts, Science and Commerce, Kamptee. His research areas are *Biogenic Synthesis*, *Phytosynthesis*, *Metal oxide/Graphene-based nanohybrids*, *Toxic Dyes Degradations* etc. He has been awarded two times with *'Rajiv Gandhi National Fellowship Award'* as JRF by UGC, New Delhi for pursuing MPhil and PhD degrees. He has published 5 books, 24 book chapters, and 102 regular articles, and 12 review articles in peer-reviewed SCI/Scopus indexing journals having Total Citations of 1728 with h_index 23. He is a recipient of several awards like 'Best Researcher Award', 'Young Scientist Award', 'Outstanding Reviewer Award by IOP-VAST', Mahatma Jyotirao Fule: State Level Best Researcher Award, etc. He worked as a Guest Editor for Scopus indexed journals *Materials Today: Proceeding*, Elsevier, and *Current Pharmaceutical Biotechnology, Current Pharmaceutical Design,* and *Current Nanosciences,* Bentham Science.

www.ingramcontent.com/pod-product-compliance
Lightning Source LLC
Chambersburg PA
CBHW071323210326
41597CB00015B/1321